1997

THE BIOSYNTHESIS OF THE TETRAPYRROLE PIGMENTS

The Ciba Foundation is an international scientific and educational charity (Registered Charity No. 313574). It was established in 1947 by the Swiss chemical and pharmaceutical company of CIBA Limited —now Ciba-Geigy Limited. The Foundation operates independently in London under English trust law.

The Ciba Foundation exists to promote international cooperation in biological, medical and chemical research. It organizes about eight international multidisciplinary symposia each year on topics that seem ready for discussion by a small group of research workers. The papers and discussions are published in the Ciba Foundation symposium series. The Foundation also holds many shorter meetings (not published), organized by the Foundation itself or by outside scientific organizations. The staff always welcome suggestions for future meetings.

The Foundation's house at 41 Portland Place, London W1N 4BN, provides facilities for meetings of all kinds. Its Media Resource Service supplies information to journalists on all scientific and technological topics. The library, open five days a week to any graduate in science or medicine, also provides information on scientific meetings throughout the world and answers general enquiries on biomedical and chemical subjects. Scientists from any part of the world may stay in the house during working visits to London.

Ciba Foundation Symposium 180

THE BIOSYNTHESIS OF THE TETRAPYRROLE PIGMENTS

1994

JOHN WILEY & SONS

Chichester · New York · Brisbane · Toronto · Singapore

©Ciba Foundation 1994

Published in 1994 by John Wiley & Sons Ltd
Baffins Lane, Chichester
West Sussex PO19 1UD, England
Telephone (+44) (243) 779777

Other Wiley Editorial Offices

John Wiley & Sons, Inc., 605 Third Avenue,
New York, NY 10158-0012, USA

Jacaranda Wiley Ltd, 33 Park Road, Milton,
Queensland 4064, Australia

John Wiley & Sons (Canada) Ltd, 22 Worcester Road,
Rexdale, Ontario M9W 1L1, Canada

John Wiley & Sons (SEA) Pte Ltd, 37 Jalan Pemimpin #05-04,
Block B, Union Industrial Building, Singapore 2057

Suggested series entry for library catalogues:
Ciba Foundation Symposia

Ciba Foundation Symposium 180
xiii + 350 pages, 94 figures, 5 tables, 19 schemes, 32 structures

Library of Congress Cataloging-in-Publication Data
A catalog record for this book is available
from the Library of Congress

British Library Cataloguing in Publication Data
A catalogue record for this book is
available from the British Library

ISBN 0 471 93947 1

Phototypeset by Dobbie Typesetting Limited, Tavistock, Devon.
Printed and bound in Great Britain by Biddles Ltd, Guildford.

This symposium was held in honour of
Sir Alan Battersby FRS

Sir Alan Battersby FRS

Contents

Participants

C. Abell University Chemical Laboratory, Lensfield Road, Cambridge CB2 1EW, UK

M. Akhtar Department of Biochemistry, School of Biological Sciences, University of Southampton, Bassett Crescent East, Southampton SO9 3TU, UK

D. Arigoni (*Chairman*) Laboratorium für Organische Chemie, ETH-Zentrum, Universitätstrasse 16, CH-8092 Zürich, Switzerland

A. R. Battersby University Chemical Laboratory, Lensfield Road, Cambridge CB2 1EW, UK

S. I. Beale Division of Biology and Medicine, Brown University, Providence, RI 02912, USA

P. A. Castelfranco Division of Biological Sciences, Section of Botany, University of California, Davis, CA 95616, USA

C. K. Chang Department of Chemistry, Michigan State University, East Lansing, MI 48824, USA

G. H. Elder Department of Medical Biochemistry, University of Wales, College of Medicine, Heath Park, Cardiff CF4 4XN, UK

A. Eschenmoser Laboratorium für Organische Chemie, ETH-Zentrum, Universitätstrasse 16, CH-8092 Zürich, Switzerland

W. T. Griffiths Department of Biochemistry, School of Medical Sciences, University of Bristol, University Walk, Bristol BS8 1TD, UK

A. Hädener Institut für Organische Chemie der Universität Basel, St Johanns-Ring 19, CH-4056 Basel, Switzerland

L. Ilag (*Bursar*) Department of Molecular Biophysics & Biochemistry, Yale University, 260 Whitney Avenue, PO Box 6666, New Haven, CT 06511, USA

P. M. Jordan   Department of Biochemistry, School of Biological Sciences, University of Southampton, Bassett Crescent East, Southampton SO9 3TU, UK

C. G. Kannangara   Department of Physiology, Carlsberg Laboratory, Gamle Carlsberg Vej 10, DK-2500 Copenhagen Valby, Denmark

B. Kräutler   Institut für Organische und Pharmazeutische Chemie, Innrain 52A, A-6020 Innsbruck, Austria

F. J. Leeper   University Chemical Laboratory, Lensfield Road, Cambridge CB2 1EW, UK

G. V. Louie   ICRF Structural Molecular Biology Unit, Department of Crystallography, Birkbeck College, University of London, Malet Street, London WC1E 7HY, UK

A. R. Pitt   University of Strathclyde, Department of Pure and Applied Chemistry, Thomas Graham Building, 295 Cathedral Street, Glasgow G1 1XL, UK

C. A. Rebeiz   Laboratory of Plant Pigment Biochemistry & Photobiology, University of Illinois, 1201 West Gregory Street, Urbana, IL 61801-3838, USA

A. I. Scott   Center for Biological NMR, Department of Chemistry, Texas A&M University, College Station, TX 77843-3255, USA

A. G. Smith   Department of Plant Sciences, University of Cambridge, Downing Street, Cambridge CB2 3EA, UK

K. M. Smith   Department of Chemistry, University of California, Davis, CA 95616, USA

P. Spencer   Department of Biochemistry, School of Biological Sciences, University of Southampton, Bassett Crescent East, Southampton SO9 3TU, UK

N. P. J. Stamford   University Chemical Laboratory, Lensfield Road, Cambridge CB2 1EW, UK

R. K. Thauer   Laboratorium für Mikrobiologie des Fachbereichs Biologie der Philipps-Universität Marburg, Karl-von-Frisch-Strasse, Postfach 1929, D-35043 Marburg/Lahn, Germany

R. Timkovich   Department of Chemistry, PO Box 870336, University of Alabama, Tuscaloosa, AL 35487-0336, USA

M. J. Warren   School of Biological Sciences, Queen Mary and Westfield College, University of London, Mile End Road, London E1 4NS, UK

S. P. Wood   ICRF Structural Molecular Biology Unit, Department of Crystallography, Birkbeck College, University of London, Malet Street, London WC1E 7HY, UK

Preface

The Ciba Foundation's 311th symposium (number 180 in the New Series of publications), on the biosynthesis of the tetrapyrrole pigments, was organized in honour of Professor Sir Alan Battersby FRS. Professor Battersby joined the Executive Council of the Ciba Foundation in 1977 but his association with the Foundation predates this by some two years as a member of the Scientific Advisory Panel. He was Chairman of the Council from 1983 to 1990 and became a Trustee in 1993. His seminal contributions to bio-organic chemistry and, in particular, to the elucidation of the biosynthesis of alkaloids and, more recently, vitamin B_{12}, are well known, and his personal distinction has been marked worldwide by a multitude of awards, honorary degrees and other accolades. The Foundation is deeply indebted to him for the support and help he has unstintingly given us during almost 20 years and we offer this book as a public expression of our appreciation and thanks. It is also a testament to an exquisitely successful collaboration between chemists and biologists which is the kind of cooperation that the Ciba Foundation is committed to fostering.

Derek J. Chadwick
Director, The Ciba Foundation

Introduction

Duilio Arigoni

Laboratorium für Organische Chemie, ETH-Zentrum, Universitätstrasse 16, CH-8092 Zürich, Switzerland

I welcome you to this meeting on The Biosynthesis of the Tetrapyrrole Pigments. Albert Eschenmoser and I happen to be two old-timers of the Ciba Foundation, having participated in the symposium on Biosynthesis of Terpenes and Sterols held here in 1958 (Ciba Foundation 1959); as I was reminiscing about the old days, I remembered that during this meeting Konrad Bloch had drawn my attention to the common origin of studies on the biosynthesis of sterols and terpenes and on the biosynthesis of the tetrapyrrole pigments. Everything started in Rittenberg's laboratories when it was found that feeding deuterated acetic acid to rats resulted in time-dependent incorporation of deuterium into fatty acids. It was soon realised that the labelled precursor was a magnificent new tool for biosynthetic investigations and a decision was taken to focus attention on cholesterol and haem. As junior members of Rittenberg's group, David Shemin and Konrad Bloch drew straws to establish who would tackle which area. Bloch got the cholesterol and Shemin got the haem.

One may wonder, in retrospect, why Shemin's contribution seemed for quite a while to have less impact on the chemical community than Bloch's work on steroid biosynthesis. Steroids were more fashionable than the tetrapyrrole pigments in the 1950s and this is probably why the discovery of mevalonic acid in Folkers' group and Woodward's new mechanistic interpretation of the old Robinson postulate linking the structures of squalene and cholesterol appealed to the organic chemists more than Shemin's discovery that 5-aminolaevulinic acid and porphobilinogen were intermediates on the pathway to tetrapyrroles. In my opinion, Shemin's work stands out as equally revolutionary and seminal and we would not have been able to gather together here today had he not made his pioneering contributions.

Although Shemin's main interest was in the biosynthesis of haem, his results were later extrapolated to the biosynthesis of chlorophyll, but it was not until the late 1970s that organic chemists started to be attracted to the area. It is fair to say that the structure of vitamin B_{12} was largely responsible for this change in mood. The B_{12} coenzyme had already taught organic chemists several unexpected lessons when it was discovered that it can act as a reservoir of biological radicals playing an important role in hitherto unprecedented biological reactions. Vitamin B_{12} has been referred to as the Everest of organic synthesis and was soon to play a similar role in biosynthetic investigations. Tackling the problem posed by the biosynthesis of this complex molecule was not an easy task in the early 1970s. The soluble enzyme cocktail responsible for the generation of the compound produced the material in disappointingly low yields and the problem of the degradation of the biosynthetic material was not helped by the scanty knowledge of the compound's chemistry. The development of ^{13}C NMR spectroscopy provided the necessary breakthrough and paved the way for studies by different groups. By the end of the 1970s, these studies had led to the identification of three different intermediates beyond uroporphyrinogen III. In the words of one of the protagonists of such studies, development in this area has been 'agonizingly slow' ever since.

In the mean time, the pathway to uroporphyrinogen III has been refined in remarkable detail and it is now possible to discuss at the molecular level the mode of action of selected enzymes.

A revitalization of the biosynthetic studies has come about in more recent years as a result of the efforts of geneticists and microbiologists who have succeeded in isolating the entire set of genes encoding all the proteins involved in B_{12} biosynthesis. This caused a somewhat paradoxical situation in which there was access to the enzymes but no specific knowledge of the reactions they were catalysing!

This situation is now changing rapidly and there has been spectacular recent progress in identifying long-sought new intermediates. All of this makes it appropriate to refresh the subject last discussed here at the Ciba Foundation in 1955 at the meeting on Porphyrin Biosynthesis and Metabolism (Ciba Foundation 1955), to review the new results, to discuss them critically and, if possible, to outline directions for future research.

It is a pleasure for me to be here as your chairman, and it adds to my pleasure that this meeting is also intended to be a tribute to the unfailing leadership of one of the scientists in this group, my good friend Alan Battersby. I am looking forward to three exciting and rewarding days.

References

Ciba Foundation 1955 Porphyrin biosynthesis and metabolism. Churchill, London
Ciba Foundation 1959 Biosynthesis of terpenes and sterols. Churchill, London

Enzymic and mechanistic studies on the conversion of glutamate to 5-aminolaevulinate

C. Gamini Kannangara, Rolf V. Andersen, Bo Pontoppidan, Robert Willows and Diter von Wettstein

Department of Physiology, Carlsberg Laboratory, Gamle Carlsberg Vej 10, DK-2500 Copenhagen Valby, Denmark

Abstract. Higher plants, algae, cyanobacteria and several other photosynthetic and non-photosynthetic bacteria synthesize 5-aminolaevulinate by a $tRNA^{Glu}$-mediated pathway. Glutamate is activated at the α-caboxyl by ligation to $tRNA^{Glu}$ with an aminoacyl-tRNA synthetase. An NADPH-dependent reductase converts glutamyl-$tRNA^{Glu}$ to glutamate 1-semialdehyde, which is finally converted to 5-aminolaevulinate by an aminotransferase. These components are soluble and in plants and algae are located in the chloroplast stroma. In plants and algae the $tRNA^{Glu}$ is encoded in chloroplast DNA whereas the enzymes are encoded in nuclear DNA. The $tRNA^{Glu}$ has a hypermodified 5-methylaminomethyl-2-thiouridine-pseudouridine-C anticodon and probably plays a role in the light–dark regulation of 5-aminolaevulinate synthesis. Ligation of glutamate to $tRNA^{Glu}$ requires ATP and Mg^{2+} and proceeds via a ternary intermediate. Glutamyl-$tRNA^{Glu}$ reduction appears to involve formation of a complex. Glutamate 1-semialdehyde non-enzymically synthesized by reductive ozonolysis from γ-vinyl GABA is used as substrate by the last enzyme. Glutamate-1-semialdehyde aminotransferase contains pyridoxal phosphate as a prosthetic group. The enzyme is converted to spectrally different forms by treatment with 4,5-diaminovalerate or 4,5-dioxovalerate. The pyridoxamine 5′-phosphate form of the enzyme converts (S)-glutamate 1-semialdehyde to 5-aminolaevulinate via 4,5-diaminovalerate through a bi-bi ping-pong mechanism.

1994 The biosynthesis of the tetrapyrrole pigments. Wiley, Chichester (Ciba Foundation Symposium 180) p 3–25

Natural tetrapyrrole pigments are chlorophylls, haems, phycobilins (including the phytochrome chromophore) and corrinoids. The pyrrole rings in these pigments are synthesized from 5-aminolaevulinate, a non-protein amino acid produced primarily for this purpose. After studying tetrapyrrole pigments for nearly 20 years, Sir Alan Battersby rated chlorophyll as the fundamental pigment of life and placed haem, the red pigment of blood, in second place (Battersby 1988). Interestingly, 5-aminolaevulinate is biosynthesized in two completely different ways for these two major tetrapyrrole pigments. Succinyl-CoA and glycine are condensed in a single step catalysed by the enzyme-5-aminolaevulinate synthase (EC 2.3.1.37) to produce 5-aminolaevulinate for biosynthesis of mammalian haem. For the synthesis of chlorophyll in plants, glutamate is converted to 5-aminolaevulinate by a multi-enzyme pathway.

In higher plants, 5-aminolaevulinate synthesis is stimulated by light. Only small amounts of 5-aminolaevulinate are made in darkness to provide for the synthesis of haem and trace amounts of protochlorophyllide, a chlorophyll precursor. Rapid synthesis of chlorophyll occurs when etiolated seedlings are placed in light. Treatment of such greening leaves with laevulinate inhibits chlorophyll synthesis, leading to accumulation of 5-aminolaevulinate. Laevulinate is a competitive inhibitor of 5-aminolaevulinate dehydratase (porphobilinogen synthase, EC 4.2.1.24) and prevents further metabolism of 5-aminolaevulinate. Beale and Castelfranco treated greening etiolated plant tissues with laevulinate and ^{14}C-labelled glutamate, glutamine, α-ketoglutarate, succinate or glycine, and found that the accumulated 5-aminolaevulinate was poorly labelled from glycine or succinate in comparison with ^{14}C incorporation from the five-carbon compounds (Beale & Castelfranco 1974). Using specifically labelled [^{14}C] glutamate together with laevulinate, by analysing the distribution of ^{14}C in the 5-aminolaevulinate, Beale et al (1975) and Meller et al (1975) demonstrated that the entire carbon skeleton of the five-carbon precursor is incorporated intact into 5-aminolaevulinate. The enzymes, cofactors and intermediates involved in the conversion of glutamate to 5-aminolaevulinate have been isolated and studied at the Carlsberg Laboratory (von Wettstein 1991).

Conversion of glutamate to 5-aminolaevulinate involves several interesting reactions (Fig. 1). Glutamate is first activated at the α-carboxyl by ligation to tRNA^{Glu}. An aminoacyl-tRNA synthetase catalyses the ligation in the presence of ATP and Mg^{2+}. Activated glutamate is then reduced to glutamate 1-semialdehyde in an NADPH-dependent reaction. Finally, glutamate 1-semialdehyde is converted to 5-aminolaevulinate in a reaction catalysed by the enzyme glutamate-1-semialdehyde aminotransferase (glutamate-1-semialdehyde 2,1-aminomutase, EC 5.4.3.8). All higher plants, algae and cyanobacteria that have been analysed synthesize 5-aminolaevulinate for their tetrapyrrole pigments by the tRNA^{Glu}-mediated pathway from glutamate, as do several photosynthetic and non-photosynthetic bacteria.

Preparation of tRNA^{Glu} and the enzymes involved in 5-aminolaevulinate biosynthesis

All the components involved in 5-aminolaevulinate synthesis in greening barley are soluble and localized in the plastid stroma. The 5-aminolaevulinate-synthesizing activity of crude preparations is extremely labile, the enzyme involved in the conversion of glutamyl-tRNA^{Glu} to glutamate 1-semialdehyde

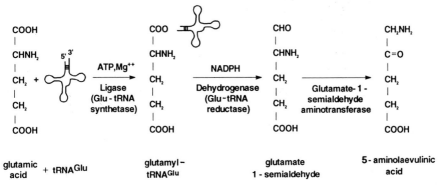

glutamic acid + tRNA^{Glu}   glutamyl–tRNA^{Glu}   glutamate 1-semialdehyde   5-aminolaevulinic acid

FIG. 1. The pathway of biosynthesis of 5-aminolaevulinic acid from glutamic acid. Glutamic acid combines with tRNA^{Glu} in a ligation reaction catalysed by glutamyl-tRNA synthetase (glutamyl-tRNA ligase, EC 6.1.1.17). The activated glutamate is then reduced to glutamate 1-semialdehyde, which is then converted to 5-aminolaevulinic acid in a reaction catalysed by glutamate-1-semialdehyde aminotransferase (glutamate-1-semi aldehyde 2,1-aminomutase, EC 5.4.3.8).

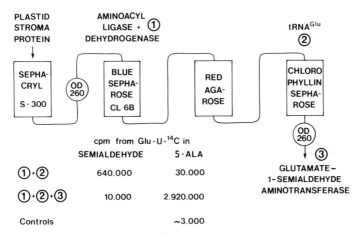

FIG. 2. The enzymes and tRNA^{Glu} required for converting glutamate to 5-aminolaevulinate (5-ALA) in the chloroplast stroma can be partially separated and purified by gel filtration and serial affinity chromatography. (1) Cibacron Blue–Sepharose binds glutamate–tRNA^{Glu} ligase and glutamyl-tRNA^{Glu} reductase (dehydrogenase). (2) Chlorophyllin–Sepharose binds the tRNA complement of the chloroplast. Glutamate-1-semialdehyde aminotransferase is collected in the run-off fraction (3).

being the most unstable component. In greening barley the components are more stable when kept at pH 9–10, rather than at pH 8 where the enzymes show their optimal activities. Rapid partial purification of active preparations by gel filtration and affinity chromatography in glycerol-containing Tricine buffers at pH 9 is a convenient way to obtain the tRNA^{Glu}, and the enzymes retain high activities (Fig. 2).

The components required for conversion of glutamate to 5-aminolaevulinate, and subsequently to uroporphyrinogen, are routinely partially purified from the stroma of greening barley plastids by Sephacryl S-300 gel filtration and affinity chromatography with, sequentially, Cibacron Blue–Sepharose, Procion Red–agarose and chlorophyllin–Sepharose or haem–Sepharose (Wang et al 1981). This procedure separates the components into three fractions which must be combined for the conversion of glutamate to 5-aminolaevulinate. Blue Sepharose binds the ligase and the reductase, and Procion Red binds several unwanted proteins. Chlorophyllin or haem Sepharose binds the tRNA complement of the chloroplasts (Kannangara et al 1984). Glutamate-1-semialdehyde aminotransferase is not retained by the affinity columns and is collected in the run-off fraction. When the aminotransferase fraction is omitted from the reconstituted mixture, the product formed is glutamate 1-semialdehyde and not 5-aminolaevulinate.

The role of tRNA^{Glu}

The intriguing feature of the conversion of glutamate to 5-aminolaevulinate is the involvement of tRNA^{Glu}. Our experiments, and those of others, indicate that all organisms investigated which use glutamate to synthesize 5-aminolaevulinate do so via the tRNA^{Glu}-mediated pathway (Avissar et al 1989, Beale & Weinstein 1990, Huang & Wang 1986, Jahn et al 1992, Jordan 1990, Kannangara 1991, Kannangara et al 1988, O'Neill et al 1991). The involvement of tRNA^{Glu} in the biosynthesis of 5-aminolaevulinate has been demonstrated in several ways: (1) ribonuclease A or snake venom phosphodiesterase sensitivity of 5-aminolaevulinate synthesis (see below); (2) reconstitution of 5-aminolaevulinate synthesis by the addition of tRNA^{Glu}; (3) conversion of glutamyl-tRNA^{Glu} to 5-aminolaevulinate with purified reductase.

The nucleotide sequences of the tRNAs involved in 5-aminolaevulinate synthesis can be fitted into the clover leaf structure. At their 3′ ends they have the C-C-A sequence characteristic of all tRNAs. The anticodon sequence U-U-C identifies them as glutamate-specific tRNAs. The 3′ C-C-A, as well as the anticodon sequence, is required for the ligation. Removal of the C-C-A by digestion with snake venom phosphodiesterase leads to complete loss of the tRNA's ability to ligate glutamate. Replacing the C-C-A sequence with nucleotidyl transferase (polyribonucleotide nucleotidyltransferase, EC 2.7.7.8) restores this ability (Schön et al 1986).

The first position of the anticodon of barley chloroplast tRNA^{Glu} is occupied by 5-methylaminomethyl-2-thiouridine (Fig. 3). This hypermodified nucleotide can be oxidized under mild conditions with iodine. The oxidized tRNA^{Glu} is unable to function in the ligase reaction but its activity can be restored by reduction with thiosulphate. This reversible oxidative inactivation of tRNA^{Glu} suggests a mechanism for the photostimulation of 5-aminolaevulinate and

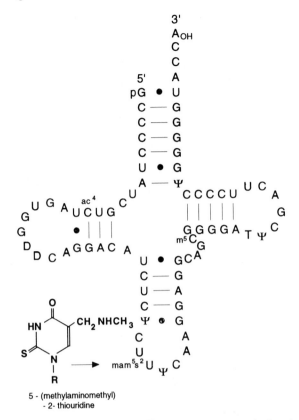

FIG. 3. Nucleotide structure of barley tRNA^{Glu}, which participates in 5-aminolaevulinate biosynthesis, and the structure of the hypermodified nucleotide, 5-methylamino-methyl-2-thiouridine, in the first position of its anticodon. ψ, pseudouridine.

protein synthesis observed in higher plants and algae. Oxidation of tRNA^{Glu} markedly curtails 5-aminolaevulinate and protein synthesis in plastids in darkness, but in light the inactive tRNA^{Glu} is rapidly reduced to its active form. The oxidation of tRNA^{Glu} probably involves a conversion of the mercapto group of the 5-methylaminomethyl-2-thiouridine at the anticodon to a disulphide form. The loss of the ability to regulate 5-aminolaevulinate synthesis by light could be lethal to the plant, as it is in some *tigrina* mutants of barley. These mutants overproduce 5-aminolaevulinate in the dark, leading to accumulation of excessive amounts of protochlorophyllide. When these mutants are grown in day and night conditions they develop a primary leaf with alternating bleached/necrotic and green bands, the *tigrina* phenotype, because the regions of leaf that grow and differentiate at night are bleached by light because of the excess of protochlorophyllide whereas those which grow and differentiate in the day accumulate chlorophyll instead of proto-chlorophyllide.

In higher plants, the tRNA^{Glu} involved in 5-aminolaevulinate synthesis is encoded by the chloroplast DNA. The barley gene is closely linked to the genes encoding tRNA^{Tyr} and tRNA^{Asp} on a 1.6 kb HindIII-EcoRI restriction fragment. The three genes are transcribed as a single precursor RNA which is processed and modified to give the functional tRNA molecules. The tRNA^{Glu} gene is transcribed both in light and in darkness, but, in cucumber, the levels of expression of this gene and other plastid tRNA genes are higher in greening cotyledons than in etiolated cotyledons (Masuda et al 1992). Application of the hormone benzyladenine stimulates the synthesis of 5-aminolaevulinate in etiolated cucumber cytoledons *in vivo*, apparently through a general increase in production of plastid RNA.

Glutamate–tRNA^{Glu} ligase (glutamyl-tRNA synthetase)

There are two glutamate–tRNA ligases in higher plants and algae, one located in the cytoplasm and the other inside the chloroplast. The latter enzyme is responsible for activating glutamate for 5-aminolaevulinate and plastid protein synthesis. The chloroplast enzyme also catalyses the ligation of glutamate to tRNA^{Gln} for its subsequent conversion to glutamine by amidophosphoribosyl-transferase (Schön et al 1988). There must be a mechanism for balancing the needs of protein and chlorophyll synthesis for glutamyl-tRNA^{Glu} to ensure the orderly development of the chloroplast. Glutamate–tRNA^{Glu} ligase probably participates in this regulatory process by forming a multi-enzyme complex which directs the path of glutamyl-tRNA^{Glu}.

The barley chloroplast tRNA^{Glu} ligase is encoded in the nuclear DNA. The protein comprises two identical subunits, both synthesized in the cytoplasm, each with a 34 amino acid pre-sequence and molecular mass of 58 kDa. The subunits are then transported into the chloroplast, processed to give the mature subunits of 526 amino acids (molecular mass 54 kDa) and assembled to give the functional enzyme. The barley chloroplast enzyme shows 40% amino acid sequence identity to other known glutamate–tRNA^{Glu} ligases (Andersen 1992).

Aminoacyl-tRNA synthetases ligate their amino acids to their cognate tRNAs in two steps. In the first step, the enzyme catalyses a reaction between the amino acid and Mg^{2+}–ATP to give an aminoacyl-adenylate and pyrophosphate. The aminoacyl moiety is then transferred from aminoacyl-adenylate to either the 2′ or the 3′ OH of the tRNA, and AMP is released. The release of pyrophosphate is not observed with glutamate–tRNA^{Glu} ligase in the absence of tRNA^{Glu}, in contrast to most other aminoacyl-tRNA synthetases; the synthetase's mechanism of catalysis is therefore likely to involve the formation of a ternary intermediate, as illustrated in Fig. 4.

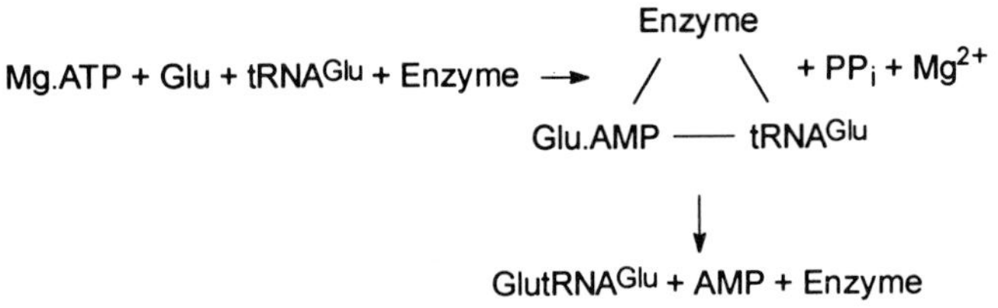

$$\text{Mg.ATP} + \text{Glu} + \text{tRNA}^{\text{Glu}} + \text{Enzyme} \longrightarrow \begin{array}{c} \text{Enzyme} \\ \diagup \quad \diagdown \\ \text{Glu.AMP} \text{—} \text{tRNA}^{\text{Glu}} \end{array} + PP_i + Mg^{2+}$$

$$\downarrow$$

$$\text{GlutRNA}^{\text{Glu}} + \text{AMP} + \text{Enzyme}$$

FIG. 4. The mechanism of glutamate–tRNA^{Glu} ligase (EC 6.1.1.17, glutamyl-tRNA synthetase) involves the formation of a ternary complex as an intermediate.

The properties of glutamyl-tRNA^{Glu} reductase

The enzyme responsible for the conversion of glutamyl-tRNA^{Glu} to glutamate 1-semialdehyde appears to differ in subunit composition and molecular mass between different organisms (Jahn et al 1992). A monomeric protein of 130 kDa purified from *Chlamydomonas reinhardtii* forms a stable complex with glutamyl-tRNA^{Glu} and reduces the bound glutamate 1-semialdehyde in the presence of NADPH (Jahn 1992). Two monomeric glutamyl-tRNA^{Glu} reductases have been found in *Escherichia coli*, with molecular masses of 45 kDa and 85 kDa. The glutamyl-tRNA^{Glu} reductase purified from *Synechocystis* 6803 is a multimeric protein of 350 kDa composed of 39 kDa subunits. An oligomeric glutamyl-tRNA^{Glu} reductase made up of 50 kDa subunits has also been indicated in *Bacillus subtilis*. The enzyme in greening barley appears to be a dimer of two identical 60 kDa subunits and carries a 450 nm light-absorbing chromophore. Glutamyl-tRNA^{Glu} is unstable at neutral pH, which makes some of its kinetic parameters difficult to measure. Haem, at 3 μM and above, inhibits the activity of glutamyl-tRNA^{Glu} reductase purified to homogeneity from *Synechocystis*. In *E. coli* cell extracts glutamyl-tRNA^{Glu} reductase activity is inhibited by haem, but both *E. coli* enzymes are insensitive to haem when purified (cf. Jahn et al 1992). Other factors are probably required for haem to inhibit the *E. coli* reductases. Partially purified enzyme preparations from greening barley convert over 90% of the added glutamate into glutamate 1-semialdehyde in coupled assays with the ligase, tRNA^{Glu}, NADPH, ATP, GTP and Mg^{2+} (Kannangara et al 1988). Increasing concentrations of haem, up to 50 μM, progressively inhibit glutamate 1-semialdehyde formation in the coupled assay, and glutamyl-tRNA^{Glu} accumulates in the reaction mixture. Haem is therefore thought to be a feedback regulator of chlorophyll biosynthesis.

The *hemA* gene product—glutamyl-tRNA^{Glu} reductase?

The *E. coli* K12 strain, which has mutations in the *hemA* gene, requires 5-amino-laevulinate for growth. *hemA* genes have been cloned and sequenced from several bacteria and higher plants and these genes complement the *E. coli hemA*

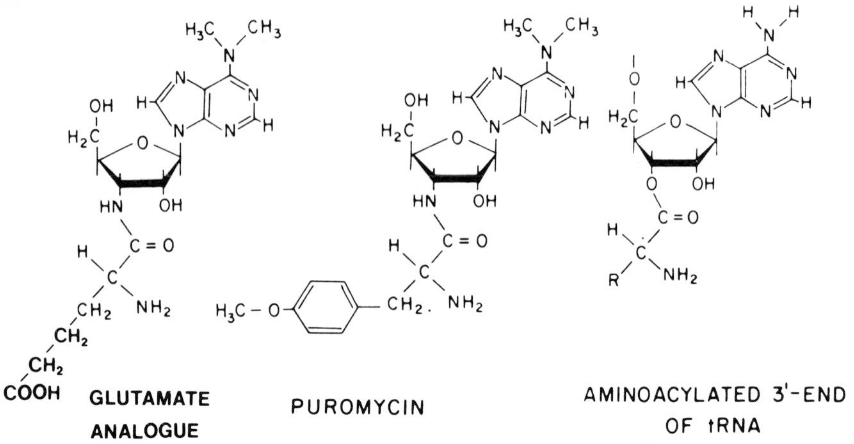

FIG. 5.   Puromycin and its glutamate analogue resemble the 3' end of an aminoacylated tRNA.

mutation. Furthermore, *E. coli* cells overproduce uroporphyrinogen when they are engineered to overexpress the *hemA* gene. Cell-free extracts from such transformed *E. coli* cells have a significantly higher capacity to convert glutamyl-tRNA^{Glu} to glutamate 1-semialdehyde than those from untransformed cells. However, it has not been possible to purify the proteins to demonstrate that reductase activity is associated with the overexpressed *hemA* protein. The *hemA* gene and the recently identified *hemM* gene (Murakami et al 1993) probably encode the proteins required for glutamyl-tRNA^{Glu} reduction.

Mechanism of conversion of glutamyl-tRNA^{Glu} to glutamate 1-semialdehyde

The conversion of glutamyl-tRNA^{Glu} to glutamate 1-semialdehyde probably occurs on a protein complex (Jahn 1992, Kannangara et al 1988). Glutamyl-tRNA^{Glu} forms stable complexes in the presence of GTP, glutamate–tRNA^{Glu} ligase and the proteins involved in glutamyl-tRNA^{Glu} reduction. A somewhat similar complex is formed during the initial step of polypeptide chain elongation, where GTP, the elongation factor EF-Tu and aminoacylated tRNA combine prior to participation of the ribosome. Puromycin, which inhibits protein synthesis and causes premature termination of polypeptide chain synthesis, and the glutamate analogue of puromycin resemble the 3' end of an aminoacylated tRNA (Fig. 5). The glutamate analogue inhibits the conversion of glutamyl-tRNA^{Glu} to glutamate 1-semialdehyde, whereas puromycin itself has no effect. On the basis of this and other observations, a likely mechanism is nucleophilic attack by the α-amino group of glutamate on the esterified carboxyl carbon of glutamyl-tRNA^{Glu} to give azeridinone propionic acid which is then reduced and hydrolysed to glutamate 1-semialdehyde (Fig. 6).

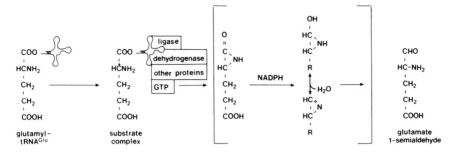

FIG. 6. The conversion of glutamyl-tRNA^{Glu} to glutamate 1-semialdehyde probably involves the formation of a substrate–enzyme complex and proceeds in several steps.

Glutamate-1-semialdehyde aminotransferase

Glutamate-1-semialdehyde aminotransferase (glutamate-1-semialdehyde 2,1-aminomutase, EC 5.4.3.8) is an abundant protein in the stroma of greening barley chloroplasts. It is encoded by nuclear DNA and has a molecular mass of 92 kDa. It is made up of two identical subunits, each of which carries a pyridoxamine 5′-phosphate cofactor. Glutamate-1-semialdehyde aminotransferase catalyses the last step of the C$_5$ pathway of 5-aminolaevulinate biosynthesis. Studies of this enzyme have been facilitated by the chemical synthesis of glutamate 1-semialdehyde.

Chemical synthesis of glutamate 1-semialdehyde

Three methods are available for synthesis of glutamate 1-semialdehyde. Two of these methods use N-carbobenzoxy-L-glutamate γ-benzyl ester as starting material; these two methods are quite cumbersome and give poor yields (5–10%) of (S)-glutamate 1-semialdehyde. The most convenient method starts from 4-amino-5-hexenoic acid (Gough et al 1989), also called γ-vinyl GABA, which is available as a mixture of R and S enantiomers from Merrell Dow (Cincinnati, USA). An aqueous solution containing equimolar amounts of 4-amino-5-hexenoic acid and HCl is treated with ozone to give the ozonide (Fig. 7), which is then reduced by passage through a Dowex 50X8 column in its H$^+$ form. This method gives pure (RS)-glutamate 1-semialdehyde in yields of up to 90%.

Glutamate 1-semialdehyde hydrochloride is stable, as a solid. We at the Carlsberg Laboratory, and Peter Jordan and co-workers at the University of London, have analysed the structure of glutamate 1-semialdehyde by NMR and infrared spectroscopy (Gough et al 1989, Hoober et al 1988, Jordan 1990). The two groups obtained identical spectra but interpreted them differently. Our interpretation is that glutamate 1-semialdehyde is a linear molecule existing as

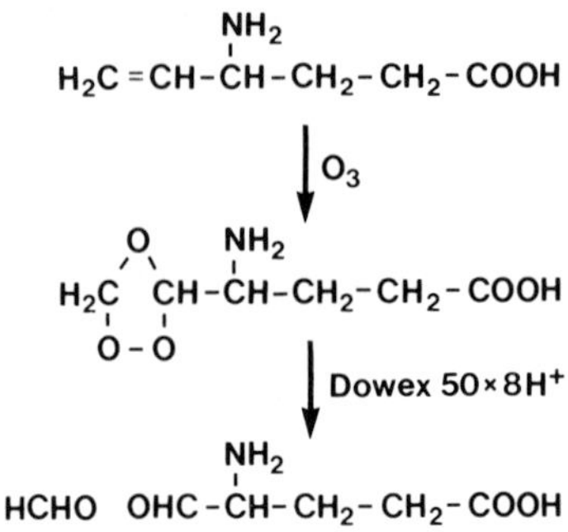

$$NH_2$$
$$H_2C = CH-CH-CH_2-CH_2-COOH$$

$$\downarrow O_3$$

$$H_2C \quad CH-CH-CH_2-CH_2-COOH$$

$$O-O$$

$$\downarrow \text{Dowex } 50 \times 8H^+$$

$$NH_2$$
$$HCHO \quad OHC-CH-CH_2-CH_2-COOH$$

FIG. 7. Chemical synthesis of glutamate 1-semialdehyde by reductive ozonolysis of γ-vinyl-γ-aminobutyric acid (γ-vinyl GABA, 4-amino-5-hexenoic acid).

a hydrate in solution. Peter Jordan's group, however, think that the spectra represent a cyclic molecule, hydroxyaminotetrahydropyranone. Stock solutions (50 to 100 mM) of glutamate 1-semialdehyde when neutralized turn yellow/brown owing to the formation of the dihydropyrazine and pyrazine (Fig. 8). A measurable amount of glutamate 1-semialdehyde is converted spontaneously to 5-aminolaevulinate at pH 8.0. Titration of (RS)-glutamate 1-semialdehyde hydrochloride with NaOH indicates the presence of a free carboxyl group (Fig. 9). Glutamate 1-semialdehyde condenses with ethyl acetoacetate or acetylacetone to give a pyrrole (Fig. 10). A vitamin B_6 enzyme converts glutamate 1-semialdehyde to 5-aminolaevulinate. These properties lead us to conclude that glutamate 1-semialdehyde has a linear structure and to discount the cyclic hydroxyaminotetrahydropyranone as a substrate for glutamate 1-semialdehyde aminotransferase.

Properties of glutamate-1-semialdehyde aminotransferase

Glutamate-1-semialdehyde aminotransferase has an exceptionally high affinity for (S)-glutamate 1-semialdehyde ($K_m = 12 \, \mu M$). The enzyme is irreversibly inhibited by gabaculine (3-amino-2,3-dihydrobenzoic acid), methylgabaculine, 2-hydroxy-3-amino-3,5-cyclohexadiene-1-carboxylic acid, 4-amino-5-hexynoic acid (acetylenic GABA), 4-amino-5-fluoropentanoic acid and glutamic acid 5-monohydroxamate (Gough et al 1992). These compounds inhibit chlorophyll synthesis in greening barley leaves causing accumulation of glutamate 1-semialdehyde. A significant reduction in the enzyme activity and a parallel reduction in chlorophyll accumulation are observed in transgenic tobacco plants carrying a DNA sequence antisense to the glutamate-1-semialdehyde aminotransferase gene.

The structural gene for glutamate-1-semialdehyde aminotransferase has been isolated and characterized from barley (Grimm 1990), E. coli, B. subtilis,

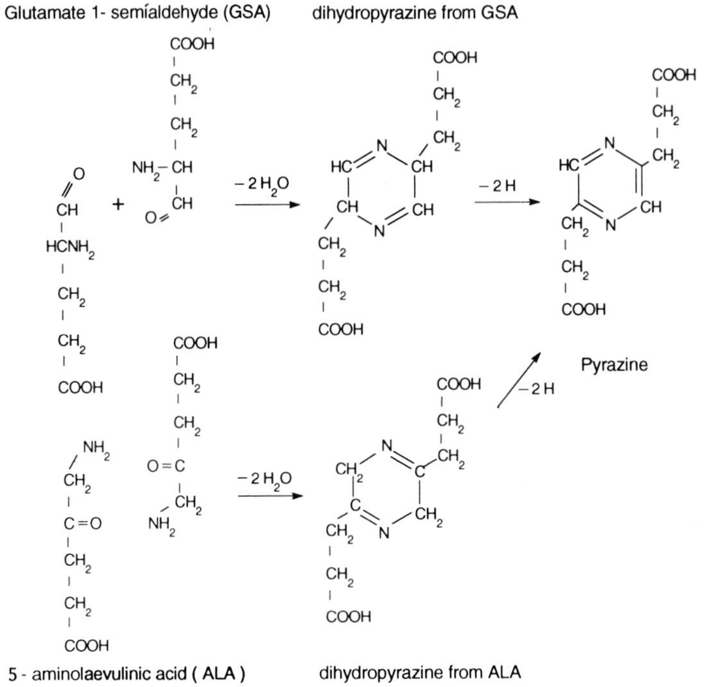

FIG. 8. Two molecules of glutamate 1-semialdehyde (GSA) or 5-aminolaevulinate (ALA) reversibly condense at alkaline pH to form yellow dihydropyrazines. Oxidation of these dihydropyrazines in air produces a stable yellow/brown-coloured pyrazine.

Salmonella, Xanthomonas (cf. Jahn et al 1992, Murakami et al 1993), *Synechococcus*, tobacco and *Arabidopsis*. In higher plants, the gene is encoded in nuclear DNA. The barley gene has two introns whereas the corresponding bacterial genes have no introns. At the nucleotide level, coding regions of barley glutamate-1-semialdehyde aminotransferase genes show 64% identity to the corresponding synechococcal gene and 40–50% identity to the *E. coli, B. subtilis* and *Salmonella* genes. In barley, and probably also in tobacco and *Arabidopsis*, glutamate-1-semialdehyde aminotransferase is made in the cytoplasm as a precursor protein of 49 540 Da and then transferred into the chloroplast. During this process a transit sequence of 34 amino acids is cleaved from the N-terminus. At the level of deduced amino acids, the primary structures of known plant and bacterial enzymes are remarkably similar, showing greater than 55% identity. *Synechococcus* glutamate-1-semialdehyde aminotransferase has 72% sequence identity with the barley enzyme. A stretch of nine amino acids has been identified as the putative pyridoxamine 5′-phosphate-binding site. There are also similarities in the distribution of the domains that are likely to have α-helices, β-sheets and hydrophobic regions.

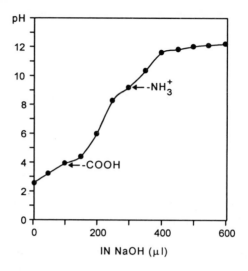

FIG. 9. Titration of (*RS*)-glutamate 1-semialdehyde hydrochloride with NaOH indicates the presence of a free carboxyl and a free amino group in the molecule.

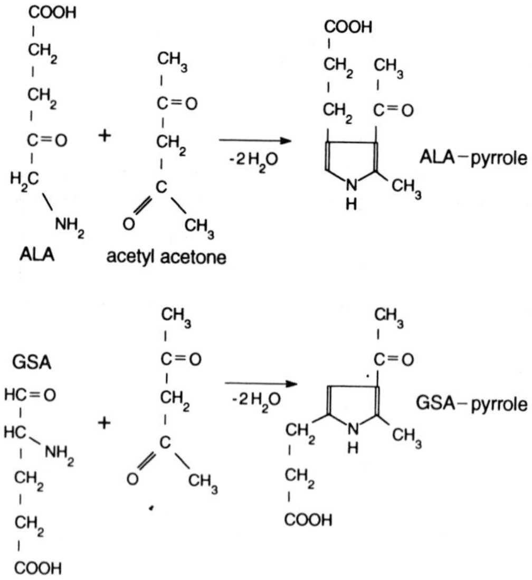

FIG. 10. Glutamate 1-semialdehyde (GSA) and 5-aminolaevulinic acid (ALA) condense with acetylacetone to give pyrroles.

Large amounts of soluble and active barley glutamate-1-semialdehyde aminotransferase have been produced in *E. coli* cells by co-expression of the barley gene (in pDS56/RBS11) with the GroEL gene (in pTG10) (Berry-Lowe et al 1992). Without overexpression of the GroEL gene, most of the barley protein produced is insoluble and inactive. The GroEL gene product probably acts as a chaperone in the folding of the barley enzyme. The glutamate-1-semialdehyde aminotransferase genes from *Synechococcus* and *E. coli* have also been overexpressed in *E. coli* cells, using pDS56/RBS11 and pET3 plasmid vectors, respectively (Grimm et al 1991, Ilag & Jahn 1992). Co-expression of the GroEL gene was not required for the expression of these bacterial genes, and large amounts of soluble and active enzyme were produced in *E. coli* cells, even though these genes show high sequence identity to the barley gene. Through this recombinant technology sufficient amounts of enzyme have been produced to elucidate its reaction mechanism by spectral and kinetic analysis (Smith et al 1991a,b).

Several mutants of *Synechococcus* 6301 have been isolated by adapting these cyanobacterial cells to grow in increasing concentrations of gabaculine (up to 100 μM) (Bull et al 1990). One of these mutants, referred to as GR 6, has a gabaculine-resistant glutamate-1-semialdehyde aminotransferase which is changed in two ways from the wild-type protein—a deletion of a tripeptide close to the N-terminus and substitution of isoleucine for methionine at position 248 (Grimm et al 1991). Two mutant genes were constructed, one with the deletion and the other with the Met→Ile substitution. Expression of the wild-type and mutant genes in *E. coli* cells followed by purification and analysis of the recombinant enzymes established that a single change, the Met-248→Ile substitution, confers gabaculine resistance to synechococcal glutamate-1-semialdehyde aminotransferase.

The sequence T^{268}TLGKIIGG^{276} of synechococcal glutamate-1-semialdehyde aminotransferase is involved in the binding of the pyridoxamine 5'-phosphate cofactor. When Lys-272 is replaced by arginine, glutamate or isoleucine, the enzyme is no longer capable of converting glutamate 1-semialdehyde to 5-aminolaevulinate (Grimm et al 1992, Ilag & Jahn 1992). This exchange is achieved by site-directed mutagenesis of the wild-type gene and subsequent expression of the mutant genes in *E. coli* cells. The mutant enzymes with arginine or isoleucine at position 272 retained bound pyridoxamine 5'-phosphate whereas the form with glutamate at position 272 had no cofactor bound, but could bind pyridoxamine 5'-phosphate when exposed to high concentrations. Lysine-272 is probably involved in the catalytic mechanism of this enzyme.

Glutamate-1-semialdehyde aminotransferase has a characteristic absorption spectrum which changes during enzyme catalysis (Berry-Lowe et al 1992, Ilag & Jahn 1992, Pugh et al 1992, Smith et al 1991a,b) (Fig. 11). It has an absorption peak at 338 nm corresponding predominantly to the pyridoxamine 5'-phosphate form of the enzyme and a peak at 418 nm corresponding to the pyridoxal

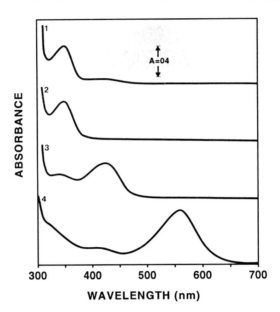

FIG. 11. Absorption spectra of synechococcal glutamate-1-semialdehyde amino-transferase and three colour-stable forms of the enzyme produced by chemical reaction with substrate and substrate analogues. (1) Purified recombinant enzyme is probably a 6:1 mixture of the pyridoxamine 5'-phosphate and pyridoxal 5'-phosphate forms. (2) The pyridoxamine 5'-phosphate form of the enzyme, generated by addition of 4,5-diaminovalerate. (3) The pyridoxal 5'-phosphate form of the enzyme, generated by addition of 4,5-dioxovalerate. (4) Inactivated enzyme with irreversibly bound acetylenic GABA (4-amino-5-hexynoic acid).

5'-phosphate form. The purified enzyme reacts with 4,5-diaminovalerate and loses its absorption peak at 418 nm. It also reacts with 4,5-dioxovalerate, which reduces the absorption at 338 nm and enhances that at 418 nm. The spectra of the treated proteins remain altered after removal of diaminovalerate or dioxovalerate by Sephadex G-50 gel filtration. If the 4,5-dioxovalerate-treated enzyme is further treated with gabaculine the peak at 418 nm disappears and that at 338 increases. The dioxovalerate-treated enzyme also reacts with acetylenic GABA, giving a new absorption peak at 560 nm. The pyridoxamine 5'-phosphate form of the enzyme is insensitive to both gabaculine and acetylenic GABA.

When (RS)-glutamate 1-semialdehyde is added to glutamate-1-semialdehyde aminotransferase, the peak at 338 nm decreases and that at 418 nm increases, with an isobestic point at 368 nm. The S enantiomer of glutamate 1-semialdehyde is used rapidly by the enzyme to give 5-aminolaevulinate (Smith et al 1991a, 1992), and addition of (S)-4,5-diaminovalerate stimulates this reaction

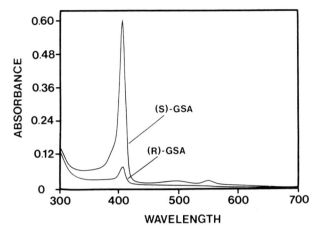

FIG. 12. Glutamate-1-semialdehyde aminotransferase utilizes (S)-glutamate 1-semi-aldehyde (GSA) rapidly and the R enantiomer slowly. An enzyme mixture containing purified aminotransferase, 5-aminolaevulinate dehydratase (porphobilinogen synthase, EC 4.2.1.24) and porphobilinogen deaminase (hydroxymethylbilane synthase, EC 4.3.1.8) was incubated with either (S)- or (R)-glutamate 1-semialdehyde. Uroporphyrinogen formed was oxidized to uroporphyrin and measured.

(Friedmann et al 1992). The enzyme also slowly converts (R)-glutamate 1-semialdehyde to 5-aminolaevulinate, probably via (R)-4,5-diaminovalerate in an anomalous reaction (Fig. 12).

Transamination of glutamate 1-semialdehyde

The transamination of glutamate 1-semialdehyde to 5-aminolaevulinate occurs via a ping-pong bi-bi mechanism, as with other pyridoxal phosphate-containing amino group-transferring enzymes. The following details have been prepared on the basis of spectral and kinetic investigations and studies on the mode of inactivation of the enzyme by gabaculine and acetylenic GABA (Fig. 13). Only the pyridoxamine 5'-phosphate form of glutamate-1-semialdehyde amino-transferase is active in catalysing the conversion of glutamate 1-semialdehyde to 5-aminolaevulinate, and the reaction proceeds via the intermediate 4,5-diaminovalerate (Hoober et al 1988, Pugh et al 1992, Smith et al 1991a). The aldehyde group of glutamate 1-semialdehyde first forms a Schiff's base with the amino group of pyridoxamine 5'-phosphate. Thereafter, a proton at the C(4') position of pyridoxamine 5'-phosphate is transferred to the C-1 of glutamate 1-semialdehyde. 4,5-Diaminovalerate and pyridoxal 5'-phosphate are thus bound to each other by a Schiff's base between the C-5 amino group of 4,5-diaminovalerate and the CHO group of pyridoxal 5'-phosphate. The ϵ amino group of Lys-272 attacks this Schiff's base to liberate 4,5-diaminovalerate, which

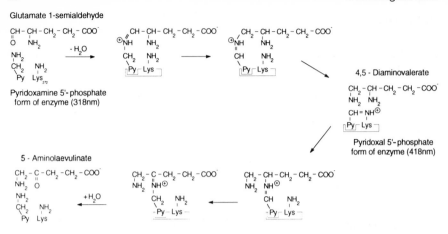

FIG. 13. A likely mechanism for the enzymic conversion of glutamate 1-semialdehyde to 5-aminolaevulinate by glutamate-1-semialdehyde aminotransferase.

immediately binds by its C-4 amino group to the CHO group of pyridoxal 5′-phosphate (a Schiff's base). The removal of a proton from C-4 of 4,5-diaminovalerate and protonation of C(4′) of pyridoxal 5′-phosphate generates 5-aminolaevulinate bound by its keto group to the amino group of pyridoxamine 5′-phosphate (a Schiff's base). Hydrolysis of this Schiff's base liberates 5-aminolaevulinate from the pyridoxamine 5′-phosphate form of the enzyme. In experiments in which the enzyme was first incubated with [14C]glutamate 1-semialdehyde and then subjected to gel filtration to remove the labelled substrate and product, no radioactivity was found associated with the enzyme. Therefore it is unlikely that a stable five-membered ring compound is formed as an intermediate during the catalysis. The participation of a short-lived intermediate with a resonance-stabilized five-membered ring is not excluded. The vitamin B_6 coenzyme and the Lys-272 of one of the subunits of glutamate-1-semialdehyde aminotransferase may participate in the complete turnover of a substrate molecule to a product molecule, but it is more likely that contributions from both subunits of the enzyme are needed to achieve the complete cycle. At present, it is not clear how the dimeric feature of glutamate-1-semialdehyde aminotransferase functions in the catalytic mechanism.

Acknowledgements

The authors thank Nina Rasmussen for the illustrations and Inge Sommer for typing the manuscript. The authors' work was supported by a grant to D. von Wettstein from the Plant Biotechnology Center of the Danish Biotechnology Programme 1991–1995.

References

Andersen RV 1992 Characterization of a barley tRNA synthetase involved in chlorophyll biosynthesis. In: Murata N (ed) Research in photosynthesis. Kluwer, The Hague (IX Int Congr Photosyn) vol 3:27–30

Avissar YJ, Ormerod JG, Beale SI 1989 Distribution of δ-aminolevulinic acid biosynthetic pathways among phototrophic bacterial groups. Arch Microbiol 151:513–519

Battersby AR 1988 Biosynthesis of the pigments of life. J Nat Prod 51:629–642

Beale SI, Castelfranco PA 1974 Biosynthesis of δ-aminolevulinic acid in higher plants. II. Formation of ^{14}C δ-aminolevulinic acid from labeled precursors in greening plant tissues. Plant Physiol (Bethesda) 53:297–303

Beale SI, Weinstein JD 1990 Tetrapyrrole metabolism in photosynthetic organisms. In: Dailey HA (ed) Biosynthesis of heme and chlorophyll. McGraw-Hill, New York, p 287–391

Beale SI, Gough SP, Granick S 1975 Biosynthesis of δ-aminolevulinic acid from the intact carbon skeleton of glutamic acid in greening barley. Proc Natl Acad Sci USA 72:2719–2723

Berry-Lowe SL, Grimm B, Smith MA, Kannangara CG 1992 Purification and characterization of glutamate 1-semialdehyde aminotransferase from barley expressed in *Escherichia coli*. Plant Physiol (Bethesda) 99:1597–1603

Bull AD, Breu V, Kannangara CG, Rogers LJ, Smith AJ 1990 Cyanobacterial glutamate 1-semialdehyde aminotransferase. Requirement of pyridoxamine phosphate. Arch Microbiol 154:56–59

Friedmann HC, Duban ME, Valasinas A, Frydman B 1992 The enantioselective participation of (S)- and (R)- diaminovaleric acids in the formation of δ-aminolevulinic acid in cyanobacteria. Biochem Biophys Res Commun 185:60–68

Gough SP, Kannangara CG, Bock K 1989 A new method for the synthesis of glutamate 1-semialdehyde. Characterization of its structure in solution by NMR. Carlsberg Res Commun 54:99–108

Gough SP, Kannangara CG, von Wettstein D 1992 Glutamate 1-semialdehyde aminotransferase as a target for herbicides. In: Böger P, Sandmann G (eds) Target assays for modern herbicides and related phytotoxic compounds. Lewis Publishers, Chelsea, MI, p 21–27

Grimm B 1990 Primary structure of a key enzyme in plant tetrapyrrole synthesis. Glutamate 1-semialdehyde aminotransferase. Proc Natl Acad Sci USA 87:4169–4173

Grimm B, Smith AJ, Kannangara CG, Smith M 1991 Gabaculine-resistant glutamate 1-semialdehyde aminotransferase of *Synechococcus*. Deletion of a tripeptide close to the NH_2 terminus and an internal amino acid substitution. J Biol Chem 266: 12495–12501

Grimm B, Smith MA, von Wettstein D 1992 The role of lys 272 in the pyridoxal 5-phosphate active site of *Synechococcus* glutamate 1-semialdehyde aminotransferase. Eur J Biochem 206:579–585

Hoober JK, Kahn A, Ash DE, Gough SP, Kannangara CG 1988 Biosynthesis of δ-aminolevulinate in greening barley leaves. IX. Structure of the substrate, mode of gabaculine inhibition and the catalytic mechanism of glutamate 1-semialdehyde aminotransferase. Carlsberg Res Commun 53:11–25

Huang D-D, Wang W-Y 1986 Chlorophyll synthesis in *Chlamydomonas* starts with the formation of glutamyl tRNA. J Biol Chem 261:13451–13455

Ilag LL, Jahn D 1992 Activity and spectroscopic properties of the *Escherichia coli* glutamate 1-semialdehyde aminotransferase and the putative active site mutant K265R. Biochemistry 31:7143–7151

Jahn D 1992 Complex formation between glutamyl tRNA synthetase and glutamyl tRNA reductase during the tRNA dependent synthesis of 5-aminolevulinic acid in *Chlamydomonas reinhardtii*. FEBS (Fed Eur Biochem Soc) Lett 314:77–80

Jahn D, Verkamp E, Soll D 1992 Glutamyl-transfer RNA: a precursor of heme and chlorophyll biosynthesis. Trends Biochem Sci 17:215–218

Jordan PM 1990 Biosynthesis of 5-aminolevulinic acid and its transformation into coproporphyrinogen in animals and bacteria. In: Dailey HA (ed) Biosynthesis of heme and chlorophyll. McGraw-Hill, New York, p 55–121

Kannangara CG 1991 Biochemistry and molecular biology of chlorophyll synthesis. In: Bogorad L, Vasil IK (eds) The photosynthetic apparatus: molecular biology and operation. Academic Press, New York (Cell Cult Somatic Cell Genet Plants 7B) p 301–329

Kannangara CG, Gough SP, Oliver RP, Rasmussen SK 1984 Biosynthesis of δ-aminolevulinate in greening barley leaves. VI. Activation of glutamate by ligation to RNA. Carlsberg Res Commun 49:417–437

Kannangara CG, Gough SP, Bruyant P, Hoober JK, Kahn A, von Wettstein D 1988 tRNA^{Glu} as a cofactor in δ-aminolevulinate biosynthesis: steps that regulate chlorophyll synthesis. Trends Biochem Sci 13:139–143

Masuda T, Komine Y, Inokuchi H, Kannangara CG, Tsuji H 1992 Sequence and expression of tRNA^{Glu} gene of cucumber chloroplast genome. Plant Physiol Biochem (Paris) 30:235–243

Meller E, Belkin S, Harel E 1975 The biosynthesis of δ-aminolevulinic acid in greening maize leaves. Phytochemistry 14:2399–2402

Murakami K, Korbsrisate S, Asahara N, Hashimoto Y, Murooka Y 1993 Cloning and characterization of the glutamate 1-semialdehyde aminomutase gene from *Xanthomonas campestris* pv. *phaseoli*. Appl Microbiol Biotechnol 38:502–506

O'Neill GP, Jahn D, Soll D 1991 Transfer RNA involvement in chlorophyll biosynthesis. In: Biswas BB, Harris JR (eds) Plant genetic engineering. Plenum, New York (Subcell Biochem 17) p 235–264

Pugh CE, Harwood JL, John RA 1992 Mechanism of glutamate semialdehyde aminotransferase. Roles of diamino-intermediates and dioxo-intermediates in the synthesis of aminolevulinate. J Biol Chem 267:1584–1588

Schön A, Krupp G, Gough S, Berry-Lowe S, Kannangara CG, Soll D 1986 The RNA required in the first step of chlorophyll biosynthesis is a chloroplast tRNA. Nature 322:281–284

Schön A, Kannangara CG, Gough SP, Soll D 1988 Protein biosynthesis in organelles requires misacylation of transfer RNA. Nature 331:187–190

Smith MA, Grimm B, Kannangara CG, von Wettstein D 1991a Spectral kinetics of glutamate 1-semialdehyde aminomutase of *Synechococcus*. Proc Natl Acad Sci USA 88:9775–9779

Smith MA, Kannangara CG, Grimm B, von Wettstein D 1991b Characterization of glutamate 1-semialdehyde aminotransferase of *Synechococcus*. Steady state kinetics. Eur J Biochem 202:749–757

Smith MA, Kannangara CG, Grimm B 1992 Glutamate 1-semialdehyde aminotransferase: anomalous enantiomeric reaction and enzyme mechanism. Biochemistry 31:11249–11254

von Wettstein D 1991 Chlorophyll biosynthesis. In: Herrmann RG, Larkins B (eds) Plant molecular biology. Plenum, New York, vol 2:449–459

Wang W-Y, Gough SP, Kannangara CG 1981 Biosynthesis of δ-aminolevulinate in greening barley leaves. IV. Isolation of three soluble enzymes required for the conversion of glutamate to δ-aminolevulinate. Carlsberg Res Commun 46:243–257

DISCUSSION

K. Smith: You are of the opinion that glutamate 1-semialdehyde exists as a hydrated linear molecule in solution. Having the cyclic form would not require chemistry different from that you described for the open form.

Kannangara: It is easier to understand how the linear molecule is used. I cannot understand how the cyclic form can convert the yellow form of the aminotransferase to a colourless form.

Arigoni: For the reaction to occur the linear and the cyclic forms must be in equilibrium with the aldehydo form, because this is the one which undergoes the reaction. In a way, it is irrelevant whether glutamate 1-semialdehyde is in the cyclic or in the linear form, but one would nevertheless like to answer this question. The two forms differ in a number of ways, not least in their molecular masses. What happens if you submit the compound to mass spectrometry?

Jordan: You get the molecular mass of the aldehyde, which is the same as that of the cyclic form, the tetrahydropyranone.

Arigoni: This is probably because you wait until the current is constant. You should insert the specimen and look at it right away.

Jordan: With electrospray mass spectrometry you get the molecular mass of 132, which is either the aldehyde or the tetrahydropyranone ring form. Of course, in the NMR spectrum you don't see an aldehyde proton. This suggests that the structure is cyclic under the acidic conditions used.

Arigoni: I am still convinced that you could track the hydrate form by normal mass spectrometric techniques. We have done this with victorin C, which is the hydrate of an ester of glyoxylic acid (Wolpert et al 1985). Initially in the spectrum you see the peak of the hydrate, then after a while you can detect a metastable peak corresponding to the transition into the aldehydo form.

K. Smith: Dr Kannangara, have you tried elemental combustion analysis?

Kannangara: There was no evidence for the hydrated molecule in elemental combustion analysis of freeze-dried glutamate 1-semialdehyde.

Arigoni: Does glutamate 1-semialdehyde crystallize?

Kannangara: It can be crystallized, but the crystals are not good enough for X-ray analysis.

Rebeiz: Three-dimensional modelling would be an ideal tool for structure–function analysis. With the powerful computers now available, it would be easy to determine the physico-chemical parameters of hydrated and unhydrated structures and relate them to function.

Castelfranco: The real question is not what the molecule is in itself, but what form it takes when it is on the enzyme. In the last step the aldehyde should react in the linear form. I too cannot visualize the cyclic form going through that third reduction step.

Jordan: In glucose biochemistry no one really worries about whether glucose is in the hydrated aldehyde or the cyclic hemiacetal form. The enzymes obviously use the free aldehyde form, but the molecule may well bind initially in the cyclic form which opens on the enzyme. It does not worry me particularly which form binds. What worries me is the presence of the free reactive aldehyde in solution.

Arigoni: This question should worry you. If you want to design inhibitors you need to know whether the substrate is linear or cyclic.

Scott: It should be possible to devise a simple means of distinction. With $H_2^{18}O$, the hydrate at C-5 will have a double isotope shift on the ^{13}C signal, whereas the cyclic form, depending on how you do the experiment (i.e., on whether you get the pH right), will experience a single isotope shift. We have found this possible in other cases (Ortiz et al 1991).

Arigoni: There is a cheaper way, to look at the NMR spectrum of the compound in methanol solution. Only the open form would convert spontaneously into a methyl hemiacetal (Wolpert et al 1985).

Leeper: We've got all the evidence we need already. The pH titration (Fig. 9) showed an ionization about pH 5, which must be due to a carboxylic acid, and there is a carboxylic acid only in the open form. At pH 7, above pH 5, it would be in the open form. It may be cyclic below pH 5, but that's not the physiological pH.

Kräutler: You could determine the isoelectric point of glutamate 1-semialdehyde. If it has a cyclic structure it should behave essentially as a basic amino acid, whereas if it has an open structure it should behave like a neutral amino acid.

Arigoni: That is not quite true. You must correct the experimental pK value with the constant for the equilibrium between the open and closed forms.

Kräutler: By measuring the isoelectric point you would get information on that.

Castelfranco: Dr Kannangara, in your opening remarks you said that 5-aminolaevulinate (ALA) is made by two different pathways for the biosynthesis of haem and chlorophyll. What you mean is that plants and certain bacteria have the C_5 pathway, whether or not they have chlorophyll, whereas other organisms such as yeast have the Shemin pathway. The distinction is not concerned with the product of the synthetic pathway but with the systematics of the organism.

You suggested that regulation of ALA synthesis by light has something to do with the 5-methylaminomethyl-2-thiouridine of the tRNA[Glu] anticodon. Have action spectra been done? The two things we usually hear about are a protochlorophyllide feedback mechanism and phytochrome regulation of some sort.

Kannangara: It is difficult to do an action spectrum for tRNA[Glu], as has been done for phytochrome, because UV light does not stimulate ALA synthesis in plants.

Thauer: The regulation of ALA synthesis via oxidative reduction of the thiouridine in the tRNA[Glu] anticodon is an intriguing idea, but the same tRNA[Glu] is present in enterobacteria, which don't interact with light and in which the regulation of ALA biosynthesis is completely different. I don't think regulation at the level of the tRNA is likely, because the tRNA concentrations in the cells are relatively high.

Kannangara: You are assuming that the regulation in bacteria and in different plants is similar, but even two different plants can be regulated in different ways. For example, recent studies in our laboratory show that the glutamate–tRNA^{Glu} ligase of tobacco is different from that of barley; the barley chloroplast ligase is a dimer whereas that of tobacco chloroplasts is a monomer. Also, glutamyl-tRNA^{Glu} reductases from different plants are different.

Thauer: Regulation at the level of the ligase does not make sense, because the tRNA^{Glu} generated is also required for protein biosynthesis. Regulation of ALA synthesis needs to be at the reductase level, not at the ligase level.

Spencer: If the hypermodified mam^{5}S^{2}UψC-tRNA is in the oxidized form in the complex, is the reductase prevented from carrying out the reduction? Can the hypermodified mam^{5}S$^2\psi$C-tRNA be oxidized after the complex has been made?

Kannangara: Glutamate falls off when glutamyl-tRNA^{Glu} is oxidized and we cannot ligate glutamate to oxidized tRNA^{Glu}.

A. Smith: There is evidence that the dehydrogenase works in the absence of the ligase. Isn't your idea of a complex difficult to reconcile with that?

Kannangara: I'm not sure that the reductase will continue to work in the absence of the ligase. The reductase utilizes tRNA^{Glu} in many cycles when ligase, ATP and NADPH are present. One may also get some non-enzymic conversion as with the aminotransferase, with the colourless form of the glutamate-1-semialdehyde aminotransferase converting dioxovalerate to ALA in a one-step reaction.

Beale: The concentration of reduced glutathione in chloroplasts is in the mM range, which would probably keep the thiol group on the tRNA^{Glu} in a reduced form under all physiological conditions. That is not compatible with oxidation–reduction of this group being a regulatory mechanism.

The multi-enzyme ALA synthesis pathway is remarkably widespread. The ALA synthase pathway is restricted to non-phototrophic eukaryotes and one small group of bacteria, the α-purples. It's intriguing that these bacteria are postulated for other reasons to be the progenitor of mitochondria. It's also remarkable that the people who worked on ALA synthase in *Rhodobacter sphaeroides* were working on the only group of bacteria to have this enzyme.

Arigoni: Franck reported some time ago that glutamate fed to duck blood is incorporated into haem (Franck et al 1984).

Beale: We tried that in 1972, when we used the avian (turkey) blood system as a positive control for ALA synthase. We did not detect conversion of glutamate to ALA in blood cells.

Arigoni: What technique were you using?

Beale: ^{14}C labelling. Only in the plant systems was there labelling.

Arigoni: Have Franck's results been reproduced?

Beale: Not as far as I know. There have also been reports of ALA synthase in plant cells which have not been reproduced. Perhaps if enough labs are looking for something, some of them will find it whether it exists or not!

Arigoni: Franck et al (1984) reported fairly low incorporation values. It should be possible to take a fresh look at this with ^{13}C NMR, to see whether the C_5 pathway operates in vertebrate systems.

Beale: The earlier experiments to discern whether plants had one or two pathways used techniques that were not powerful enough to give a definitive answer. The internal carbons of glutamate can appear in ALA regardless of which pathway is taken. It's only incorporation of the C-1 of glutamate that indicates the C_5 pathway has been used.

Thauer: Have there been studies on ALA synthesis in root tissue culture? One might expect root mitochondria to have the Shemin pathway.

A. Smith: Danielle Werck-Reichardt (personal communication) has found no ALA synthase activity in mitochondria isolated from a special organ of *Arum*, the spadix.

Beale: We have looked specifically at the incorporation of precursors into mitochondrial cytochrome oxidase haem *a* from several plant systems, including etiolated maize epicotyls, the one system in which we could get sufficient incorporation of exogenous label. All haems, including mitochondrial haem *a*, appeared to be made from glutamate and not glycine.

Hädener: Has the exchange of an amino group from the aminotransferase to the glutamate 1-semialdehyde been proven by labelling, ^{15}N labelling, for example?

Kannangara: I don't know.

Arigoni: This cannot be done in a normal experiment because the coenzyme is acting as a catalyst.

Hädener: You could use enzyme in substrate quantities.

Beale: In a similar experiment we showed that the amino group on the ALA produced is not derived from the same glutamate 1-semialdehyde molecule as the carbon atoms (Mayer et al 1993).

Leeper: Is that exclusive, or could there be, say, 10% provided intramolecularly and 90% intermolecularly?

Beale: The data are compatible with 100% intermolecular transfer, but of course there could be a small amount of intramolecular transfer, less than 10%.

Arigoni: Dr Kannangara, you assigned a specific role to the α-amino group in the reduction of the ester group of glutamyl-tRNA^{Glu}. Is there any other example of an ester group being reduced enzymically by NADPH? This works well with thioesters, but I am not aware of any esters that are reduced in this way.

Kannangara: The bond between glutamate and tRNA^{Glu} is not a real ester bond, but an active ester which is unstable, and glutamate falls off quickly above pH 7.0.

Akhtar: The ΔG of hydrolysis of the ester bond in aminoacyl-tRNA is about 2 kcal more negative than that of ordinary esters (Berg et al 1961, Wolfenden 1963).

Arigoni: My question remains unanswered.

Eschenmoser: Dr Kannangara, in Fig. 6 why did you show the α-lactam to be an intermediate?

Kannangara: For several reasons. You can isolate a glutamyl-tRNA^{Glu}–reductase–GTP complex with a molecular mass of about 1.2×10^6 Da by Sephacryl S-300 gel filtration. This complex can act as precursor to produce glutamate 1-semialdehyde when NADPH is provided.

Eschenmoser: Professor Arigoni's last question was an important one. In non-enzymic chemistry, sodium borohydride will not reduce esters, but it will sometimes reduce esters which are α-substituted by electronegative substituents. Here, we have to remember, the α-substituent is an ammonium ion and not an amino group. The α-ammonium ester is an activated ester, and we should simply take note of Dr Kannangara's finding that NADPH can reduce such activated esters.

Arigoni: I would express this differently. If sodium borohydride can do it, it is not surprising that NADPH should work as well.

References

Berg P, Bergmann FH, Ofengaud EJ, Dieckmann M 1961 The enzymic synthesis of aminoacyl derivatives of RNA. J Biol Chem 236:1726–1734

Franck B, Bruse M, Dahmer J 1984 Häm-Biosynthese aus L-Glutaminsäure. Angew Chem 96:1000–1001

Mayer SM, Gawlita E, Avissar YJ, Anderson V, Beale SI 1993 Intermolecular nitrogen transfer in the enzymatic conversion of glutamate to δ-aminolevulinic acid by extracts of *Chlorella vulgaris*. Plant Physiol 101:1029–1038

Ortiz C, Tellier C, Williams HJ, Stolowich NJ, Scott AI 1991 Diastereotopic covalent binding of the natural inhibitor leupeptin to trypsin: detection of two interconverting hemiacetals by solution and solid-state NMR spectroscopy. Biochemistry 30:10026–10034

Wolfenden R 1963 The mechanism of hydrolysis of aminoacyl RNA. Biochemistry 2:1090–1092

Wolpert TJ, Macko V, Acklin W et al 1985 Structure of victorin C, the major host-selective toxin from *Cochliobolus victoriae*. Experientia (Basel) 41:1524–1529

5-Aminolaevulinic acid synthase and uroporphyrinogen methylase: two key control enzymes of tetrapyrrole biosynthesis and modification

Martin J. Warren, Edward Bolt and Sarah C. Woodcock

School of Biological Sciences, Queen Mary and Westfield College, University of London, Mile End Road, London E1 4NS, UK

Abstract. Two enzymes which play an important role in regulation and flux control through the tetapyrrole biosynthetic pathway are considered. The *Rhodobacter sphaeroides* 5-aminolaevulinic acid synthase isoenzymes are discussed and the progress being made on their recombinant expression and isolation is reported. The *Escherichia coli* uroporphyrinogen methylase, which is encoded by the *cysG* gene, is also examined. In this case evidence is provided which demonstrates that the gene product is responsible for the complete synthesis of sirohaem from uroporphyrinogen III. The enzyme is thus capable of performing two *S*-adenosylmethionine-dependent methylation reactions, an $NADP^+$-dependent dehydrogenation and iron chelation. The uroporphyrinogen methylase is thus a small multifunctional enzyme.

1994 The biosynthesis of the tetrapyrrole pigments. Wiley, Chichester (Ciba Foundation Symposium 180) p 26–49

The enzymes 5-aminolaevulinic acid (ALA) synthase (EC 2.3.1.37) and uroporphyrinogen methylase not only play respective roles in the biosynthesis of ALA and sirohaem but also play key roles in regulating flux through the tetrapyrrolic biosynthetic pathway. When these proteins are overexpressed in *Escherichia coli* they give rise to fluorescent recombinant strains which are easily identifiable under UV light owing to the build up of porphyrinoid material. Here we shall be discussing some of our current research into these enzymes. In the case of ALA synthase we shall look at the progress being made now that we can produce the two isoenzymes from the photosynthetic bacterium *Rhodobacter sphaeroides* in a recombinant expression system. With the *E. coli* uroporphyrinogen methylase, which is encoded by the *cysG* gene, we shall describe how a combination of gene dissection, comparative protein studies and enzymology have demonstrated that this particular protein is capable of synthesizing sirohaem from uroporphyrinogen III.

ALA synthase and the C_4 pathway

The biosynthesis of ALA occurs by one of two routes, either by transformation of the intact carbon skeleton of glutamate (the C_5 pathway) or via the condensation of succinyl-CoA and glycine (the Shemin or C_4 pathway). In evolutionary terms the C_5 pathway probably represents the primordial route of synthesis of ALA, but the succinyl-CoA route was elucidated experimentally many years before the C_5 pathway was discovered. Here we shall be discussing some recent work on the enzyme responsible for the condensation of glycine and succinyl-CoA into ALA, ALA synthase.

The biosynthesis of ALA acid via glycine and succinyl-CoA occurs in photosynthetic bacteria and most eukaryotes, the major exceptions in this latter class being higher plants and algae in which the C_5 pathway is prevalent. The enzyme was first described more than 35 years ago (Gibson et al 1958, Kikuchi et al 1958) and has now been isolated from many sources including *R. sphaeroides* (Warnick & Burnham 1971), rat and chicken liver (Ohashi & Kikuchi 1979, Borthwick et al 1983) and yeast (Volland & Felix 1984), although the enzyme would appear to be quite unstable in most of these cases and has been isolated in only modest amounts. The nucleotide sequence encoding the enzyme has not escaped the interest of molecular biologists and the sequence of either the gene or cDNA has been published from several sources including *Bradyrhizobium japonicum* (McClung et al 1987), *R. sphaeroides* (Neidle & Kaplan 1993), yeast (Urban-Grimal et al 1986), and human and chicken liver (Bawden et al 1987, Maguire et al 1986). The enzyme's role in the control of flux through the pathway has attracted considerable attention. It has been known for some time that the biosynthesis of ALA is the rate-limiting step in tetrapyrrole biosynthesis. The activity of ALA synthase is regulated by feedback inhibition, its production is regulated at the nucleic acid level and, in eukaryotes, the transport of the enzyme precursor into the mitochondria is also controlled (Jordan 1991).

The mechanism

The reaction stoichiometry is shown in Fig. 1; ALA is produced along with CoASH and CO_2. Pyridoxal 5'-phosphate is required as a coenzyme and the reaction stereochemistry has been elucidated. The two substrates, glycine and succinyl-CoA, bind to the enzyme prior to the release of CoASH, CO_2 and ALA. The enzyme reaction proceeds with overall inversion of configuration, with stereospecific loss of the pro-R hydrogen of glycine. This has allowed the mechanism to be worked out in detail, as summarized in Fig. 1, and has granted the mechanism a three-dimensional character because the stereochemistry provides information concerning the side of the substrate molecule from which protons are removed and transferred. However, the mechanism lacks a protein scaffold: how does the reaction relate to the structure of the enzyme, and what is the nature of the catalytic groups involved in the catalytic cycle?

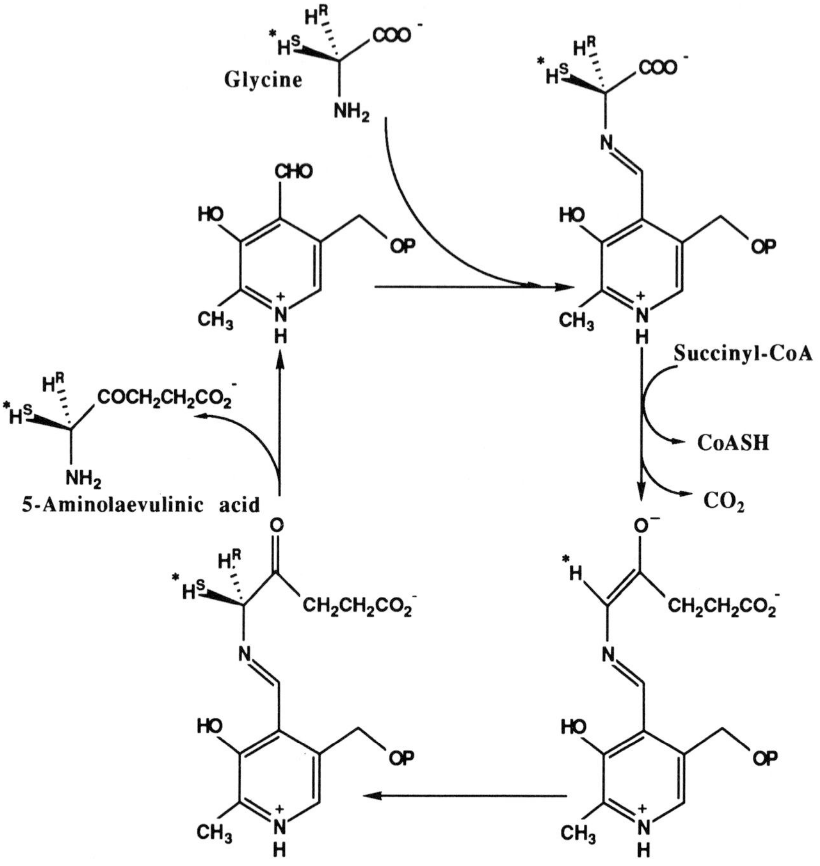

FIG. 1.   The reaction mechanism of 5-aminolaevulinic acid synthase (EC 2.3.1.37), as proposed by Jordan (1991). The catalytic cycle involves the formation of a Schiff's base between the glycine and the enzyme-associated pyridoxal phosphate coenzyme. Abstraction of the pro-*R* hydrogen as a proton generates a carbanion species which readily reacts with the second substrate, succinyl CoA. The ensuing chemistry leads to release of CoASH and CO_2 and finally 5-aminolaevulinic acid.

To try to answer some of these questions we have turned our attention to recombinant molecular biology to aid us in our quest for a complete catalytic portrait.

ALA synthases from *R. sphaeroides*

We have been studying the ALA synthase system from *R. sphaeroides* for several reasons. Firstly, the enzyme from *R. sphaeroides* has been well characterized already, and most of the mechanistic and stereochemical work has been done on this enzyme. Secondly, and interestingly, the isolation and sequencing of the ALA synthase gene from *R. sphaeroides* revealed that there are two genes, each encoding a different isoenzyme (Tai et al 1988, Neidle & Kaplan 1993). The two genes, termed the *hemA* and the *hemT* genes, are located on different chromosomes (Tai et al 1988). The genome of *R. sphaeroides* is made up of a large and a small chromosome and was the first prokaryotic genome shown to consist of more than a single chromosome (Suwanto & Kaplan 1989). The *hemA* gene is located on the large chromosome while the *hemT* gene is located on the smaller one. The two genes appear to be under different control systems. Two transcriptional products, possibly a single and a polycistronic message, are associated with the *hemA* gene, which is regulated by anaerobic metabolic activity (Neidle & Kaplan 1993); under photosynthetic conditions levels of the *hemA* transcript are three-fold higher than under aerobic conditions. The *hemT* gene does not seem to be expressed under any physiological conditions so far tested. Why *R. sphaeroides* requires two isoenzymes for ALA synthase is therefore unclear. However, the existence of the two genes and their recombinant expression does provide us with the opportunity to compare the kinetic activities of the two enzymes.

The *hemA* and *hemT* genes encode isoenzymes of 406 amino acids which have about 50% amino acid sequence identity. The *hemA* gene of *R. sphaeroides* is more like the *hemA* gene of *Bradrhizobium japonicum* than it is like the *hemT* gene (Neidle & Kaplan 1993). With the close cooperation of Professor S. Kaplan (Houston) and Professor P. M. Jordan (London), we have been studying the two isoenzymes produced as recombinant proteins in *E. coli*. The *hemA* gene is expressed reasonably well in a pUC plasmid to give a level of about 1 mg enzyme per litre of culture, but the *hemT* gene is expressed only at very low levels owing, in part, to the presence of a weak ribosome-binding site. To overcome this deficiency we have now modified the upstream sequence of the *hemT* gene and incorporated a much stronger Shine–Dalgarno sequence. This has had the desired effect of increasing the expression of ALA synthase in transformed *E. coli*.

The HemA isoenzyme has been purified essentially to homogeneity by a combination of different chromatographic steps. The purified protein has a molecular mass on PAGE of 46 000 Da (Fig. 2), in close aggreement to that predicted from the gene sequence. This approach allows us to compare the activities of the two isoenzymes in full and will, we hope, enable us to put more detail on the mechanism proposed for the enzyme as well as to investigate why two isoenzymes exist.

Uroporphyrinogen methylase

The *cysG* gene was named thus because of its involvement in the biosynthesis of cysteine. *cysG* mutants were unable to make a competent sulphite reductase,

Protein
standards
$(10^{-3} \times M_r)$

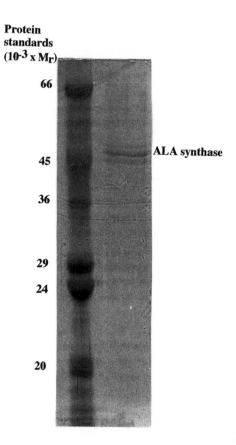

FIG. 2. SDS–polyacrylamide gel of purified 5-aminolaevulinic acid (ALA) synthase from a recombinant *Escherichia coli* strain harbouring the *Rhodobacter sphaeroides hemA* plasmid. The enzyme was purified by a modification of the procedure described by Jordan & Laghai-Newton (1986).

suggesting that the gene may be involved in the biosynthesis of sirohaem, the prosthetic group of both sulphite and nitrite reductases (Cole et al 1980). Sirohaem is a modified tetrapyrrole belonging to the same family of natural products as haem, chlorophyll, factor F_{430} and the corrin macrocycle of vitamin B_{12} (Warren & Scott 1990). In common with these prosthetic groups, sirohaem's structure is based on uroporphyrinogen III. Three steps are required to convert uroporphyrinogen III into sirohaem: *S*-adenosyl-L-methionine (SAM)-dependent methylation at positions 2 and 7 (UPAC–IUB JCBN nomenclature) to produce dihydrosirohydrochlorin (precorrin-2); oxidation to produce sirohydrochlorin; and iron chelation to yield sirohaem (Fig. 3) (Warren & Scott

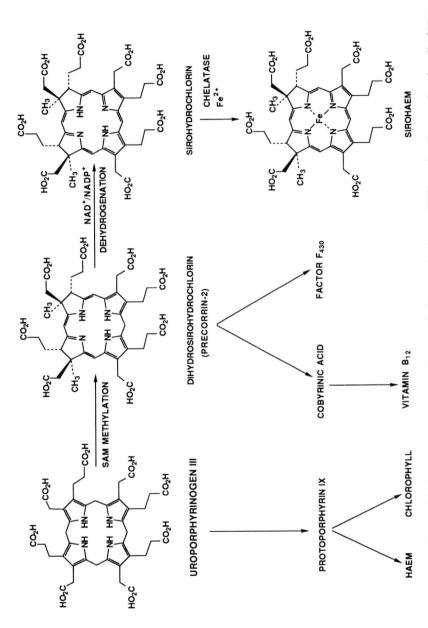

FIG. 3. The biosynthesis of sirohaem from uroporphyrinogen III. Uroporphyrinogen III is used as the template for all of Nature's tetrapyrrolic prosthetic groups. The biosynthesis of sirohaem requires methylation of uroporphyrinogen III, oxidation and chelation.

1990). On the basis of the phenotype of the *cysG* mutants it was proposed that the product of this gene was involved in the SAM-dependent methylation (Cole et al 1980).

The sub-cloning and overexpression of the *cysG* gene enabled us to test this hypothesis (Warren et al 1990a). Incubation of the purified protein with uroporphyrinogen III and SAM clearly demonstrated that the protein encoded by *cysG*, CysG, was capable of methylating uroporphyrinogen III at positions 2 and 7. The product under rigorous anaerobic conditions was dihydrosirohydrochlorin (Warren et al 1990a,b) which is also referred to as precorrin-2 because of its intermediary role in the biosynthesis of vitamin B_{12}. Dihydrosirohydrochlorin is remarkably unstable, oxidizing rapidly if conditions are not kept rigorously anaerobic. A detailed account of the absolute structure of this compound as derived by NMR spectroscopy and its importance in the biosynthesis of vitamin B_{12} is given elsewhere (Warren et al 1992). How, then, is dihydrosirohydrochlorin converted into sirohaem? It had been thought that two other enzymes must be involved, one for the oxidation and one for the iron chelation. However, no mutants lacking in such activities have been found. We can now demonstrate that CysG alone is responsible for the biosynthesis of sirohaem from uroporphyrinogen III, as explained in the following sections.

Comparative protein studies

The sequencing of *E. coli cysG* revealed that it encoded a protein consisting of 457 amino acids with a predicted molecular mass of about 50 000 Da (Peakman et al 1990). Other uroporphyrinogen methylases from *Pseudomonas denitrificans* (encoded by *cobA*; Crouzet et al 1990a), *Bacillus megaterium* (*corA*; Robin et al 1991), *Methanobacterium ivanovii* (*cobA*; Blanche et al 1991) and *Anacystis nidulans R2* (*cobA*; Dr Alison Smith, Cambridge, personal communication) have now also been sequenced. These nucleotide sequences encode proteins of predicted molecular masses of around 27 000 Da, about half the size of CysG. Protein comparison studies reveal that there is similarity between these sequences and the C-terminal portion of CysG, between amino acids 203 and 457. This similarity was particularly apparent in three areas of the C-terminus of CysG, as shown in Fig. 4a. All these proteins catalyse the SAM-dependent methylation of uroporphyrinogen III as observed with CysG. The three conserved areas may be the SAM- and uroporphyrinogen-binding sites and a catalytic site.

Dissection of the *cysG* gene: evidence for a multifunctional protein

What is the function of the first 202 N-terminal amino acids of CysG? Are they somehow involved in the SAM-dependent methylation reaction, perhaps acting as a regulatory element, or is the N-terminus of the protein involved in the

(a)

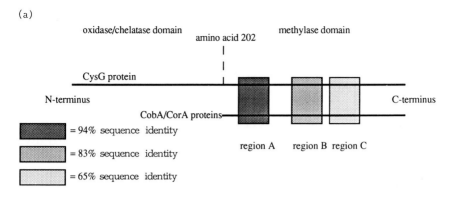

(b)

Uroporphyrinogen methyltransferase

Escherichia coli CysG	V	V	L	V	G	A	G	P	G	D	A	G	L	L	T	L	K
Pseudomonas denitrificans CobA	V	W	L	V	G	A	G	P	G	D	P	G	L	L	T	L	H
Bacillus megaterium CorA	V	Y	L	V	G	A	G	P	G	D	P	D	L	I	T	L	K
Methanobacterium ivanovii CobA	V	Y	L	V	G	A	G	P	G	D	P	E	L	I	T	L	K
Anacystis nidulans CobA	V	Y	L	V	G	A	G	P	G	D	P	E	Y	L	T	L	Q

B$_{12}$ methyltransferases

P. denitrificans C-20 methylase	L	I	G	V	G	T	G	P	G	D	P	E	L	L	T	V	K
P. denitrificans CobM (C-11?)	V	H	F	I	G	A	G	P	G	A	A	D	L	I	T	V	R

Other methylases

Rat guanidinoacetate methylase	V	L	E	V	G	F	G	M	A	I	A	A	S	R	V	Q	Q
Saccharopolyspora erythraea EryG	V	L	D	V	G	F	G	L	G	A	Q	D	F	F	W	R	K
Human D-Asp/L-isoAsp	A	L	D	V	G	S	G	S	G	I	L	T	A	C	F	A	R

FIG. 4. (a) Similarity between the complete *E. coli* CysG protein and other uroporphyrinogen methylases encoded by genes *cobA* and *corA*. The similarity is in the C-terminus of CysG, largely concentrated in three areas. Area A represents the putative *S*-adenosyl-L-methionine (SAM)-binding site (expanded in b) and regions B and C may represent the uroporphyrinogen-binding site and a catalytic site. (b) Expansion of region A, amino acids 218–234, containing the likely SAM-binding site. The sequences of SAM-binding sites from other vitamin B$_{12}$ methyltransferases (sequences from Crouzet et al 1990b) and a selection of other SAM-dependent methyltransferases are given for comparison (rat guanidoacetate methyltransferase taken from Takata & Fujioka 1992; EryG [erythromycin *O*-methyltransferase] sequence taken from Haydock et al 1991; human aspartyl/isoaspartyl methyltransferase [protein-L-isoaspartate (D-aspartate) *O*-methyltransferase, EC 2.1.1.77] sequence taken from Ingrosso et al 1989).

conversion of dihydrosirohydrochlorin into sirohaem? To discriminate between these possibilities we dissected the *cysG* gene in half, to produce one half consisting of the nucleotide sequence encoding amino acids 1 to 202 and the other that encoding amino acids 203 to 437, and cloned the two independently. The C-terminal portion of the protein was then expressed to high levels, producing highly fluorescent bacteria. The accumulated porphyrinoid material within the bacteria transformed with this C-terminal truncated *cysG* gene contained sirohydrochlorin. This demonstrated immediately not only that the C-terminus of CysG was catalytically active on its own, but also that it was indeed involved in the methylation reaction, as predicted from comparison with other methyltransferases (Fig. 4). The truncated protein was purified and found to have a molecular mass of about 29 000 Da, in close agreement with that predicted from the cloned partial gene (28 000 Da) (Fig. 5). The properties of purified truncated CysG were apparently very similar to those of the complete protein. The truncated protein has a high affinity for SAM (data not shown), and, under the appropriate conditions, will convert uroporphyrinogen III into dihydrosirohydrochlorin. However, the truncated C-terminal protein is unable to complement *E. coli cysG⁻* strains, indicating that the complete protein is required to carry out the transformations beyond the two methylations. An

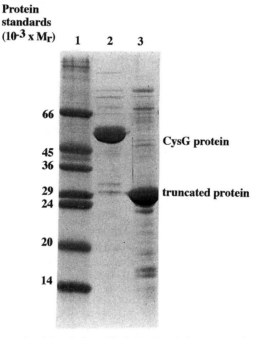

FIG. 5.   SDS–polyacrylamide gel of purified CysG and the truncated C-terminal portion of CysG which is involved in SAM-dependent methylation of uroporphyrinogen III. Lane 1, molecular mass standards; lane 2, purified CysG; lane 3, purified truncated protein.

idea of the further reactions which can be catalysed by the complete protein can be gained from the strains used to overproduce the full and truncated proteins. The strain harbouring *cysG* accumulates large quantities of sirohaem, whereas the strain harbouring truncated C-terminal *cysG* accumulates sirohydrochlorin, as well as 'overmethylated' (C-12-methylated) trimethyl-pyrrocorphin (Warren et al 1990b, Scott et al 1990). Although these findings do not enable us to discern whether truncated CysG is deficient in an oxidase activity, because dihydrosirohydrochlorin is extremely prone to oxidation, they do suggest that truncated CysG is deficient in ferrochelatase activity, which must therefore be mediated from within the N-terminus of the complete CysG protein.

The SAM-binding site and the discovery of a putative NADP⁺-binding site

SAM-binding sites are quite often identifiable by their high glycine content, as are other nucleotide coenzyme-binding sites such as those for NAD⁺, NADP⁺ and FAD. The putative SAM-binding site in CysG, the sequence GXGXG, lies between amino acids 221 and 227. This sequence is conserved between all identified uroporphyrinogen methylases as well as in many other methyltransferases (Fig. 4b) (Haydock et al 1991). While comparing the CysG SAM-binding site with other nucleotide-binding structures, we noticed that the N-terminal portion of CysG contained another glycine-rich sequence. This glycine-rich sequence was found in the centre of a predicted βαβ fold, a structure which is found in many nicotinamide adenine dinucleotide-dependent enzymes (Hanukoglu & Gutfinger 1989). A comparison of consensus NAD⁺ and NADP⁺ sites with the putative binding site on the N-terminus of CysG was therefore made (Fig. 6). The putative site showed strong similarity to the fingerprint region of a typical NADP⁺ site (Hanukoglu & Gutfinger 1989, Scrutton et al 1990). It occurred to us that an NADP⁺ coenzyme could be

Binding site	sequence
consensus NADP	X X X X X G X G X X A X X X A X X X X X X G X X X X X X
consensus NAD	X X X X X G X G X X G X X X G X X X X X X G X X X X X X
consensus FAD	X X X X X G X G X X G X X X A X X X X X X G X X X X X X
cysG sequence (14–41)	D C L I V G G G D V A E P K A R L L L D A G A R L T V N
predicted secondary structure	T β β β β β T T T · α α α α α α α α α α α α β β β β β β

FIG. 6. Comparison of the putative nicotinamide adenine dinucleotide-binding site found on the N-terminus of the CysG protein with consensus binding sites of other nucleotide coenzymes. The pattern found on CysG closely resembles the pattern observed in the consensus NADP⁺ site. Conserved hydrophobic amino acids are marked with a bold X.

utilized in the enzymic dehydrogenation of dihydrosirohydrochlorin into sirohydrochlorin. To test whether CysG uses NADP$^+$ in the biosynthesis of sirohydrochlorin, we incubated NADP$^+$ or NADPH with CysG. The results clearly demonstrated that NADP$^+$ promoted the efficient conversion of dihydrosirohydrochlorin into sirohydrochlorin (Fig. 7); only in the presence of NADP$^+$ did the spectrum of the incubation mixture change from that of a dipyrrocorphin (dihydrosirohydrochlorin) to that of an isobacteriochlorin (sirohydrochlorin) with a new absorption maximum at 378 nm. When truncated CysG was incubated under similar conditions with NADP$^+$ no sirohydrochlorin was formed. Because the truncated protein does not contain the putative nucleotide-binding site, this result supports the idea that this enzymic dehydrogenation activity is mediated by the N-terminal region of CysG. The NADP$^+$ coenzyme thus appears to be required in catalysis of the oxidation of dihydrosirohydrochlorin by transfer of a hydride and a proton from the macrocyclic ring. However, NADP$^+$ seems not to be the preferred substrate for the reaction; NAD$^+$ functions considerably faster (A. I. Scott & P. Spencer, personal communication).

The evidence that we have presented here demonstrates that CysG is a multifunctional enzyme which is capable of catalysing three separate reactions. Firstly, it catalyses two SAM-dependent *C*-methylations at positions 2 and 7 of uroporphyrinogen III, to produce dihydrosirohydrochlorin. Secondly, using a nicotinamide adenine dinucleotide coenzyme which binds to the N-terminal

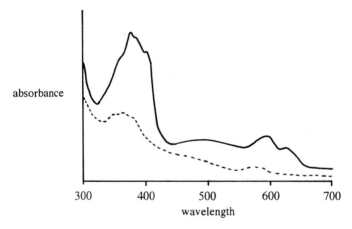

FIG. 7. Absorption spectrum of the reaction carried out by CysG in the presence of uroporphyrinogen III and SAM (----) and the spectrum after the addition of NADP$^+$ (——). The spectrum of the reaction in the absence of NADP$^+$ is characteristic of that of a dipyrrocorphin, as would be expected during the formation of dihydrosirohydrochlorin (Warren et al 1990b), whereas that in the presence of NADP$^+$ with the absorption maximum at 378 nm is characteristic of an isobacteriochlorin, as expected during the formation of sirohydrochlorin (Warren et al 1990a).

half of the protein, it catalyses the dehydrogenation of dihydrosirohydrochlorin into sirohydrochlorin. Thirdly, it inserts iron into the centre of sirohydrochlorin to produce sirohaem. This final catalytic function also resides within the N-terminal portion of the protein. With the discovery of these new properties, CysG should perhaps be called sirohaem synthase. Interestingly, the protein encoded by *cysG* in *Salmonella*, which is highly similar to the *E. coli cysG* gene, is not only responsible for the biosynthesis of sirohaem but is also required for the biosynthesis of vitamin B_{12} (*Salmonella*, unlike *E. coli*, is able to biosynthesize vitamin B_{12} *de novo*); CysG must therefore also be able to catalyse the synthesis of dihydrosirohydrochlorin (precorrin-2) for the biosynthesis of the corrin macrocycle (Jeter et al 1984).

Evolutionary origin: gene fusion?

Questions about the number of active sites and substrate-binding sites and the nature of the catalytic groups are currently being addressed. One question which we have pondered concerns the evolutionary origin of the enzyme and how CysG (sirohaem synthase) evolved to be this multifunctional enzyme. The answer to this probably has something to do with the fact that CysG's major role is in the biosynthesis of sirohaem. *Escherichia coli* methylates uroporphyrinogen only for the biosynthesis of sirohaem. The other uroporphyrinogen methylases all make precorrin-2 (dihydrosirohydrochlorin) as an intermediate largely for the biosynthesis of either vitamin B_{12} or factor F_{430}, the prosthetic group required by methanogenic bacteria (Thauer & Bonacker 1994, this volume). To make sirohaem, these organisms either have a separate *cysG* locus, as in the case of *Pseudomonas denitrificans* (Crouzet et al 1990a), or have a separate oxidase/chelatase enzyme to complement the activity of the uroporphyrinogen methylase, as in the case of *Bacillus megaterium* (Robin et al 1991). As *E. coli* lost the ability to make vitamin B_{12}, so the necessity for a separate and independent methylase and oxidase/chelatase function was also lost. The nucleotide sequences encoding these separate proteins might then have been fused to give a single gene encoding the multifunctional enzyme. The process of gene fusion may well have given rise to the current-day sirohaem synthase. The advantage of a multifunctional enzyme is that it can channel its highly labile substrates from one active site to the next within the confines of one protein. Why then does *Salmonella* use the *cysG* locus for vitamin B_{12} biosynthesis? The reason may be that *Salmonella* can make vitamin B_{12} only under anaerobic conditions. The *cysG* gene is expressed at quite low levels, and therefore the concentration of precorrin-2 is always low. Only under anaerobic conditions is precorrin-2 stable enough to be methylated further by the other methylases in the pathway to construct vitamin B_{12}. Thus CysG in *Salmonella* is able to meet the demand for the small quantities of precorrin-2 required for B_{12} synthesis, but only when conditions are suitable for precorrin-2 stability. There is therefore no requirement for a separate uroporphyrinogen methylase, but there is a requirement for anaerobiosis.

Acknowledgements

The authors wish to thank Professor S. Kaplan for the generous gift of the *hemA* and *hemT* clones and for helpful discussions. Likewise, the helpfulness of Professor A. I. Scott is also acknowledged, especially for the disclosure of results prior to publication. We should also like to thank Professor P. M. Jordan for his assistance and encouragement with this research. Finally, financial support from SERC, AFRC and NATO is greatly appreciated.

References

Bawden MJ, Borthwick IA, Healy HM, Morris CP, May BK, Elliott WH 1987 Sequence of human 5-aminolaevulinate synthase cDNA. Nucleic Acids Res 15:8563

Blanche F, Robin C, Couder M et al 1991 Purification, characterisation, and molecular cloning of S-adenosyl-L-methionine:uroporphyrinogen III methyltransferase from *Methanobacterium ivanovii*. J Bacteriol 173:4637–4645

Borthwick IA, Srivastava G, Brooker JD, May BK, Elliott WH 1983 Purification of 5-aminolaevulinic acid synthase from liver mitochondria of chick embryo. Eur J Biochem 129:615–620

Cole JA, Newman BM, White P 1980 Biochemical and genetic characterisation of *nirB* mutants of *Escherichia coli* K12 pleiotropically defective in nitrite and sulphite reduction. J Gen Microbiol 120:475–483

Crouzet J, Cauchois L, Blanche F et al 1990a Nucleotide sequence of a *Pseudomonas denitrificans* 5.4-kilobase DNA fragment containing five *cob* genes and identification of structural genes encoding S-adenosyl-L-methionine:uroporphyrinogen III methyltransferase and cobyrinic acid a,c-diamide synthase. J Bacteriol 172:5968–5979

Crouzet J, Cameron B, Cauchois L et al 1990b Genetic and sequence analysis of an 8.7-kilobase *Pseudomonas denitrificans* fragment carrying eight genes involved in transformation of precorrin-2 to cobyrinic acid. J Bacteriol 172:5980–5990

Gibson KD, Laver WG, Neuberger A 1958 Initial stages in the biosynthesis of porphyrins. The formation of δ-aminolaevulinic acid from glycine and succinyl-coenzyme A by particles from chicken erythrocytes. Biochem J 70:71–81

Hanukoglu I, Gutfinger T 1989 cDNA sequence of adrenodoxin reductase. Identification of NADP-binding sites in oxidoreductases. Eur J Biochem 180:479-484

Haydock SF, Dowson JA, Dhillon N, Roberts GA, Cortes J, Leadlay PF 1991 Cloning and sequence analysis of genes involved in erythromycin biosynthesis in *Saccharopolyspora erythraea*: sequence similarities between ErgG and a family of S-adenosylmethionine-dependent methyltransferases. Mol Gen Genet 230:120–128

Ingrosso D, Fowler AV, Bleibaum J, Clarke S 1989 Sequence of the D-aspartyl/L-isoaspartyl protein methyltransferase from human erythrocytes. Common sequence motifs for protein, DNA, RNA, and small molecule S-adenosylmethionine-dependent methyltransferases. J Biol Chem 264:20131–20139

Jeter RM, Olivera BM, Roth JR 1984 *Salmonella typhimurium* synthesizes cobalamin (vitamin B_{12}) *de novo* under anaerobic growth conditions. J Bacteriol 159:206–213

Jordan PM 1991 The biosynthesis of 5-aminolaevulinic acid and its transformation into uroporphyrinogen III. In: Jordan PM (ed) Biosynthesis of tetrapyrroles. Elsevier, Amsterdam, p 1–59

Jordan PM, Laghai-Newton A 1986 Purification of 5-aminolaevulinate synthase. Methods Enzymol 123:435–443

Kikuchi G, Kumar A, Talmage P, Shemin D 1958 The enzymatic synthesis of δ-aminolaevulinic acid. J Biol Chem 233:1214–1219

McClung CR, Somerville JE, Guerinot ML, Chelm BK 1987 Structure of the *Bradyrhizobium japonicum* gene *hemA* encoding 5-aminolaevulinic acid synthase. Gene 54:133–139

Maguire DJ, Day AR, Borthwick IA et al 1986 Nucleotide sequence of the chicken 5-aminolaevulinic acid synthase gene. Nucleic Acids Res 14:1379–1391

Neidle EL, Kaplan S 1993 Expression of the *Rhodobacter sphaeroides hemA* and *hemT* genes, encoding two 5-aminolevulinic acid synthase isozymes. J Bacteriol 175:2292–2303

Ohashi A, Kikuchi G 1979 Purification and some properties of two forms of δ-aminolaevulinate synthase from rat liver cytosol. J Biochem 85:239–247

Peakman T, Crouzet J, Mayaux J-F et al 1990 Nucleotide sequence, organisation and structural analysis of the products of genes in the *nirB-cysG* region of the *E. coli* K-12 chromosome. Eur J Biochem 191:315–323

Robin C, Blanche F, Cauchois L, Cameron B, Couder M, Crouzet J 1991 Primary structure, expression in *Escherichia coli*, and properties of S-adenosyl-L-methionine: uroporphyrinogen III methyltransferase from *Bacillus megaterium*. J Bacteriol 173:4893–4896

Scott AI, Warren MJ, Roessner CA, Stolowich NJ, Santander PJ 1990 Development of an 'overmethylation' strategy for corrin synthesis. Multi-enzyme preparation of pyrrocorphins. J Chem Soc Chem Comm, p 593–597

Scrutton NS, Berry A, Perham RN 1990 Redesign of the coenzyme specificity of a dehydrogenase by protein engineering. Nature 343:38–43

Suwanto A, Kaplan S 1989 Physical and genetic mapping of the *Rhodobacter sphaeroides* 2.4.1 genome: presence of two unique circular chromosomes. J Bacteriol 171:5850–5859

Tai T-N, Moore MD, Kaplan S 1988 Cloning and characterisation of the 5-aminolaevulinate synthase gene(s) from *Rhodobacter sphaeroides*. Gene 70:139–151

Takata Y, Fujioka M 1992 Identification of a tyrosine residue in rat guanidinoacetate methyltransferase that is photolabelled with S-adenosyl-L-methionine. Biochemistry 31:4369–4374

Thauer RK, Bonacker LG 1994 Biosynthesis of coenzyme F_{430}, a nickel porphinoid involved in methanogeneis. In: The biosynthesis of the tetrapyrrole pigments. Wiley, Chichester (Ciba Found Symp 180) p 210–227

Urban-Grimal D, Volland C, Garnier T, Dehoux P, Labbe-Bois R 1986 The nucleotide sequence of the HEM1 gene and evidence for a precursor form of the mitochondrial 5-aminolaevulinate synthase in *Saccharomyces cerevisiae*. Eur J Biochem 156:511–519

Volland C, Felix F 1984 Isolation and properties of 5-aminolaevulinate synthase from *Saccharomyces cerevisiae*. Eur J Biochem 142:551–557

Warnick GR, Burnham BF 1971 Regulation of porphyrin biosynthesis: purification and characterisation of δ-aminolaevulinic acid synthetase. J Biol Chem 246:6880-6885

Warren MJ, Scott AI 1990 Tetrapyrrole assembly and modification into the ligands of biologically functional cofactors. Trends Biochem Sci 180:486-491

Warren MJ, Roessner CA, Santander PJ, Scott AI 1990a The *Escherichia coli cysG* gene encodes S-adenosylmethionine-dependent uroporphyrinogen III methylase. Biochem J 265:725–729

Warren MJ, Stolowich NJ, Santander PJ, Roessner CA, Sowa BA, Scott AI 1990b
    Enzymatic synthesis of dihydrosirohydrochlorin (precorrin-2) and of a novel
    pyrrocorphin by uroporphyrinogen III methylase. FEBS (Fed Eur Biochem Soc) Lett
    261:76–80
Warren MJ, Roessner CA, Ozaki S, Stolowich NJ, Santander PJ, Scott AI 1992
    Enzymatic synthesis and structure of precorrin-3, a trimethyldipyrrocorphin
    intermediate in vitamin B_{12} biosynthesis. Biochemistry 31:603–609

DISCUSSION

Arigoni: Professor Akhtar worked out the stereochemical details of ALA
synthase some time ago, but one question remains. If my memory is correct,
the overall result corresponds to one retention and one inversion, but we do
not yet know which is which.

Akhtar: I am of the opinion that the second step in the mechanism in Scheme
1 (Abboud et al 1974) is an inversion. My opinion is based on analogy with
the work Cornforth's group and yours did on malate synthase and citrate synthase
(Cornforth et al 1969, Luthy et al 1969). In both these cases, the conversion
of acetyl-CoA into an enol intermediate and the neutralization of the latter by
reaction with a carbonyl group occurs with inversion of sterochemistry. Step
2 in Scheme 1 is reminiscent of a formal reversal of this process.

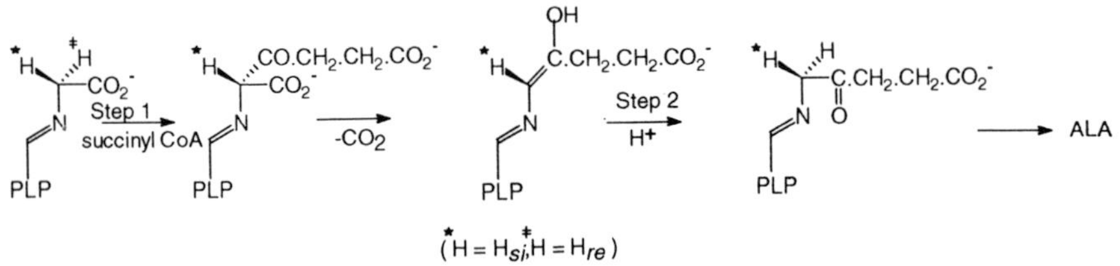

$$(\overset{*}{H} = H_{si}, \overset{+}{H} = H_{re})$$

Scheme 1 (*Akhtar*)

Arigoni: That is unconventional. Most pyridoxal phosphate-catalysed
decarboxylations display a retention mechanism.

Akhtar: After the β-keto acid intermediate, we have *not* used pyridoxal
phosphate to promote the decarboxylation but the carbonyl group. I was so
influenced by your remarkable work that I thought, why can't I take advantage
of the enol to force the inversion of stereochemistry?

Jordan: It is easier to explain the overall inversion if pyridoxal phosphate
is not mechanistically involved in the decarboxylation reaction. The condensation
would proceed by retention in the normal way, but the newly formed ketone
would promote the decarboxylation rather than the protonated pyridine ring.

Beale: I am puzzled by the existence of two genes for ALA synthase, *hemA*
and *hemT*, in *R. sphaeroides*. Results from S. Kaplan's laboratory show that
if either gene is knocked out, induction of bacteriochlorophyll synthesis is
inhibited, even though when both are intact mRNA was detected from only

one gene (Neidle & Kaplan 1993a,b). The second gene is not a pseudogene or a silent gene; it is used somehow.

Warren: If you knock out the *hemA* gene, the *hemT* gene is used, but under normal conditions you don't see *hemT* transcripts.

Beale: If you look for mRNA by Northern analysis you can detect only *hemA* expression; but don't forget that *hemT* was discovered in a mutant strain that had *hemA* but couldn't accumulate bacteriochlorophyll.

Why does knocking out *hemT* prevent the induction of bacteriochlorophyll synthesis?

Warren: I'm not sure exactly of the results of all those mutations. You are essentially asking what the selection pressure on the *hemT* gene was—why does the bacterium need two isoenzymes? That's a good question.

Beale: One can't help drawing an analogy with mammals and birds where one gene encodes the housekeeping enzyme and the other the erythrocyte-specific enzyme, but there's no evidence for that.

Warren: The fact that we can't detect a *hemT* transcript is evidence against that.

Beale: In a closely related organism, *Rhodobacter capsulatus*, there's only one gene, and only one chromosome.

A. Smith: What is the significance of the small chromosome? What other genes does it carry?

Warren: It encodes some essential genes, but everything found on the small chromosome is also present on the large chromosome. To be certain it's essential, one would need to find an auxotrophic mutant that maps in the small chromosome.

Leeper: I'm taken by the idea that ALA synthase arose first in purple photosynthetic bacteria, and that these are the origin of mitochondria, and so the origin of ALA synthase in everything. I wondered if there is any idea of where it arose from? Does it share sequence similarity with anything else? It catalyses a fairly unique chemical reaction. Are there chemical analogies?

Warren: There are two enzymes in *E. coli* that catalyse similar reactions, aminoketobutyrate–CoA ligase, which condenses glycine and acetyl-CoA to form 2-amino-3-ketobutyrate, and aminoketopelargonic acid synthase, which condenses alanine and pimelic acid–CoA in the biosynthesis of biotin. These have some sequence identity with *hemA*.

Arigoni: Are there any organisms that can make ALA both ways, by the Shemin pathway and by the C_5 pathway?

Beale: As far as I know, this has been shown only in *Euglena gracilis*, in which the plastid haems and chlorophyll are derived exclusively from glutamate whereas mitochondrial haems are derived from glycine only. The two pathways can operate simultaneously in wild-type cells in the light, but in the dark, and in mutant cells in which plastids are unable to develop, only the glycine–succinate pathway operates. When you illuminate wild-type cells there is a large induction of the C_5 pathway in the chloroplast. We don't yet know from which compartment the haems in cytoplasmic haemoproteins are derived. I don't know of any other organism in which both pathways have been measured simultaneously at the enzyme level *in vitro* as well as by *in vivo* labelling.

Arigoni: Do the intermediates in the pathway leak from one system to the other, or are they completely separate?

A. Smith: Sam Beale's work implies that there are two separate pathways for the conversion of ALA to haem, because labelled glycine is incorporated only into haem *a* (Weinstein & Beale 1983). We have not been able to find non-plastidic forms of ALA dehydratase and porphobilinogen deaminase in *Euglena gracilis*. The fact that we can't find them doesn't mean they're not there. At the moment I'm agnostic; I don't think there's another set of enzymes, but I am prepared to be shown otherwise.

Rebeiz: We've also been unable to detect biosynthesis of protoporphyrin IX from ALA in mitochondria from higher plants, although we can detect it in insect mitochondria.

Castelfranco: Dr Warren, you said that in the conversion of uroporphyrinogen III into sirohaem, the methylation and the oxidation of the macrocycles are coupled. Is it correct that the uroporphyrinogen is methylated but that you cannot isolate the methylated product, or is it the case that it tends to go over to the oxidized state?

Warren: You can isolate precorrin-2 (dihydrosirohydrochlorin), but NAD or NADP is needed to continue the cycle.

Arigoni: Is the activity of the enzyme the same in the presence and absence of the oxidizing agent?

Warren: It seems to be more or less the same. The slow step is the methylation.

Thauer: Is it possible to stop the reaction in some way to isolate the monomethylated form? I understand that the enzyme has a low turnover number. Do you think that there is only one binding site for SAM and that the tetrapyrrole molecule has to move before it is methylated for the second time, or do you think there are two SAM-binding sites? If there were two SAM-binding sites, increasing the concentration of SAM might increase the turnover number.

Warren: It's easier to cope with one SAM-binding site. We haven't observed much monomethylated material during the reaction. Adding Cu^{2+} inhibits the second methylation reaction, for some reason, leading to an increase in the accumulation of the monomethylated material, Factor I.

Arigoni: As I understand Blanche and Crouzet's work, the first methylation product comes off the enzyme and then goes back on again for the second methylation.

Battersby: The evidence for this came from *in vitro* studies with the first methylase which always yielded a mixture of mono- and dimethylated products. Also, precorrin-1 is as good a substrate for this methylase as uroporphyrinogen III.

Arigoni: This seems to suggest that a readjustment of the substrate is needed after the first methylation step so that the second methylation can occur with a similar topology.

Warren: By varying the amount of enzyme you can vary the amount of Factor I and precorrin-2 produced. We tend to have an excess of the enzymes in our systems, because we are trying to produce precorrin-2. With low enzyme concentrations there is a greater build-up of the monomethylated material.

Thauer: If you plot rates and SAM concentration to derive the K_m, do you see Michaelis–Menten-type kinetics or do you see a sigmoidal dependence with a Hill coefficient of 2?

Warren: We haven't done kinetic studies on CysG.

Beale: I gather that the turnover of substrate by CysG is about 1/min. That seems awfully slow for an enzyme. I wonder if there might be a build up of inhibitors. Is it possible that the trimethylated compound, or some other aberrant product, could be accumulating in your *in vitro* systems and poisoning the enzymes?

Warren: There are several possible explanations. There could be inactive enzyme present, for example. The tight binding of the substrates and some of the products of the reaction, the *S*-adenosylhomocysteine, for example, could also contribute. In the cell, when you induce these enzymes, you quickly see a build up of porphyrinoid material. I suspect that we are losing enzyme activity during purification.

Pitt: Most SAM-dependent methylases are strongly inhibited by low concentrations of *S*-adenosylhomocysteine. This is likely to be inhibitory unless you can get rid of it quickly.

Rebeiz: Given the low turnover number, is the enzyme's activity sufficient to account for the production of the end product under normal conditions?

Warren: I haven't worked out how much sirohaem is actually required by *E. coli*, but it's not a huge amount. The activity must suffice, because we can complement *cysG⁻* strains with quite low expression plasmids.

Beale: Do aerobic organisms that make B_{12} have the *cysG* gene, or do they rely on another chelatase?

Warren: As far as I can gather, Blanche and Crouzet's groups seem to think that in *Pseudomonas denitrificans* there is a separate *cysG* locus. In *Bacillus megatarium*, their evidence suggests there is a separate chelatase/oxidase-type of protein which is independent of the uroporphyrinogen methylase.

Scott: We know that CysG is necessary for sirohaem synthesis in *Salmonella typhimurium*. Goldman & Roth (1993) have made 66 auxotrophic *cysG* mutants which are unable to make either sirohaem or B_{12}. I don't know where the lesions are, because the sequences are not known yet, but it appears that *cysG*, at least in anaerobes such as *Salmonella*, is essential for both the B_{12} and the sirohaem pathways.

Warren: Mutations in *cysG* in *E. coli* seem to make the SAM-dependent methylation even slower, or inhibit the oxidation process. We've also identified a few point mutations that seem to cause the protein encoded to fall out of solution.

Beale: I would be a little worried about using *Salmonella typhimurium* as an example because it's an extremely poor producer of B_{12}.

Scott: It is poor in comparison with industrial production strains such as *Propionibacterium shermanii* and *Pseudomonas denitrificans*, making only about five molecules per cell.

Beale: And only under anaerobic conditions.

Scott: That's true, but it does make it, so all the genes are there! Production can be increased 100-fold by adding dimethylbenzimidazole (J. Roth, personal communication). That is still a small amount but at least it is detectable by isolation rather than bioassay.

A. Smith: You described expression of the C-terminal domain of *cysG*; have you expressed the N-terminal domain?

Warren: We have cloned the N-terminal domain but the protein does not seem to be overexpressed. The two clones, the C-terminal and N-terminal domains, put together on the same plasmid, do not appear to complement $cysG^-$ strains, so we can't draw any conclusions about the N-terminus of CysG. The N-terminal protein might be being produced but not folded properly, or rapidly degraded. We've tried to express this protein using several strategies, but without success.

A. Smith: Is there a change in the phenotype of the cells? You can't detect protein, but do you see formation of any strange compounds? Do the bacteria change colour?

Warren: There is no significant change that we can see.

Eschenmoser: The 'over-methylation', in which a methyl group is added in ring C, is very striking from a chemical point of view. It occurs at C-12, which is exactly the position one would expect by chemical reasoning, provided that the endocyclic double bond of ring B (position C-8–C-9) has moved into the macrocyclic ring (position C-9–C-10) before the methylation in ring C takes place. An analogous argument applies to the site of methylation in ring B, where the endocyclic double bond of ring A (position C-3–C-4) would have to move into position C-4–C-5 before methylation in ring B occurred. Such double bond shifts preceding methylations could be a factor determining the selection of the methylation site (Leumann et al 1983). If such pre-methylation shifts do not occur in enzymic methylations, one would have to conclude that the sites are selected by the enzymes according to constitutional correspondence of positions C-2, C-7 and C-12.

Is anything known about the movement of double bonds in the macrocyclic ring after or before methylation steps?

Warren: We have done some quite detailed NMR studies on precorrin-2 which indicated a tautomerization of the double bonds in the northern half of the molecule, but there was no evidence for the movement of double bonds at the level of precorrin-1 (Warren et al 1992).

Scott: There was no direct evidence. Perhaps one could set up the enzyme and substrate without the SAM to see, by NMR, whether the molecule prepares itself for this kind of movement before methylation.

Eschenmoser: Where is that double bond in precorrin-1?

Battersby: That is not certain. It was clearly shown by synthesis that the oxidation level of precorrin-1 is the same as that of uroporphyrinogen III (see Brunt et al 1989). Two double-bond isomers around ring A were obtained from our synthesis and they were used in admixture for the incorporation experiment, which demonstrated an efficient and specific conversion into cobyrinic acid. So, although one of these isomers is precorrin-1, we do not know which one.

Arigoni: Professor Eschenmoser suggested that methylation at position 7, in ring B, is chemically reasonable after the first methylation at position 2, but this still leaves the first question unanswered. Why does the first methylation occur at position 2, in ring A, if there is almost no difference between positions 2 and 3?

Eschenmoser: That is a quite different question, because my arguments referred to rings B and C. However, one could imagine that methylation of ring A could be preceded by a double bond shift from C-18–C-19 to C-19–C-20; through such a shift position C-2 would be strongly and specifically activated towards methylation.

Warren: The enzyme must have a preference for its new substrate, precorrin-1, if the methylation is to proceed in this particular manner. It is programmed to catalyse two methylations, going round in a 'clockwise' direction, recognizing the order of acetate and propionate side chains.

Spencer: Is it known how copper inhibits the second (C-7) methylation? Might it prevent the double bond rearrangement proposed to precede the second methylation?

Warren: I'm not sure. It must somehow affect the character of the substrate.

Chang: In a fully aromatic porphyrin system, the ring C double bond would be the preferred site of attack. Consider the hydrogenation or attack by osmium oxide of these porphyrins. If you remove one double bond, the second saturation step, in a free-base porphyrin, is always in ring C, forming a bacteriochlorin. However, when there is a metal inside, the preferred site of attack changes. Instead of bacteriochlorin being formed, ring B becomes the preferred site of attack.

Leeper: We have some results from incorporation of deuterated porphobilinogen into sirohydrochlorin which are relevant to the question of the location of the double bond in precorrin-1 and precorrin-2 (Weaver et al 1990). Deuterium from position 11 of porphobilinogen is totally washed out at both C-5 and C-20 of sirohydrochlorin. The easiest explanation is that at the precorrin-1 stage the bonds move around rather easily. I suspect that it will be impossible to pin down the tautomer of precorrin-1 produced by the methylase.

Scott: There is evidence about stereochemistry at C-15 in the conversion of precorrin-2 to Factor II, for example. You did experiments with chiral porphobilinogen, with ^2H at C-11, in the early days. Can these experiments now be re-interpreted in the light of today's knowledge of the dehydrogenase activity of CysG, i.e., O_2^- versus NAD-mediated oxidation?

Leeper: In our system, C-15 was, we believe, oxidized non-enzymically in air. However, we have not done any experiments to test whether there is any significant dehydrogenase activity. It is true that a surprisingly high degree of stereoselectivity was observed for this oxidation.

Battersby: As far as I know, there's no evidence about movement of the double bonds other than that Finian Leeper has just described.

Eschenmoser: For a perfect description of the enzymic process one would need to know which tautomeric forms of the chromophore are methylated.

Battersby: I agree, but this is extremely difficult to establish.

Eschenmoser: Interestingly, in that famous compound with the *C*-methyl at position 11, the ring B double bond is in the original position again.

Battersby: Remember that the materials are isolated as esters. One has already treated these materials under strongly acidic conditions before they are isolated.

Akhtar: Professor Eschenmoser, you suggested essentially that position 7 becomes much more susceptible to attack by *S*-adenosylmethionine if there is a double bond between positions 4 and 5. Is this a theoretical deduction, or do you say this on the basis of your experience?

Eschenmoser: This is both what one expects theoretically and what has been shown to be the case in studies on the methylation of synthetic pyrrocorphins and derivatives.

Akhtar: From this, one would expect the C-5 to undergo exchange with protons of H_2O during the reversible transfer of the double bond from C-3–C-4 to C-4–C-5.

Kräutler: Movement of the double bond from the 3–4 position to the 4–5 position is not achieved easily, kinetically. C-20 deprotonation should be achieved relatively easily.

Arigoni: Movement of the double bond from C-3–C-4 to C-4–C-5 would be expected to occur as shown in Scheme 2.

Kräutler: This would explain the C-20 isotopic equilibration.

Scott: The *C*-methylation situation is even more interesting when there's a methyl group already at C-20, such as in the Factor III or precorrin-3 series, because, at first glance, there is no obviously preferable site for the insertion of a new methyl group. However, Albert Eschenmoser's *in vitro* chemistry has shown that there is regiospecific control over C-12 for example, as against C-17, depending on the absence or presence of coordinating metal. Thus, the methyl group at C-20 perturbs the substrate in a way that Albert might well have

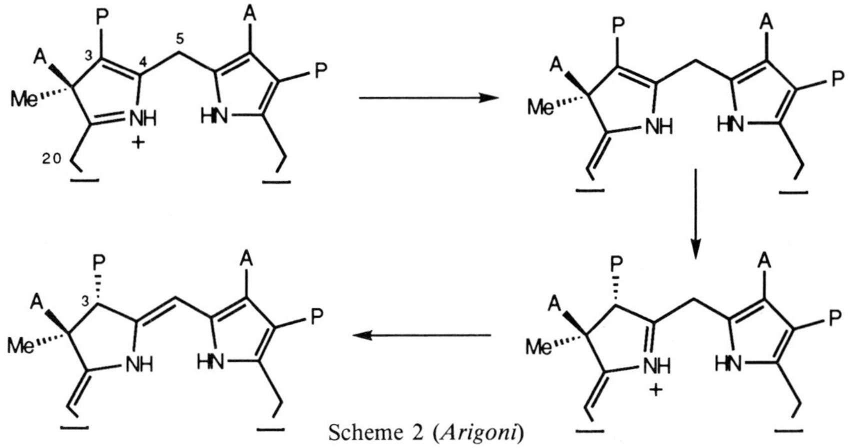

Scheme 2 (*Arigoni*)

predicted but which I find quite remarkable because of the stereoelectronic effects on the rest of the system.

Akhtar: Professor Eschenmoser raised an interesting point. Is the second methylation governed by constitutional similarity with the first ring, or is it governed by the electronic factors, which require the double bond to be first placed in the 4–5 position? He almost convinced me that having the 4–5 double bond would be a good idea, but then the point was made that the C-4–C-5 double bond is difficult to introduce.

Arigoni: No one said this was difficult.

Akhtar: Not easy, then; but it would become easy if constitutional factors determined that ring B is methylated first. Then one can see how wonderfully the system is set up for migration of the double bond. Position 5 is now next to the C=N, beautifully set up to undergo the otherwise difficult double bond rearrangement. If, however, rearrangement occurs first, one needs to put a proton at C-3, which would be an anti-Markownikoff process. This particular position, C-3, cannot be sufficiently electron-rich to undergo an electrophilic attack. However, when you use the constitutional argument, and put the methyl group at C-7 first, all of a sudden new factors come into play and the fact that the C-5 hydrogen is now α to the C=N is important, turning a difficult double bond migration into an easy one (Scheme 3).

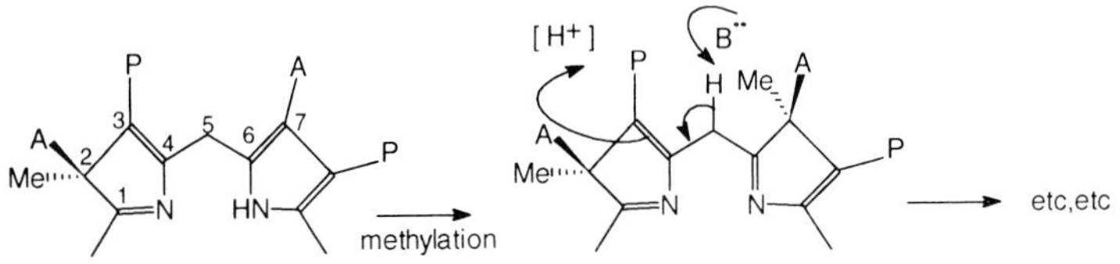

Scheme 3 (*Akhtar*)

Eschenmoser: You would say, then, that from a chemical point of view, the second methylation is more likely to happen under topological control.

Akhtar: Absolutely.

Eschenmoser: This might be more economical and there is no chemical argument against it.

Arigoni: I still favour the protonation–deprotonation sequence outlined in Scheme 2.

Akhtar: Then you are involving the C-20 position in the migration of the double bond. I was assuming, perhaps mistakenly, that until the second methylation, C-20 is not touched.

Thauer: One could determine the rate-limiting step by adding the mono-methylated compound to the enzyme and determining whether the rate of the second methylation is of the same order of magnitude as the rate of the overall reaction, or higher.

Warren: Blanche and Crouzet's groups have reduced Factor I and find that Factor II is produced ultimately.

Battersby: We have shown (Brunt et al 1989) that reduction of Factor I octamethyl ester with sodium amalgam yields a mixture of isomers. The non-pyrrolic double bonds of the major isomer lie at C-20–C-1 and N_A–C-4 and those of the minor isomer at C-1–N_A and C-4–C-5.

Warren: Uroporphyrinogen III is always methylated at positions 2 and 7, obviously, but even the uroporphyrinogen II isomer, in which the side chains are APPAAPPA, offering no molecular 'handles' to guide the next methylation, is still sequentially methylated in rings A and B at positions 2 and 7.

Castelfranco: Forget about the enzyme for a moment. What would the order of methylation be if you shook the starting material in aqueous solution with methyl iodide?

Eschenmoser: We have done this on dipyrroles, but not on tetrapyrroles.

K. Smith: Using methyl iodide would give products which might be *N*-methylated anyway.

Battersby: The point about the importance of C-15 of precorrin-2 in stereochemical studies (p 46) poses an interesting analytical problem. If porphobilinogen stereospecifically labelled with ^2H at C-11 were incorporated into precorrin-2, the product should have the ^2H at C-15 above or below the general plane of the molecule. The question is how to determine which it is. Of course, we have our ideas on this problem, but it is worthy of discussion.

Arigoni: We were asking whether there is any stereospecificity in the formation of the oxidized compound.

Battersby: Actually, both questions are important and it is interesting to consider how one might tackle the stereochemical problem at the precorrin-2 stage.

Scott: Could one ozonolyse to get the malonate? If the malonate were unsymmetrical, i.e., labelled only in one carboxyl with ^{13}C, the configuration of ^2H at C-2 could be determined.

Battersby: That's one possible way. In our view, appropriate labelling followed by degradation is probably the best way forward.

Scott: If the dihydro form could be crystallized, with deuterium at C-15, you could use neutron diffraction studies.

Arigoni: That's cheating!

References

Abboud MM, Jordan PM, Akhtar M 1974 Biosynthesis of 5-amino laevulinic acid: involvement of a retention-inversion mechanism. J Chem Soc Chem Commun, p 643–644

Brunt D, Leeper FJ, Grugurina I, Battersby AR 1989 Biosynthesis of vitamin B_{12}: synthesis of (+)-[5-^{13}C]faktor 1 ester; determination of the oxidation state of precorrin-1. J Chem Soc Chem Commun, p 428–431

Cornforth JW, Redmond JW, Eggerer H, Buckle W, Gutschow C 1969 Asymmetric methyl groups and the mechanism of malate synthase. Nature 221:1212–1213

Goldman BS, Roth JR 1993 Genetic structure and regulation of the *cysG* gene in *Salmonella typhimurium*. J Bacteriol 175:1457–1466

Leumann C, Hilpert K, Schreiber J, Eschenmoser A 1983 Chemistry of pyrrocorphins: C-methylations at the periphery or pyrrocorphins and related corphinoid ligand systems. J Chem Soc Chem Commun, p 1404–1407

Luthy J, Retey J, Arigoni D 1969 Preparation and detection of chiral methyl groups. Nature 221:1213–1215

Neidle EL, Kaplan S 1993a Expression of the *Rhodobacter sphaeroides hemA* and *hemT* genes, encoding two 5-aminolevulinic acid synthase isozymes. J Bacteriol 175: 2292–2303

Neidle EL, Kaplan S 1993b 5-aminolaevulinic acid availability and control of spectral complex formation in HemA and HemT mutants of *Rhodobacter sphaeroides*. J Bacteriol 175:2304–2313

Warren MJ, Roessner CA, Ozaki S, Stolowich NJ, Santander PJ, Scott AI 1992 Enzymatic synthesis and structure of precorrin-3, a trimethyldipyrrocorphin intermediate in vitamin B_{12} biosynthesis. Biochemistry 31:603–609

Weaver GW, Blanche F, Thibaut D, Debussche L, Leeper FJ, Battersby AR 1990 Biosynthesis of vitamin B_{12}: incorporation of (11S)-[11-$^2H_1$]- and (11R)-[11-$^2H_1$]- porphobilinogen into sirohydrochlorin and 2,7,20-trimethylisobacteriochlorin. J Chem Soc Chem Commun, p 1125–1127

Weinstein JD, Beale SI 1983 Separate physiological roles and subcellular compartments for two tetrapyrrole biosynthetic pathways in *Euglena gracilis*. J Biol Chem 258:6799–6807

5-Aminolaevulinic acid dehydratase: characterization of the α and β metal-binding sites of the *Escherichia coli* enzyme

P. Spencer* and P. M. Jordan*

School of Biological Sciences, Queen Mary and Westfield College, University of London, Mile End Road, London E1 4NS, UK

Abstract. The α and β metal-binding sites of 5-aminolaevulinic acid dehydratase (ALAD) (porphobilinogen synthase, EC 4.2.1.24) from *Escherichia coli* were investigated to determine the function of each metal ion and the role of the reactive cysteines in metal binding. Occupancy of the α site by Zn^{2+} restored virtually all catalytic activity to the inactive metal-depleted ALAD (apoALAD). Occupancy of the α site by Co^{2+} also yielded an active enzyme and resulted in a charge-transfer band indicative of a single cysteine amongst the metal ligands. Subsequent labelling of this cysteine residue with ^{14}C-labelled N-ethylmaleimide, followed by peptide analysis, indicated the involvement of Cys-130. The metal ion at the α site is thought to be essential for binding of the second molecule of substrate at the A substrate-binding site that forms the acetic acid side of the product, porphobilinogen. Binding of Zn^{2+} to the β site restored little activity if the α site was unfilled. Metal ion binding to the β site could be monitored by following the change in protein fluorescence with Zn^{2+} titration of apoALAD at pH 6. A conformational change upon β site occupancy may explain why binding of Mg^{2+} at the α site can occur only if Zn^{2+} is bound at the β site. The binding of Co^{2+} at the β site produced an inactive enzyme that exhibited a charge-transfer band indicative of at least three cysteine ligands.

1994 The biosynthesis of the tetrapyrrole pigments. Wiley, Chichester (Ciba Foundation Symposium 180) p 50–69

The enzyme 5-aminolaevulinic acid dehydratase (ALAD, EC 4.2.1.24), also called porphobilinogen synthase, catalyses the formation of porphobilinogen by the dimerization of two molecules of 5-aminolaevulinic acid (ALA) (1). The binding sites for the two substrate molecules have been designated as the A and P sites, according to which side of the product porphobilinogen they form. The

---

Present address: Department of Biochemistry, School of Biological Sciences, University of Southampton, Bassett Crescent East, Southampton SO9 3TU, UK.

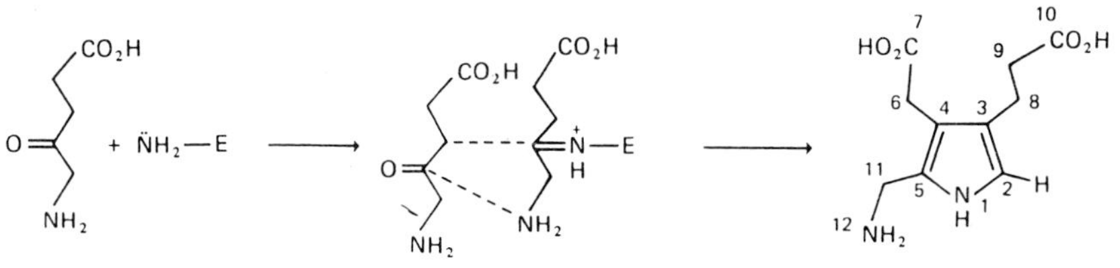

Scheme 1

substrate initially bound to the dehydratase forms a Schiff's base (2) at the P site and gives rise to the propionic side chain of porphobilinogen (3) (Jordan & Seehra 1980; Scheme 1). The second molecule of ALA to bind to the enzyme interacts with the A site (Jordan & Seehra 1980), giving rise to the acetic acid side chain of porphobilinogen. The A site is thought to require a divalent metal; removal of the metal ion prevents the overall enzymic reaction but does not affect the ability of the enzyme to bind substrate at the P site (Chaudhry et al 1976, Jaffé et al 1984). Despite several proposals, the precise mechanism of catalysis still remains to be determined.

The enzyme has been purified from a variety of sources including human erythrocytes (Anderson & Desnick 1979), bovine liver (Bevan et al 1980, Jordan & Seehra 1986), spinach (Liedgens et al 1983), pea (P. Spencer & P. M. Jordan, unpublished results), *Escherichia coli* (Spencer & Jordan 1993), *Rhodobacter sphaeroides* (Nandi et al 1968) and yeast (Borralho et al 1990). All ALADs studied to date have shown a requirement for either Zn^{2+} (mammalian and some bacterial ALADs) or Mg^{2+} (plant and some bacterial ALADs). The properties of 5-aminolaevulinic acid dehydratases have been reviewed by Shemin (1972), Cheh & Neilands (1976) and Jordan (1991). Although ALADs were at first classified on the basis of their prokaryotic or eukaryotic origin, a division on the basis of whether they require Zn^{2+} or Mg^{2+} now appears to be more informative.

The mammalian and avian enzymes are zinc metalloenzymes with pH optima in the range 6.3–7.0 that are inhibited by EDTA or exposure to oxidative environments. These eukaryotic dehydratases have a molecular mass of about 280 kDa and are composed of eight identical subunits each of 35 kDa. The enzymes isolated from spinach (Liedgens et al 1983) and pea (P. Spencer & P. M. Jordan, unpublished results) require magnesium ions for maximal activity and have a high pH optimum (above pH 8). Spinach ALAD is a hexamer of 50 kDa subunits (Liedgens et al 1983). The yeast dehydratase (Borralho et al 1990) appears to be a zinc metalloenzyme with a high pH optimum. The bacterial ALADs are more variable in their metal ion requirement. The ALAD from *R. sphaeroides*, although not immediately inhibited by EDTA, loses activity on dialysis; activity can be restored by the addition of magnesium ions (Shemin 1972). Potassium ions increase further the activity of the *R. sphaeroides* enzyme as a result of protein aggregation (Nandi & Shemin 1968). In contrast, the *E. coli* ALAD, which is an octamer composed of 35.5 kDa subunits, can use Zn^{2+} alone or Zn^{2+} and Mg^{2+} to support activity; the metal-depleted apoALAD is inactive (Spencer & Jordan 1993).

Those ALADs requiring Zn^{2+} are also highly sensitive to oxygen and thiol reagents such as Ellman's reagent (5,5'-dithiobis[2-nitrobenzoic acid]; DTNB), suggesting that free cysteines are involved in metal ligation (Spencer & Jordan 1993). ALADs of plant origin, which require Mg^{2+}, are not sensitive to oxygen. The ALAD gene sequences available, which include human (Wetmur et al 1986), *E. coli* (Li et al 1989; Echelard et al 1988), rat (Bishop et al 1986), pea (Boese et al 1991) and spinach (Schaumburg et al 1991), indicate that the oxygen insensitivity of plant ALADs arises from replacement of several cysteines with aspartyl residues; this might assist the functioning of ALAD in the plant cell, which will at times be a more oxidizing environment than its mammalian counterpart because of the evolution of oxygen from photosynthesis. The replacement of the probable 'soft' sulphur metal ligands with the 'hard' oxygen metal ligands of the aspartic acid residues may also explain the preference for Mg^{2+} ions (Boese et al 1991).

More recent research has centred on defining the nature of the metal-binding sites of ALAD (Dent et al 1990, Jaffé et al 1992), which are most probably located in a region of the protein between residues 118 and 140 (*E. coli* numbering). The aim of our studies on ALAD from *E. coli* has been to define the role of the oxygen-sensitive thiol groups in metal ion binding. Of the six cysteines present in native *E. coli* ALAD, four are accessible for modification without prior denaturation; when ALAD from *E. coli* is reacted with DTNB the cysteines form two intramolecular disulphide bonds (Spencer & Jordan 1993). ALAD from *E. coli* has been shown to bind two Zn^{2+} ions, one of which is lost on the formation of one equivalent of disulphide bonds per subunit with the concomitant loss of virtually all enzymic activity (Spencer & Jordan 1993). Below, we describe our attempts to identify the nature of the two metal-binding sites in *E. coli* ALAD and their possible function.

Materials and methods

Sources of chemicals were as in Spencer & Jordan (1993), as were methods for assay. The number of sulphydryl groups was determined as previously (Spencer & Jordan 1993), to confirm which enzyme species was being investigated. ApoALAD was generated from the holoenzyme by overnight incubation with 10 mM EDTA in 50 mM potassium phosphate buffer (pH 6) containing 20 mM β-mercaptoethanol followed by gel filtration under nitrogen in potassium phosphate buffer (pH 6 or 8) to remove EDTA and β-mercaptoethanol. Removal of thiols was essential for Co^{2+} binding; a brown precipitate was formed otherwise. Samples for metal analysis in atomic absorption studies were

determined as in Spencer & Jordan (1993). For fluorimetry, apoALAD
(8–50 μM) was titrated with Zn^{2+} ions at 15 °C with excitation at 280 nm (slit
10 nm) and emission at 330 nm (slit 10 nm) in a Hitachi F-2000 fluorimeter.

Effect of pH on the relationship between Zn^{2+} content and ALAD activity

Metal-depleted *E. coli* ALAD (apoALAD) is inactive (Spencer & Jordan 1993).
Readdition of Zn^{2+} ions in 50 mM potassium phosphate buffer, pH 6,
containing 10 mM β-mercaptoethanol restored full activity only on the binding
of two Zn^{2+} ions per subunit (Fig. 1). In contrast, at pH 8 the binding of only
one mol Zn^{2+} per mol ALAD subunits was sufficient to restore near maximum
(81%) specific activity (Fig. 1). A possible explanation for this is that there are
two distinct Zn^{2+}-binding sites, one of which, termed the α site, must be filled
for catalytic activity to be restored. This α site may be occupied preferentially
by Zn^{2+} at pH 8 whereas the other site, termed the β-site, is preferentially filled
at pH 6 but without restoration of activity. Alternatively, binding of metal ion
to the β site at pH 8 may itself be sufficient to restore activity. It has been
reported that on addition of half the stoichiometric amount of Zn^{2+} ions to
bovine apoALAD at pH 7 full activity was restored (Bevan et al 1980, Jaffé

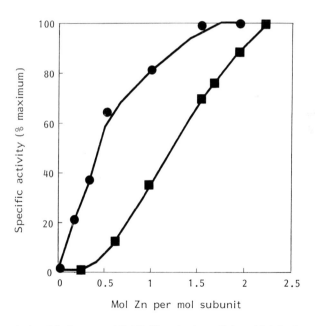

FIG. 1.   The relationship between ALAD (5-aminolaevulinic acid dehydratase) activity
and Zn^{2+} content at pH 6 and pH 8. ALAD from *Escherichia coli* was titrated with Zn^{2+}
ions (0–750 μM) in 50 mM potassium phosphate buffer, pH 8 (●) or pH 6 (■). Samples
were assayed for enzymic activity and Zn^{2+} content after gel filtration in the same buffer.

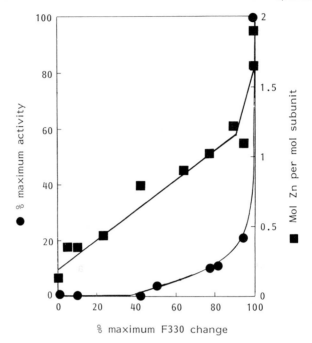

FIG. 2.   ALAD activity (% maximum, ●) and Zn²⁺ content (mol Zn²⁺/mol subunit, ■) against fluorescence (as a percentage of the maximum change) at 300 nm (F330) (with excitation at 280 nm). The enzyme was in 50 mM potassium phosphate buffer, pH 6, 10 mM β-mercaptoethanol at 15 °C throughout.

et al 1984). In contrast, a linear relationship between metal content and activity was reported for human ALAD at pH 7 (Gibbs 1984).

Decrease of fluorescence on binding of Zn²⁺ to apoALAD at pH 6

On addition of Zn²⁺ ions to apoALAD in 50 mM potassium phosphate buffer, pH 6, containing 10 mM β-mercaptoethanol at 15 °C protein fluorescence monitored at 330 nm from excitation at 280 nm decreased. Klotz analysis (Stinson & Holbrook 1973) indicated the presence of one binding site with a K_d of 6 μM (±2 μM) (data not shown). To investigate whether this site is responsible for the Zn²⁺ binding which gives little increase in activity, we established the relationships between enzymic activity, fluorescence change, and Zn²⁺ content (Fig. 2). The binding of 1.1 equivalent of Zn²⁺ was found to restore only 22% of the activity, yet it gave nearly the maximum reduction (94%) of the fluorescence signal at 330 nm, indicating that a Zn²⁺ ion is binding to a site at which its binding is unable to restore activity (the β site).

The exact cause of the fluorescence change has not been established, although

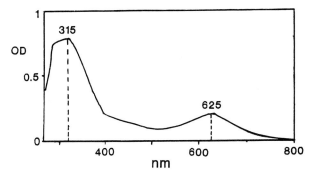

FIG. 3. Absorption spectrum of Co^{2+}-substituted ALAD. The spectrum is a difference scan of Co^{2+}–ALAD (100 μM) against Zn^{2+}–ALAD (100 μM) in 50 mM potassium phosphate buffer, pH 7.8, showing the absorbance peaks at 625 nm ($\varepsilon = 2000\,M^{-1}\,cm^{-1}$) and 315 nm ($\varepsilon = 8000\,M^{-1}\,cm^{-1}$).

similar results were obtained on excitation at 295 nm, a wavelength at which only tryptophan residues are excited. This appears to rule out any direct interaction of the bound Zn^{2+} with tyrosine residues, known to be potential metal ligands. An actual conformational change of the protein may be responsible for the fluorescence decrease. The binding of Mg^{2+} may be reliant on this conformational change, which would account for the inability of apoALAD to bind Mg^{2+} in the absence of bound Zn^{2+} (Spencer & Jordan 1993).

Incorporation of Co^{2+}

Co^{2+} ions ligated to sulphydryl groups in cysteine residues have distinct absorption spectra (Garbett et al 1972). To investigate the possibility that cysteine may be acting as a metal ion ligand in ALAD, we added Co^{2+} (1 mM) to apoALAD (100 μM) in 50 mM potassium phosphate, pH 7.8, in the absence of β-mercaptoethanol but under anaerobic conditions. The resulting protein was green and showed two major peaks on difference spectroscopy against Zn^{2+}–holoALAD (Fig. 3). The absorbance maxima ($\lambda_{max} = 315$ nm and 625 nm) and extinction coefficients ($\varepsilon = 8000$ and $1900\,M^{-1}\,cm^{-1}$ respectively) observed in Co^{2+}–ALAD are suggestive of a charge-transfer band cysteine–S^{-}–Co^{2+} in a tetrahedral environment (Garbett et al 1972). The spectrum is similar to that of fully Co^{2+}-substituted alcohol dehydrogenase ($\lambda_{max} = 340$ nm, $\varepsilon = 15\,500\,M^{-1}\,cm^{-1}$; $\lambda_{max} = 655$ nm, $\varepsilon = 2050\,M^{-1}\,cm^{-1}$), as described by Maret et al (1979). Cysteine residues are known to be ligands for the bound Co^{2+} ions in Co^{2+}-substituted alcohol dehydrogenase.

Subsequent gel filtration of Co^{2+}-substituted ALAD in 50 mM potassium phosphate buffer, pH 7.8, to remove excess Co^{2+}, did not alter the observed spectrum, and atomic absorption spectroscopy showed that 2.1 mol Co^{2+} were

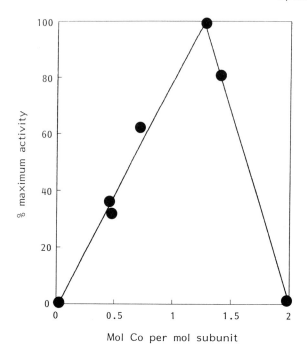

FIG. 4.   ALAD activity against Co^{2+} content. *E. coli* apoALAD (150 μM) was titrated with Co^{2+} ions (0–500 μM) in 50 mM potassium phosphate buffer, pH 8. Activity and Co^{2+} content were determined after gel filtration in the same buffer under nitrogen.

bound per mol subunit and a maximum of only 0.17 mol Zn^{2+} per mol subunit. All the cysteines remained in the reduced form in the Co^{2+}-substituted ALAD.

Relationship between ALAD activity at pH 8 and Co^{2+} content and absorbance

The full spectral characteristic of ALAD with two Co^{2+} bound developed gradually after exposure of apoALAD to 1 mM Co^{2+}. During this time the catalytic activity, although decreasing, was still detectable. A more detailed investigation of Co^{2+} content as related to activity indicated that maximum specific activity (96 U ml^{-1}, similar to that achieved with addition of Zn^{2+}, 80 U ml^{-1} at pH 8) could be restored by the binding of 1.2 mol Co^{2+} per mol subunit, predominantly at the α site (Fig. 4). The subsequent binding of the second mol of Co^{2+} per mol subunit to the β site resulted in inactivation of the enzyme (Fig. 4). Determination of the absorbance spectrum of ALAD containing between 0 and 2 mol Co^{2+} per mol subunit indicated that cysteine

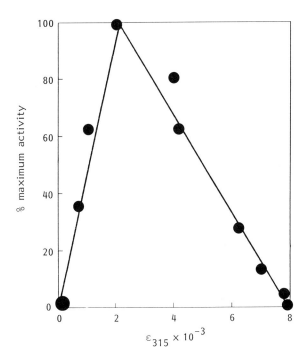

FIG. 5. Co^{2+}-ALAD activity against ε_{315} at pH 8. *E. coli* apoALAD (150 μM) was titrated with Co^{2+} ions (0–500 μM) in 50 mM potassium phosphate buffer, pH 8, then A$_{315}$ and activity were determined.

ligands may be associated with both the α and the β sites. The magnitude of the extinction coefficients at 315 and 625 nm increased markedly from 1600 M^{-1} cm^{-1} and 200 M^{-1} cm^{-1} respectively with one molar equivalent of Co^{2+} bound to 8000 M^{-1} cm^{-1} and 2000 M^{-1} cm^{-1} respectively on the binding of the second molar equivalent of Co^{2+} to the β site, which resulted in the inactivation of ALAD (Fig. 5). The disproportionate increase in the extinction coefficients at 315 and 625 nm indicates there are several cysteine ligands involved in the binding of the second molar equivalent of Co^{2+} to ALAD at the β site. Replacement of Co^{2+} into proteins with known cysteine ligands indicates an extinction coefficient at 310–350 nm of between 1500 and 2000 M^{-1} cm^{-1} per cysteine ligand (Maret et al 1979, Garbett et al 1972). According to this value, the extinction coefficient of 1600 M^{-1} cm^{-1} observed for Co^{2+} binding to the α site of *E. coli* ALAD would implicate the involvement of a single cysteine residue. The further increase in the extinction coefficient at 315 nm on the binding of the second mole of Co^{2+} at the β site suggests the involvement of three or four more cysteine ligands.

The effect of pH on the relationship between Co²⁺–ALAD absorption and activity

Initial binding of Co^{2+} to ALAD was predominantly at the α site at both pH 6 and pH 8, although the affinity for Co^{2+} at pH 6 was lower, as apparent from the need for exogenous Co^{2+} to maintain activity. The increase in activity observed on addition of Co^{2+} ions to ALAD was associated with an increase in absorbance at 315 nm and 625 nm, to give extinction coefficients of around $1600 \, M^{-1} cm^{-1}$ and $300 \, M^{-1} cm^{-1}$, respectively, at both pH 6 and 8. However, although addition of more Co^{2+} at pH 8 led to a further increase in the extinction coefficient, it did not at pH 6. Lowering the pH from 8 to 6 of ALAD with two molar equivalents of Co^{2+} bound, in the presence of $500 \, \mu M$ exogenous Co^{2+}, did not change the extinction coefficients significantly (ε_{315} decreased by less than $500 \, M^{-1} cm^{-1}$). The virtual pH independence of the extinction coefficients at 315 nm and 625 nm indicates that their lower value in Co^{2+}–ALAD at pH 6 is due to the enzyme's inability to bind Co^{2+} at the cysteine-rich, β site at this pH (at least in the presence of a seven-fold molar excess of Co^{2+}; 1 mM). Gel filtration under nitrogen at pH 6 of ALAD with two molar equivalents of Co^{2+} bound at pH 8 removed one molar equivalent of Co^{2+} and the extinction coefficients were reduced from $8000 \, M^{-1} cm^{-1}$ to $6400 \, M^{-1} cm^{-1}$ at 315 nm and from $2000 \, M^{-1} cm^{-1}$ to $1700 \, M^{-1} cm^{-1}$ at 625 nm, indicating that Co^{2+} had been lost from the α site. The resulting $Co^{2+}_{(1\beta)}$–ALAD was still inactive. Although Co^{2+} did not bind to the β site at pH 6, once it bound to the β site at pH 8 lowering the pH to 6 did not result in the loss of Co^{2+} from the β site (as shown by the characteristic spectrum and inactivity of this species), presumably as a result of kinetic factors.

The inability of Co^{2+} to bind to the β site at pH 6 and its slow kinetics of binding at pH 8 permits the determination of the K_d for Co^{2+} at the α site essentially without interference from metal ion binding at the β site. The amount of Co^{2+} bound to the α site could be determined by assessing the maximum percentage change in absorbance with maximum occupancy at the α site (resulting in an $\varepsilon_{315} = 1600 \, M^{-1} cm^{-1}$). The K_d for Co^{2+} for the α site fell from $260 \, \mu M$ at pH 6 to $2 \, \mu M$ at pH 8.

Determination of the number of available cysteines in native Co²⁺–ALAD

Co^{2+} ions (1 mM) were added to apoALAD ($200 \, \mu M$) in 50 mM potassium phosphate buffer, pH 8, under nitrogen, and the absorbance at 315 nm was monitored until binding was complete. On reaction of the Co^{2+}_2–ALAD species with excess DTNB (1 mM) at pH 8, one molar equivalent of TNB was released, indicating that only cysteine residue was available. This is in contrast to Zn^{2+}_2–ALAD and apoALAD where four cysteines are readily available in the native form (Spencer & Jordan 1993). Reaction of $Co^{2+}_{(1\beta)}$–ALAD, produced by gel filtration of Co^{2+}_2–ALAD under nitrogen in 50 mM potassium phosphate buffer, pH 6, also indicated one available cysteine. This reaction with DTNB could be abolished by prior exposure of $Co^{2+}_{(1\beta)}$–ALAD to one molar equivalent of *N*-ethylmaleimide (NEM) at pH 6; at this pH NEM specifically modifies cysteine residues. Subsequent removal of any unreacted NEM by gel filtration showed that only 0.2 mol free cysteine per mol subunit was available after NEM treatment.

Identification of the reactive cysteine of Co^{2+}–ALAD

A sample of Co^{2+}_2–ALAD (3.8 ml of 3.4 mg ml^{-1}, 0.36 μmol) was gel filtered at pH 6 to give $Co^{2+}_{(1\beta)}$–ALAD, which was reacted with 0.8 molar equivalents of [^{14}C] NEM (specific activity 6 μCi μmol^{-1}, 10 560 d.p.m. nmol^{-1}). After one hour excess NEM was removed by gel filtration, yielding 0.22 μmol protein with 1.6×10^6 d.p.m. incorporated (0.7 mol NEM per subunit). The sample was then denatured in 4 M guanidinium-HCl, 200 mM potassium phosphate, pH 8, containing 10 mM β-mercaptoethanol, and the remaining cysteines were blocked by the addition of 20 mM iodoacetic acid. After one hour the sample was dialysed against 2×4 l of 2 mM potassium phosphate, pH 8, then freeze-dried. The sample was redissolved in 50 mM potassium phosphate buffer, pH 8, containing 1 mM EDTA and digested with V8 proteinase for 16 h at 37 °C. Following digestion, 89% of the label remained in the soluble fraction (1×10^6 d.p.m.) which was then applied to a reverse-phase HPLC C_{18} column with a gradient of 0.5% trifluoroacetic acid (TFA) in water:0.1% TFA in acetonitrile. Two labelled peaks were detected (NEM1 and NEM2) containing 40% and 25% of the applied counts. The material in these peaks was subsequently purified further by a second passage down the C_{18} column, each original peak giving a single labelled peak. Yields of NEM1 and NEM2 during purification are shown in Table 1.

The purified peptides were sequenced by Edman degradation and residues were identified as the phenylthiohydantoin (PTH)· derivatives. In a second sequencing experiment the amino acids were modified to the anilinothiazolinone (ATZ) derivatives and then collected for radioactive counting, giving a 30–50% recovery of label (40–60% allowing for the repetitive yield). The results of the

TABLE 1 Recovery of ^{14}C-labelled ALAD-derived peptides on HPLC

	1st passage			2nd passage			
	% Total d.p.m.	Label (nmol)	% Yield	% Total d.p.m.	Label (nmol)	% Yield	Final % yield
NEM1	40	12	13	85	4	33	4
NEM2	25	11	12	80	3	27	3

TABLE 2   Sequence and labelling patterns of ^{14}C-labelled ALAD-derived peptides[a]

	NEM1[b]		NEM2[c]	
	pmol (PTH)	Radioactivity (ATZ) (%)	pmol (PTH)	Radioactivity (ATZ) (%)
Tyr-124	92	—	48	—
Thr	—	—	45	—
Ser	33	—	23	—
His	—	—	52	—
Gly	40	—	28	—
His	27	—	32	8
Cys-130	—	27	—	60
Glu	30	13	14	16
Val	16	4	5	8
Leu	22	—	6	—
Cys-134	—	30	—	—
Glu-135	4	19	—	—

[a]ALAD was labelled with N-[14C]ethylmaleimide (NEM) before digestion and subsequent purification of peptides by HPLC (Table 1). The two labelled peptide peaks derived, NEM1 and NEM2, were sequenced by Edman degradation and amino acids were analysed as the phenylhydantoin (PTH) derivatives.
[b]For peptide NEM1, 200 pmol were sequenced as the PTH derivatives. A total of 10 000 d.p.m. were sequenced, of which 40% was recovered on derivatization to the anilinothiozoline (ATZ) form. Another peptide, with the sequence RTELIGAYQVSG, co-purified with peptide NEM1.
[c]For peptide NEM2, 100 pmol were sequenced as the PTH derivatives. 2700 d.p.m. were sequenced, of which 25% was recovered in the ATZ-derivatized form. Another peptide, with the sequence DGLVARMSPICKQ, co-purified with peptide NEM2.

peptide sequencing are given in Table 2. The protein sequence determined agreed with both of the published gene sequences for *E. coli* ALAD. The counts obtained in NEM1 were equally distributed between Cys-130 and Cys-134 (Table 2). NEM1 accounted for 40% of the total counts initially incorporated, therefore each cysteine labelled accounts for 20% of the initial label incorporated. The counts obtained in NEM2 were found only at Cys-130; because NEM2 accounted for 25% of the initial counts incorporated, this site must therefore contain 25% of the initial label incorporated. Combining the total number of counts at each site indicates that the majority of the label (70%) was incorporated at Cys-130, with a smaller amount (30%) being incorporated at Cys-134.

Discussion

From the experiments described above it is clear that *E. coli* ALAD has two distinct metal-binding sites. On binding of metal ion to the α site, activity is restored to the apoALAD. The marked effect of pH on the Zn^{2+} content required for enzyme activity suggests that the metal-binding sites have different pH-dependent affinities. The relative affinity of the two sites may be affected by pH such that the α site is filled first by Zn^{2+} at pH 8 whereas the β site is filled first at pH 6. The alternative possibility, that the binding of Zn^{2+} to the β site supports activity at pH 8 but not at pH 6, does not appear to be the case, because addition of one molar equivalent of Zn^{2+} to apoALAD at pH 8 leaves the β site free and capable of binding Co^{2+}. This generates an inactive mixed metalloALAD species with an extinction coefficient at 315 nm of $6400 M^{-1}$, indicative of Co^{2+} at the β site (P. Spencer & P. M. Jordan, unpublished results). Affinity of binding of Co^{2+} to the α site shows a more marked pH dependence (see p 58), similar to that noted for the affinity of Zn^{2+} for glycerol dehydrogenase (Spencer et al 1989), where two histidines have been implicated in Zn^{2+} binding (Spencer et al 1991). The findings are also consistent with extended X-ray absorption fine structure (EXAFS) studies on bovine ALAD which indicated that two histidine ligands may be present at a site termed the A site at which binding of Zn^{2+} restored activity (Dent et al 1990).

The interaction between *E. coli* ALAD and Co^{2+} ions has been shown here to be sensitive to exogenous thiols and the pH of the buffer. Whether, or to what extent, addition of Co^{2+} ions to *E. coli* apoALAD stimulates enzymic activity depends on the amount of Co^{2+} that binds and whether it does so at the α or β site. This effect may also underlie conflicting reports on whether bovine apoALAD is active or inactive in the presence of Co^{2+} ions (Bevan et al 1980, Cheh & Neilands 1976). The failure to regenerate active bovine Co^{2+}–ALAD (Jaffé et al 1984) may also be due to the presence of β-mercaptoethanol, which gives a brown precipitate. It is interesting that studies on inhibition of human ALAD by lead ions also provided indirect evidence for two different metal-binding sites (Gibbs et al 1985).

The magnitude of the charge-transfer bands arising on Co^{2+} binding may give some indication about the actual number of cysteine ligands contributing to each metal-binding site. The data are consistent with there being one cysteine ligand at the α site and three or four cysteines at the β site. Because only four cysteines appear to be accessible in native *E. coli* ALAD it is likely that one of these is at the α site and that the remaining three are acting as ligands at the β site. These results are comparable with those from bovine ALAD, where EXAFS studies suggest one and four cysteine residues contribute to the A and B sites respectively (Dent et al 1990).

The labelling of the available cysteine in $Co^{2+}{}_{(1\beta)}$–ALAD strongly suggests that this cysteine, Cys-130, is involved in ligation of the α-site (catalytic) metal ion. The implication that Cys-130 acts as a ligand at the α site can be used to direct further experiments, although the single available cysteine residue in the enzymically inactive $Co^{2+}{}_2$–ALAD species (possibly arising from an altered ligation state) is not necessarily the same cysteine ligand as that operating in

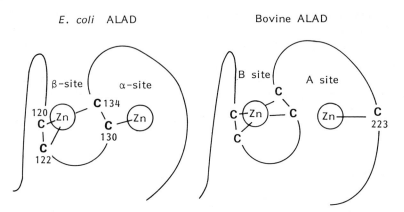

FIG. 6.   Schematic illustration of the proposed role of the cysteine residues in metal ligation at the α and β sites of *E. coli* ALAD (*left*) and, for comparison, the A and B metal ion-binding sites of bovine ALAD (*right*; Dent et al 1990, Jaffé et al 1992).

the active $Co^{2+}{}_{(1\alpha)}$–ALAD species. Chemical modification of bovine ALAD with 5-chlorolaevulinic acid (CLA), a substrate analogue, identified two possible ligands to the catalytic metal, a cysteine residue and a histidine residue (Jaffé et al 1992). However, in the absence of any reported stoichiometry for CLA incorporation it is unclear whether these two residues were equally labelled or if one or the other was the major modification site. The residues in *E. coli* ALAD equivalent to those modified in bovine ALAD by CLA are Ser-229 in place of the Cys and Gln-270 in place of the His (numbering as in Spencer & Jordan 1993). Ser-229 in *E. coli* is replaced by Thr in plant ALADs (which use Mg^{2+}) and Cys in mammalian ALADs (which use Zn^{2+}), suggesting a role in metal binding and selectivity. Because *E. coli* can utilize Zn^{2+} or Mg^{2+} at the α site, the alteration of Cys-229, a soft ligand, to Ser, a hard ligand, may facilitate the additional acceptance of Mg^{2+} at the α site. Such ligand substitutions have been proposed for the replacement of cysteine by acidic residues, hard ligands, at the proposed metal-binding domain in plant ALADs (Jordan 1991, Boese et al 1991). Whether plant ALADs also have structural and catalytic metal-binding sites remains to be established.

The proposed roles of the cysteine residues in metal ion binding at the α and β sites in *E. coli* ALAD are shown in Fig. 6; the roles of the cysteine ligands in the proposed A and B metal-binding sites of bovine ALAD suggested by EXAFS studies (Dent et al 1990) and chemical labelling studies (Jaffé et al 1992) are also shown.

The exact role of the metal ions in the functioning of ALAD is, as yet, unclear. It is known that Schiff's base formation with the substrate binding at the P site of ALAD can proceed in the absence of any bound metal ion in ALAD from *E. coli* (P. Spencer & P. M. Jordan, unpublished results) and in bovine

ALAD (Chaudhry et al 1976, Jaffé et al 1984). Therefore, restoration of activity on metal ion binding to the α site in *E. coli* ALAD (the equivalent of the A metal-binding site in bovine ALAD) would appear to relate to the binding of the second ALA substrate molecule, which gives rise to the acetic acid side of porphobilinogen. The precise role, if any, played by the metal ion in subsequent bond formation remains to be established, although it is significant that all the metal ions known to bind at this site (Zn^{2+}, Co^{2+} and Mg^{2+}) could support catalysis by acting as a Lewis acid to promote the aldol condensation.

Acknowledgements

The SERC (MRI) and AFRC (PMB) are gratefully acknowledged for financial support. Thanks are due to Dr M. Gore and Mr L. Hunt at the Protein Sequencing Unit, Southampton University, for carrying out the Edman sequencing of peptides, and to Dr C. Rossner, for providing the *E. coli hemB* strain.

References

Anderson PM, Desnick RJ 1979 Purification and properties of δ-aminolevulinate acid dehydratase from human erythrocytes. J Biol Chem 254:6924–6930

Bevan DR, Bodlaender P, Shemin D 1980 Mechanism of porphobilinogen synthase. Requirement of Zn^{2+} for enzyme activity. J Biol Chem 255:2030–2035

Bishop TR, Cohen PJ, Boyer SH, Noyes AN, Frelin LP 1986 Isolation of a rat liver δ-ALAD cDNA clone—evidence for unequal ALAD gene dosage among inbred mouse strains. Proc Natl Acad Sci USA 83:5568–5572

Boese QF, Spano AJ, Li J, Timko MP 1991 Aminolevulinic acid dehydratase in pea (*Pisum sativum* L.). Identification of an unusual metal-binding domain in the plant enzyme. J Biol Chem 266:17060–17066

Borralho LM, Ortiz CHD, Panek AD, Mattoon JR 1990 Purification of δ-aminolevulinic acid dehydratase from genetically engineered yeast. Yeast 6:319–330

Chaudhry AG, Gore MG, Jordan PM 1976 Studies on the inactivation of 5-aminolevulinic dehydratase by alkylation. Biochem Soc Trans 4:301–303

Cheh AM, Neilands JB 1976 The δ-aminolevulinic acid dehydratases. Struct Bonding 29:123–169

Dent AJ, Beyersmann D, Block C, Hasnain SS 1990 Two different zinc sites in bovine 5-aminolevulinate dehydratase distinguished by extended X-ray absorption fine structure. Biochemistry 29:7822–7828

Echelard Y, Dymetryszyn J, Drolet M, Sasarman A 1988 Nucleotide sequence of the *hemB* gene of *Escherichia coli* K12. Mol & Gen Genet 214:503–508

Garbett KG, Partridge GW, Williams RJP 1972 Models for metal thiolate complexes in enzymes. Bioinorg Chem 1:309–329

Gibbs PNB 1984 Mechanistic studies on enzyme-catalysed dimerization reactions. PhD thesis, University of Southampton, UK

Gibbs PNB, Gore MG, Jordan PM 1985 Investigation of the effect of metal ions on the reactivity of thiol groups in human 5-aminolevulinate acid dehydratase. Biochem J 225:573–850

Jaffé EK, Salowe SP, Chen NT, DeHaven PA 1984 Porphobilinogen synthase modification with methylmethanethiousulfonate. J Biol Chem 259:5032–5036

Jaffé EK, Abrams WR, Kaempfen HX, Harris KA 1992 5-Chlorolevulinate modification of porphobilinogen synthase identifies a potential role for the catalytic zinc. Biochemistry 31:2113–2123

Jordan PM 1991 The biosynthesis of 5-aminolaevulinic acid and its transformation into uroporphyrinogen III. In: Jordan PM (ed) Biosynthesis of tetrapyrroles. Elsevier, Amsterdam (New Compr Biochem 19) p 1–66

Jordan PM, Seehra JS 1980 ^{14}C NMR as a probe for the study of enzyme catalysed reactions. Mechanism of action of 5-aminolevulinic acid dehydratase. FEBS (Fed Eur Biochem Soc) Lett 114:283–286

Jordan PM, Seehra JS 1986 Purification of porphobilinogen synthase from bovine liver. Methods Enzymol 123:427–434

Li JM, Russell CS, Cosloy SD 1989 The structure of the *Escherichia coli hemB* gene. Gene 75: 177–184

Liedgens W, Lütz C, Schneider HAW 1983 Molecular properties of 5-aminolevulinic acid dehydratase from *Spinacia oleracea*. Eur J Biochem 135:75–79

Maret W, Andersson I, Dietrich H, Schneider-Bernlörh H, Einarsson R, Zeppezauer M 1979 Site-specific substituted cobalt (II) horse liver alcohol dehydrogenases. Eur J Biochem 98:501–512

Nandi DL, Shemin D 1968 δ-Aminolevulinic acid dehydratase of *Rhodopseudomonas sphaeroides*. 2. Association to polymers and dissociation to subunits. J Biol Chem 243:1231–1235

Nandi DL, Baker-Cohen KF, Shemin D 1968 δ-Aminolevulinic acid dehydratase of *Rhodopseudomonas sphaeroides*. 1. Isolation and properties. J Biol Chem 243:1224–1230

Schaumburg A, Schneider-Poetsh AAW, Eckerskorn C 1991 Nucleotide sequence of the *hemB* gene from *Spinacia oleracea*. Z Naturforsh Sect C Biosci 47:77–84

Shemin D 1972 δ-Aminolaevulinic acid dehydratase. In: Boyer PD (ed) The enzymes, 3rd edn. Academic Press, New York, vol 7:323–337

Spencer P, Jordan PM 1993 Purification and characterization of 5-aminolaevulinic acid dehydratase from *Escherichia coli* and a study of the reactive thiols at the metal-binding domain. Biochem J 290:279–287

Spencer P, Brown, KJ, Scawen MD, Atkinson T, Gore MG 1989 Isolation and characterisation of the glycerol dehydrogenase from *Bacillus stearothermophilus*. Biochim Biophys Acta 994:270–279

Spencer P, Scawen MD, Atkinson T, Gore MG 1991 The identification of a structurally important cysteine residue in the glycerol dehydrogenase from *Bacillus stearothermophilus*. Biochim Biophys Acta 1073:386–393

Stinson RA, Holbrook JJ 1973 Equilibrium binding of nicotinamide nucleotides to lactate dehydrogenase. Biochem J 131:719–728

Wetmer JG, Bishop DF, Cantelmo C, Desnick RJ 1986 Human δ-aminolevulinate dehydratase: nucleotide sequence of a full-length cDNA clone. Proc Natl Acad Sci USA 83:7703–7707

DISCUSSION

Castelfranco: You said that plant ALADs require Mg^{2+}. Do any use Zn^{2+} instead of, or as well as, Mg^{2+}?

Spencer: There are no reports of any plant enzyme utilizing zinc for ALAD activity. We are now investigating the pea enzyme, which is activated by magnesium but not zinc.

Castelfranco: Are there still two sites?

Spencer: Preliminary results with equilibrium dialysis suggest there are two sites, though I should stress that we use Co^{2+} in those experiments because it binds more tightly than Mg^{2+} to the pea enzyme.

Warren: The extracted *E. coli* enzyme can use just Zn^{2+}, or a combination of Mg^{2+} and Zn^{2+}. Physiologically, do you think the enzyme uses only Zn^{2+}, or the combination?

Spencer: The isolated enzyme is more active when it has Mg^{2+} and Zn^{2+} bound. If we extract the enzyme into zinc-containing buffer, it contains Zn^{2+} at both sites. If we extract into buffer containing Mg^{2+}, activity is stimulated. Which metal ions the enzyme uses physiologically probably depends largely on which are available. We are overexpressing the enzyme, so presumably whatever metals we put into the growth medium will be incorporated.

Warren: I wouldn't have thought the intracellular concentration of Mg^{2+} would be limiting.

Spencer: I don't know about that, but I know the zinc concentration is.

Beale: Plant ALAD has been reported to be either an octamer (Huault et al 1987) or a hexamer (Liedgens et al 1980). Is that question now resolved?

Spencer: Spinach ALAD is reported to be a hexamer (Liedgens et al 1983). Our studies suggest that the pea enzyme is an octamer.

Beale: The mammalian enzyme is a zinc enzyme, but might Mg^{2+} be involved with that enzyme too?

Spencer: The bovine enzyme seems not to work with Mg^{2+}, but this may depend on which sites are filled, on whether half the sites contain zinc which might allow the other half to bind Mg^{2+}. I don't know whether all these possibilities have been studied thoroughly.

Beale: Shemin reported some time ago that bovine liver ALAD shows half-of-the-sites reactivity, and that the dialysed enzyme is reactivated when four mol Zn^{2+} are incorporated per mol octamer (Shemin 1976).

Spencer: That appears to be the case with the bovine enzyme, but the *E. coli* enzyme loses all activity on incorporation of one ALA per subunit. We don't see half-sites reactivity with respect to ALA incorporation in the *E. coli* dehydratase.

Arigoni: I gather from what you said that the mechanistic role of the metal ions is not understood.

Spencer: That's correct. It has been proposed that they might work as Lewis acids, polarizing the carbonyl bond of the substrate.

Beale: There was a report that the bovine liver enzyme was still active, even after Zn^{2+} extraction, under anaerobic conditions in which thiol oxidation is prevented (Tsukamoto et al 1979). Is that still held to be correct?

Spencer: When handling metal-depleted metalloenzymes it is difficult to prevent the apoenzyme from acquiring exogenous metal ions from the buffer, the concentration of which are often in the 0.5–1 μM range. We assayed *E. coli* apoALAD at 100 μM, so that even if the maximum amount of metal ions had been picked up, this would represent only 1% of the maximum activity. However, Tsukamoto et al (1979) assayed bovine ALAD at 0.5 μM such that the small amounts of metal ions that might be acquired from the assay buffer could restore activity to near maximum. Their finding that bovine apoALAD maintains activity has not been replicated at higher enzyme concentrations.

Beale: So in your hands the enzyme absolutely requires zinc, even under anaerobic conditions.

Spencer: Yes.

Scott: Many years ago I saw a slide shown by David Shemin of crystals of this enzyme. What progress has been made in that direction?

Spencer: We can obtain crystals readily, and they do diffract, but their lifetime in the synchrotron beam is not great. We have run into problems in making metal derivatives, probably because exchange of metals with those already bound to the enzyme changes the conformation. For example, when we insert two Co^{2+} ions, the enzyme binds these ions but is no longer active. We have not yet succeeded in crystallizing the apoenzyme.

Leeper: 5-Chlorolaevulinic acid and 3-chlorolaevulinic acid alkylate different cysteines in the bovine enzyme. Have you tried that with *E. coli* ALAD?

Spencer: We used iodolaevulinic acid to attach the substrate analogue for crystallographic studies. I haven't done detailed studies to identify the labelling site or labelling studies with chlorolaevulinic acid. Iodolaevulinic acid inactivates the enzyme, but I don't know which cysteine it alkylates.

Akhtar: If you can get full activity with only one Co^{2+} ion bound per subunit, that ion must be fulfilling the structural role as well as the catalytic role.

Spencer: The same applies to zinc. At pH 8 one molar equivalent of Zn^{2+} will give almost fully active enzyme.

Akhtar: Do you have any instinctive feeling about whether under those conditions the metal ion is going to the so-called α site, the catalytic site?

Spencer: If we bind Zn^{2+} first to the α site and then put Co^{2+} in at the β site, we see an extinction coefficient of around $6000\,M^{-1}\,cm^{-1}$, corresponding to the β site being filled with cobalt, and the enzyme is inactive.

Jordan: The extinction coefficient wouldn't change as it does if the Co^{2+} ion were not binding to sulphur ligands; when cobalt interacts with sulphur ligands, it gives the green coloration.

Akhtar: How do you interpret the fact that only one mole of metal ions per subunit gives full activity, whereas the enzyme normally requires two metal ions, physiologically, one for the structural site and one for the catalytic site?

Jordan: I don't really know whether under physiological conditions you could say that the β site, the structural site, is important. Under the experimental conditions we use, either Co^{2+} or Zn^{2+} can fulfil the catalytic role with the β site still unfilled.

Warren: Is there much sequence identity between the ALADs from *E. coli* and pea?

Spencer: Yes, around 48%.

Warren: So one could reasonably assume that the two metals, magnesium and zinc, are playing the same kind of role in catalysis.

Spencer: They could well be.

Warren: I'm really trying to ask why the plant enzymes use magnesium rather than zinc.

Spencer: There are two possible reasons. Photosynthesis generates oxygen, so an enzyme with a lot of cysteines might not be ideal in plants because oxidation might inactivate it. The other possibility is that the Mg^{2+} is also playing a regulatory role in chlorophyll production—the system is producing chlorophyll, which binds Mg^{2+}. The K_d of the pea enzyme for Mg^{2+} is about $700\,\mu M$, which is quite poor for a metalloenzyme, so a large amount of free magnesium, not in chlorophyll, would be required to stimulate the enzyme.

A. Smith: Oxygen is unlikely to be important, because the ALAD in cyanobacteria, which evolve oxygen, has a zinc-binding site rather than a magnesium-binding site (Jones et al 1994).

Stamford: As well as the cysteines which you have indicated are necessary for metal chelation, there appear to be some histidine residues that are conserved between the various ALADs. Have you done any experiments to see if any of those residues are also involved in metal chelation?

Spencer: No.

Scott: Jaffé did some ^{13}C NMR experiments of the half reaction, i.e., with the first ALA adduct (Jaffé et al 1990). Jaffé actually observed the Schiff's base as well as the reduced Schiff's base. Using the latter, with the ^{13}C label properly placed in the ALA, you might be able to begin to use NMR spectroscopy to investigate the relationship of the metal ion to the substrate bound in the half reaction. The edited heteronuclear experiment or the cobalt NMR experiment would be very good, although the M_r is high, of course.

Akhtar: What is the M_r limit for NMR of proteins? Some say it is 15 000, while others say that new techniques will take you to 60 000.

Scott: Some years ago, although it's always dangerous to make predictions, I wrote that 30 000 was the maximum, and I was proved wrong when someone did NMR spectroscopy on a protein of M_r 50 000. We have now seen a perfectly respectable spectrum of a ^{13}C-enriched ligand bound to an enzyme of M_r 120 000. With the edited heteronuclear experiment, the ^{13}C label allows the inverse detection of a single proton, and the NOESY from that proton to the backbone of the enzyme (Ortiz et al 1991) gives a tremendous amount of information, even at such a high M_r. Jaffé's ALAD was an intact octamer of about 280 000, and she got good signals from that.

Arigoni: Could the ^{13}C content of selected amino acids in the sequence be increased, so that ^{13}C NMR could be used to probe for specific contact with the substrate?

Scott: You could grow the bacteria in minimal medium enriched with ^{13}C-labelled glucose, but it's difficult to get [^{13}C]glucose at the moment in the USA because this is a popular technique and Ad Bax at NIH has bought up almost the entire world supply. The situation will improve soon though.

Akhtar: Could you not selectively enrich with particular ^{13}C-labelled amino acids?

Scott: You can feed ^{13}C-labelled amino acids and get reasonable incorporation in some cases, but this is expensive.

Arigoni: The reaction catalysed by ALAD was always regarded as provocative by organic chemists, because if you try to mimic what goes on, by non-enzymic means, the reaction normally takes a different course. The only exception has been put on record by Ian Scott and I wonder if anyone else has taken up the problem since.

Scott: I have seen research proposals but no successful results. The best we could achieve was a rather miserable yield of less than 10% after 11 days in the presence of the resin; otherwise we got only isoporphobilinogen, the C-3 aminopyrrole.

Arigoni: Dr Spencer, there are some mismatches between the gene sequence and the protein sequence of ALAD from *E. coli*. How often does this happen, and, when it does, which does one trust? Also, what are the sources of errors?

Spencer: I would tend to trust the protein sequence more. If you miss a base in the DNA sequence, and insert another downstream, you can get sections which are completely mismatched. For example, comparison of the two gene sequences for *E. coli* ALAD of Echelard et al (1988) and Li et al (1989) revealed a mismatch from 18 to 42 resulting from a frame-shift. We also picked up several discrepancies between the predicted amino acid sequence of *E. coli* ALAD and that obtained by protein sequencing (see Spencer & Jordan 1993).

Arigoni: It is possible to check the molecular mass of proteins predicted by nucleotide and amino acid sequences. Do such data fit with expectations for *E. coli* ALAD?

Spencer: Electrospray mass spectrometry of the purified peptides I described agreed with both the gene-derived protein sequence and that obtained by protein sequencing. However, other *E. coli* ALAD peptides we have studied have given a mass consistent with that of the determined protein sequence but not the gene-derived sequence. The molecular mass of the entire protein as measured by electrospray mass spectrometry, $35\,500 \pm 3$ Da, is 46 ± 3 Da higher than that predicted from what we know after comparison of the gene sequence of Li et al (1989) with our regions of protein sequence, so a mistake may remain in the gene-derived amino acid sequence.

Abell: We have done a lot of electrospray mass spectrometry on many different proteins, and I would say that 10–15% of them come out with a mass different from that predicted; in one or two cases the discrepancy was tracked down to mistakes in the sequence, but this hasn't been checked yet in the other cases.

Arigoni: How large are the discrepancies? One amino acid, normally?

Abell: Yes, about one amino acid. Occasionally the discrepancies arise from frame-shifts during DNA sequencing.

Arigoni: Can you identify the missing amino acid by looking at the difference in the molecular mass?

Abell: No, because the differences are usually small, unless a section has been missed off the end, somewhere at the C-terminus perhaps.

Akhtar: Your group has a lot of experience with this technique. What is the largest protein for which you have used the technique successfully?

Abell: Using the quadripole detector, we can go routinely to 50 000–80 000 M_r. Using a sector machine, with some manipulation of the data afterwards, Dr Staunton at Cambridge has looked at proteins of up to 300 000 M_r.

Griffiths: What sort of resolution do you get at that molecular mass?

Abell: For proteins of around 50 000 M_r, we use of the order of 150–200 pmol of protein, i.e., a few micrograms. The mass resolution is about one part in 10 000.

References

Echelard Y, Dymetryszyn J, Drolet M, Sasarman A 1988 Nucleotide sequence of the *hemB* gene of *Escherichia coli* K12. Mol & Gen Genet 214:503–508

Huault C, Aoues A, Colin P 1987 Reconsidération de la sous-unitaire de la δ-aminolévulinate déshydratase de feuilles d'épinard. C R Acad Sci Paris Ser III Sci Vie 305:671–676

Jaffé EK, Markham GD, Rajagopalan JS 1990 ^{15}N and ^{13}C NMR studies of ligands bound to the 280,000-dalton protein porphobilinogen synthase elucidate the structures of enzyme-bound product and Schiff base intermediate. Biochemistry 29:8345–8350

Jones ME, Jenkins JM, Smith AG, Howe CJ 1994 Plant Mol Biol, in press

Li JM, Russell CS, Cosloy SD 1989 The structure of the *Escherichia coli hemB* gene. Gene 75:177–184

Liedgens W, Grützmann R, Schneider HAW 1980 Highly efficient purification of the labile plant enzyme 5-aminolevulinate dehydratase (EC 4.2.1.24) by means of monoclonal antibodies. Z Naturforsch Sect C Biosci 35:958–962

Liedgens W, Lütz C, Schneider HAW 1983 Molecular properties of 5-aminolevulinic acid dehydratase from *Spinacia oleracea*. Eur J Biochem 135:75–79

Ortiz C, Tellier C, Williams HJ, Stolowich NJ, Scott AI 1991 Diastereotopic covalent binding of the natural inhibitor leupeptin to trypsin: detection of two interconverting hemiacetals by solution and solid-state NMR spectroscopy. Biochemistry 30:10026–10034

Shemin D 1976 Structure, function, and mechanism of δ-aminolevulinic acid dehydratase. J Biochem 79:37P–39P

Spencer P, Jordan PM 1993 Purification and characterization of 5-aminolaevulinic acid dehydratase from *Escherichia coli* and a study of the reactive thiols at the metal-binding domain. Biochem J 290:279–287

Tsukamoto I, Yoshinaga T, Sano S 1979 The role of zinc with special reference to the essential thiol groups in δ-aminolevulinic acid dehydratase of bovine liver. Biochim Biophys Acta 570:167–178

The biosynthesis of uroporphyrinogen III: mechanism of action of porphobilinogen deaminase

Peter M. Jordan*

School of Biological Sciences, Queen Mary and Westfield College, University of London, Mile End Road, London E1 4NS, UK

Abstract. The biosynthesis of the uroporphyrinogen III macrocycle from porphobilinogen requires the sequential participation of two enzymes— porphobilinogen deaminase (1-hydroxymethylbilane synthase, EC 4.3.1.8) and uroporphyrinogen III synthase (cosynthase, EC 4.2.1.75). The product of the deaminase-catalysed reaction is a highly unstable 1-hydroxymethylbilane called preuroporphyrinogen which acts as the substrate for the uroporphyrinogen III synthase, resulting in the exclusive formation of uroporphyrinogen III. In the absence of the synthase, preuroporphyrinogen cyclizes spontaneously to give uroporphyrinogen I. Porphobilinogen deaminase contains a dipyrromethane cofactor that acts as a primer onto which the tetrapyrrole chain is built. The assembly process occurs in stages through enzyme–intermediate complexes, ES, ES_2, ES_3 and ES_4. The negatively charged carboxylates of the cofactor, substrate and intermediate complexes interact with positively charged amino acid side chains in the catalytic cleft. Mutagenesis of conserved arginines has dramatic effects on the assembly of the dipyrromethane cofactor and on the tetrapolymerization process. During the polymerization, the enzyme changes conformation to accommodate the elongating pyrrole chain. The structure of the deaminase from *Escherichia coli* has been determined by X-ray crystallograpy at 1.9Å resolution and gives important insight into the enzymic mechanism. Aspartate 84 plays a key role in catalysis and its substitution by glutamate reduces k_{cat} by two orders of magnitude.

1994 The biosynthesis of the tetrapyrrole pigments. Wiley, Chichester (Ciba Foundation Symposium 180) p 70–96

The biosynthesis of uroporphyrinogen III, the universal precursor of haem, chlorophyll and all other tetrapyrroles, requires the participation of two enzymes, porphobilinogen deaminase (1-hydroxymethylbilane synthase, EC 4.3.1.8) and uroporphyrinogen III synthase (EC 4.2.1.75, also known as cosynthase), as shown in Fig. 1. In the first stage, porphobilinogen deaminase catalyses the

---

Present address: Department of Biochemistry, School of Biological Sciences, University of Southampton, Bassett Crescent East, Southampton SO9 3TU, UK

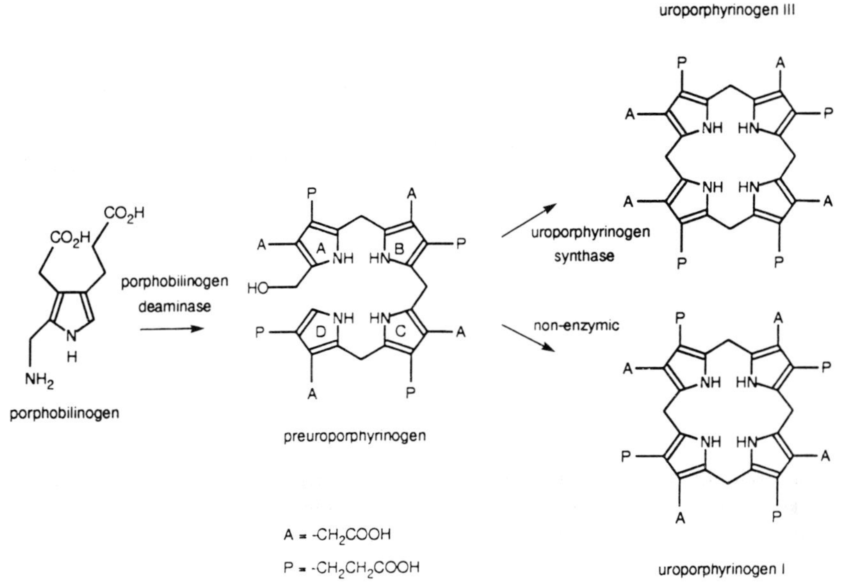

FIG. 1. The synthesis of uroporphyrinogens I and II.

deamination and step-wise polymerization of four molecules of the monopyrrole, porphobilinogen, to give a highly unstable 1-hydroxymethylbilane named preuroporphyrinogen. In the second stage, preuroporphyrinogen is transformed into uroporphyrinogen III by uroporphyrinogen III synthase in a reaction involving rearrangement of ring D and cyclization to give the asymmetric macrocyclic system characteristic of natural tetrapyrroles. In the absence of uroporphyrinogen III synthase, preuroporphyrinogen cyclizes non-enzymically without rearrangement, to give uroporphyrinogen I. The reactions catalysed by the deaminase and synthase have been reviewed recently (Jordan 1991).

Sources of porphobilinogen deaminase

Porphobilinogen deaminases have been purified to homogeneity from many sources, including *Rhodobacter sphaeroides*, spinach, human erythrocytes, *Chlorella regularis*, *Euglena gracilis* and rat liver. The enzyme has also been isolated from recombinant strains of *Escherichia coli*. All the deaminases are monomeric proteins, of 34 000–44 000 M_r, and most have similar properties including remarkable heat stability, pH optima of 8.0–8.5 and isoelectric points between pH 4 and 4.5. The K_m values for porphobilinogen are all in the low μmolar range (see Jordan 1991 for a review).

Substrate requirements for porphobilinogen deaminase

Porphobilinogen deaminase will accept only porphobilinogen as a substrate and cannot use effectively aminomethyldipyrromethanes or aminomethyltripyrranes which, instead, act as enzyme inhibitors. However, the enzyme will deaminate the linear 1-aminomethylbilane ($NH_2CH_2APAPAPAP$, where A = acetate and P = propionate). Although it is a poor substrate, its incubation with the combined deaminase–synthase system produces uroporphyrinogen III. These observations led to an earlier proposal that this 1-aminomethylbilane is the product of the deaminase-catalysed reaction and acts as the substrate for the uroporphyrinogen III synthase (Battersby et al 1978). Although this 1-aminomethylbilane is not in fact a natural intermediate of the tetrapyrrole pathway, it was used to establish that ring D of uroporphyrinogen III is rearranged at the tetrapyrrole (bilane) stage. Studies with ^{13}C NMR confirmed that a single intramolecular rearrangement of ring D occurred and that the terminal pyrrole ring carrying the aminomethyl group in 1-aminomethylbilane gives rise to ring A in uroporphyrinogen III (reviewed in Battersby 1978).

Preuroporphyrinogen, the product of porphobilinogen deaminase

The discovery of preuroporphyrinogen was a major breakthrough in our understanding of how porphobilinogen deaminase and uroporphyrinogen III synthase participate in the biosynthesis of uroporphyrinogen III. NMR signals from preuroporphyrinogen were first observed on incubation of porphobilinogen deaminase with [11-^{13}C]porphobilinogen in an NMR spectrometer when resonances at $\delta = 23$ p.p.m. and $\delta = 57$ p.p.m., integrating in the ratio of 3:1, were formed transiently (Burton et al 1979). These observations eliminated the 1-aminomethylbilane as an intermediate, and the analogous 1-hydroxy-methylbilane, termed preuroporphyrinogen, was proposed as an alternative (Burton et al 1979). Preuroporphyrinogen spontaneously forms uroporphyrinogen I with a $t_{1/2}$ of only 4.5 min at pH 8. However, most importantly, on incubation of preuroporphyrinogen with homogeneous uroporphyrinogen III synthase, without the deaminase, a rapid and quantitative formation of uroporphyrinogen III followed, establishing preuroporphyrinogen as the true substrate for the synthase (Jordan et al 1979). Thus the two enzymes were shown for the first time to function independently and sequentially in the overall transformation of porphobilinogen into uroporphyrinogen III.

Porphobilinogen deaminase was shown to catalyse the deamination of the 1-aminomethylbilane to preuroporphyrinogen, which then acted as the substrate for the synthase, explaining the earlier observation that uroporphyrinogen III can be formed from the 1-aminomethylbilane in the presence of both enzymes (Battersby et al 1978). Preuroporphyrinogen was subsequently synthesized chemically and shown unambiguously to be a 1-hydroxymethylbilane (Battersby et al 1979a), with chemical and enzymic properties identical to those of the intermediate demonstrated earlier in the enzymic experiments (Burton et al 1979; Jordan et al 1979), thus eliminating other proposed structures (Burton et al 1979). Further investigations (Jordan & Berry 1980) with uroporphyrinogen III synthases from several sources firmly established preuroporphyrinogen as the universal precursor of uroporphyrinogen III. The understanding of how the two enzymes function allowed the development of a rapid direct assay for uroporphyrinogen III synthase based on the fact that preuroporphyrinogen cyclizes to uroporphyrinogen I more slowly than the uroporphyrinogen III synthase-catalysed reaction gives uroporphyrinogen III (Jordan 1991).

The order of assembly of the four pyrrole rings

Single turnover reactions with a radiochemical approach (Jordan & Seehra 1979) or a ^{13}C NMR method (Battersby et al 1979b) established that preuroporphyrinogen is assembled by the deaminase with ring A binding first to the enzyme, followed by rings B, C and finally D.

Enzyme–intermediate complexes between the deaminase and porphobilinogen

Earlier studies on the deaminase (Pluscec & Bogorad 1970, Davies & Neuberger 1973) had demonstrated that incubation with porphobilinogen and high concentrations of NH_3 (one of the products from the deaminase reaction), NH_2OH or NH_2OCH_3 led to the formation of monopyrroles, dipyrromethanes, tripyrranes and bilanes linked to the inhibitory base. This suggested that a covalent linkage is formed between the enzyme and the bound substrates and pointed to a mechanism in which the four porphobilinogen molecules are incorporated into the tetrapyrrole in a step-wise fashion. Incubating porphobilinogen deaminase with [^{3}H]porphobilinogen generated labelled enzyme–intermediate complexes with one, two, three and four pyrrole units linked to the enzyme (Anderson & Desnick 1980). These intermediate complexes are referred to as ES, ES_2, ES_3 and ES_4, respectively. Similar behaviour was observed when substrate was added to the purified deaminases from *R. sphaeroides* (Berry et al 1981) and *E. coli* (Warren & Jordan 1988), although the ES_4 complex from the bacterial enzymes was too unstable to be isolated.

The dipyrromethane cofactor, a resident active site primer

The studies described above suggested that preuroporphyrinogen was assembled while covalently linked to the deaminase. Initial experiments invoked lysine (Hart et al 1984) as the protein residue linked covalently to the substrate molecules. However, the true nature of the linkage between the enzymes and the substrate was first indicated when a peptide isolated from a ^{14}C-labelled ES$_2$ complex from *R. sphaeroides* was found to contain a non-radioactive pink chromophore (A. Berry & P. M. Jordan, unpublished work). Treatment of the *E. coli* enzyme with acid yielded uroporphyrin I, indicating that the chromophore was composed of porphobilinogen units (Jordan & Warren 1987). The reaction of the native deaminase with Ehrlich's reagent was typical of a dipyrromethane (Pluscec & Bogorad 1970), leading us to conclude that the pink chromophore observed at acid pH is derived from a covalently bound dipyrromethane system. Most significantly, the reaction of the ES$_2$ complex with Ehrlich's reagent was typical of that of a tetrapyrrole (bilane) reaction, indicating that the two molecules of substrate were actually linked covalently to the two rings of the dipyrromethane. This dipyrromethane system was therefore named the dipyrromethane cofactor (Jordan & Warren 1987).

The dipyrromethane cofactor could be labelled specifically with ^{14}C by growing a deaminase-overproducing *hemA*$^-$ *E. coli* strain in medium containing 5-amino[5-^{14}C]laevulinic acid (Jordan & Warren 1987, Warren & Jordan 1988). When the resulting ^{14}C-labelled deaminase was isolated and incubated with non-radioactive porphobilinogen no radioactivity was incorporated into the enzymic product and, most significantly, the enzyme retained all the original label, indicating that the dipyrromethane cofactor is not subject to catalytic turnover (Jordan & Warren 1987, Warren & Jordan 1988).

The existence of a pyrromethane system as the substrate-binding group in *E. coli* porphobilinogen deaminase was supported by ^{13}C NMR studies (Hart et al 1987). The ES complex prepared from [11-^{13}C]porphobilinogen produced a ^{13}C NMR signal at $\delta = 24.6$ p.p.m. This chemical shift was assigned to a methylene ($-^{13}$CH$_2-$) group between two pyrrole rings and is about 20 p.p.m. upfield of the signal expected from a pyrrole α-methylene attached to the ϵ-amino group of lysine. A positive reaction with Ehrlich's reagent and the formation of uroporphyrinogen I in acid were also noted in these experiments.

The dipyrromethane cofactor has since been demonstrated in porphobilinogen deaminases from animals (human), dicotyledonous and monocotyledonous plants and *R. sphaeroides* (Jordan 1991). Figure 2 indicates how the polymerization process occurs while the intermediates are anchored covalently to the dipyrromethane cofactor.

Further compelling evidence that the substrate molecules are added step-wise to the dipyrromethane cofactor during the tetrapolymerization reaction came from studies with the porphobilinogen analogue α-bromoporphobilinogen (Warren & Jordan 1988). α-Bromoporphobilinogen, with its reactive α-position blocked by the bromine atom, acts as a chain terminator and suicide substrate in the polymerization reaction. When α-bromoporphobilinogen was incubated with the native deaminase not only was the enzyme inactivated but the Ehrlich's reaction was also inhibited, showing unambiguously that the inhibitor had reacted directly with the dipyrromethane cofactor. E, ES, ES$_2$ and ES$_3$ all

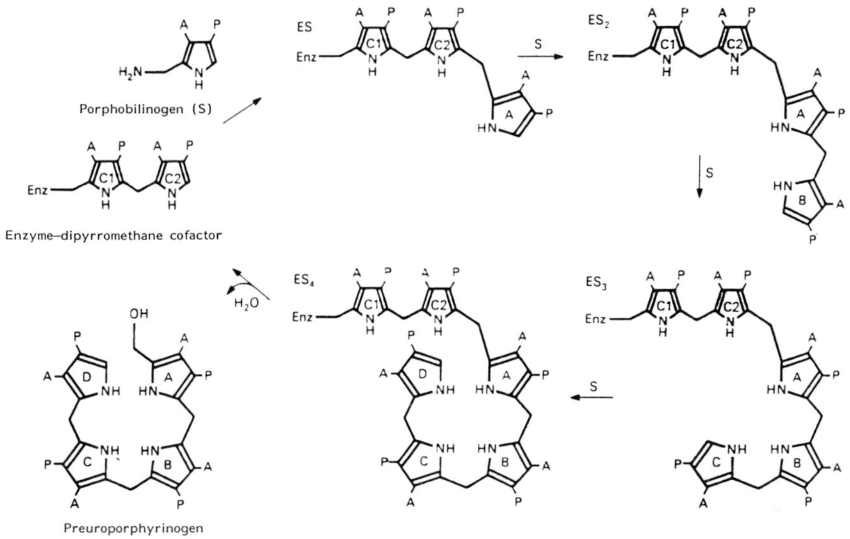

FIG. 2. Structure of the dipyrromethane cofactor and its role in the tetrapolymerization of porphobilinogen.

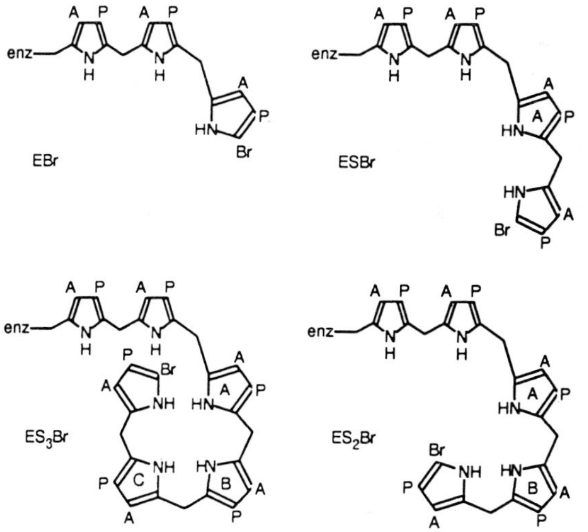

FIG. 3. Termination complexes EBr, ESBr, ES$_2$Br and ES$_3$Br from reaction of the enzyme (E) and intermediate complexes with α-bromoporphobilinogen (Br).

E. coli
Human
Rat
Mouse
Pea
Arabidopsis
Euglena
Bacillus
Yeast
Pseudomonas

```
                         195       200       205       210       215       220       225       230       235       240       245
E. coli      S L P A V G Q G A V G I E C R L D D S R T R E L L A A L N H H E T A L R V T A E R A M N T R L E G G C Q V P I
Human        C M Y A V G Q G A L G V E V R A K D Q D I L D L V G V L H D P E T L L R C I A E R A F L R H L E G G C S V P V
Rat          C M Y A V G Q G A L A V E V R A K D Q D I L D L V G V L H D P E T L L R C I A E R D F L R H L E G G C S V P V
Mouse        C M Y A V G Q G A L A V E V R A K D Q D I L D L V S V L H D P E T L L R C I A E R A F L R H L E G G C S V P V
Pea          M L P A V A Q G A I G I A C R S N D D K M A E Y L A S L N H E E T R L A I S C E R A F L T T L D G S C R T P I
Arabidopsis  M L P A V A Q G A I G I A C R T D D D K M A T Y L A S L N H E E T R L A I S C E R A F L E T L D G S C R T P I
Euglena      M C P A A G Q G A L S I E L R T N D P E I A A L L E P L H H I P D A V T V A C E R A M N R R L N G G C Q V P I
Bacillus     C L P A V G Q G A L A I E C R E S D E E L L A L F S Q F T D E Y T K R T V L A E R A F L N A M E G G C Q V P I
Yeast        M Y H A V G Q G A L G I E I R K G D T K M M K I L D E I C D L N A T I C C L S E R A L M R T L Z G G C S V P I
Pseudomonas  S L F A G G Q G A V G I E C R T A D S D L E A L L E P L H H T D T A L R V T A E R A L N K R L N G G C Q V P I

                 250            255       260       265       270       275
E. coli      G S Y A E L I — — D G E I W L R A L V G A P D G S Q I I R G E R R G A — — — — — — — — — — — — — — — — — —
Human        A V H T A H K — — D G Q L Y L T G G V W S L D G S D S I Q E T H Q A T I H V P A Q H E D G P E D D P Q L V G I
Rat          A V H T V H K — — D G Q L Y L T G G V W S L D G S D S H Q E T H Q A T I Q V P V Q Q E D G P E D D P Q L V G I
Mouse        A V H T V I K — — D G Q L Y L T G G V W S L D G S D S H Q E T M Q A T I Q V P V Q Q E D G P E D D P Q L V G I
Pea          A G Y A S R D — K D G N C L F R G L V A S P D G T R V L E T S R I G S Y T — — — — — — — — — — — — — — — —
Arabidopsis  A G Y A S K D — E E G N C I F R G L V A S P D G T K V L E T S R K G P Y V — — — — — — — — — — — — — — — —
Euglena      S G F A Q L K — — D G Q L R M E A R V G S V T G K G P L I I Q S K T F R L P W S G R T W P Q L — — — — — — —
Bacillus     A G Y S V L N — G Q D E I E M T G L V A S P D G K I I F K E T V T — — — — — — — — — — — — — — — — — — — —
Yeast        G V E S K Y N E E T K K L L L K A I V V D V E G T E A V E D E I E M L I E — — — — — — — — — — — — — — — —
Pseudomonas  A C Y A I R E — — G D Q L W L R G L V G Q P D G T Q L L R A E G R — — — — — — — — — — — — — — — — — — — —

                                  280       285       290       295       300       305       310
E. coli      — — — — — — — — — P Q D A E Q M G I S L A E E L L N N G A R E I L A E V Y N G D A P A — —
Human        T A R N I P R G P Q L A A Q N L G I S L A N L L L S K G A K N I L D V A R Q L N D A H — —
Rat          T A R N I P R G A Q L A A E N L G I S L A S L L L N K G A K N I L D V A R Q L N D V R — —
Mouse        T A R N I P R G A Q L A A E N L G I S L A S L L L N K G A K N I L D V A R Q L N D V R — —
Pea          — — — — — — — — — Y E D M M K I G K D A G E E L L S R A G P G F F N S — — — — — — — — —
Arabidopsis  — — — — — — — — — Y E D M V K M G K D A G Q E L L S R A G P G F F G N — — — — — — — — —
Euglena      — — — — — — — — — Q K E S E A L G V E V A D M L L A D G A Q A Y L D E A Y A S R T L G W A
Bacillus     — — — — — — — — — G N D P E E V G K R S A A L M A D K G A K D L S D R V K R E L D E D G K
Yeast        — — — — — — — N V K E D S M A C G K I L A E R H I A D G A K K I L D E I N L D R I K — — —
Pseudomonas  — — — — — — — A P L A D A E A L G V R V A E D L L E Q G A E A I L E A V Y G E A G H P — —
```

FIG. 4. Primary structures of porphobilinogen deaminases. The invariant residues are boxed.

reacted with α-bromoporphobilinogen (Br) to yield EBr, ESBr, ES$_2$Br and ES$_3$Br (Warren & Jordan 1988). With the exception of ES$_3$Br, these termination complexes could all be isolated; their structures are shown in Fig. 3.

The structure of the dipyrromethane cofactor was further confirmed by labelling with ^{13}C, either by growing a recombinant strain of *E. coli* in the presence of 5-amino[5-^{13}C]laevulinic acid (Jordan et al 1988a) or by incubating [11-^{13}C]porphobilinogen with the apodeaminase (Beifuss et al 1988). Difference NMR spectra revealed four signals: at $\delta = 117$ and 128 p.p.m., assigned to the two aromatic carbon atoms, a third at $\delta = 25$ p.p.m. from the bridging methylene position and a fourth at $\delta = 27-29$ p.p.m. arising from the ^{13}C methylene carbon atom directly linked to the enzyme. The chemical shift of the fourth resonance is consistent with that expected for a pyrrole methylene group attached as a thioether to cysteine. Cysteine 242 in *E. coli* deaminase was shown to be the attachment site for the dipyrromethane cofactor by protein chemistry and site-directed mutagenesis (Jordan et al 1988a). This residue lies in the sequence LEGGCQVP.

Molecular biology and primary structure of porphobilinogen deaminases

The primary structures of several porphobilinogen deaminases have been deduced from gene or cDNA sequences. Sequences are available from *E. coli* (Thomas & Jordan 1986), human (Raich et al 1986), rat (Stubnicer et al 1988), mouse (Beaumont et al 1989), *Euglena gracilis* (Sharif et al 1989), *Bacillus subtilis* (Petricek et al 1990), yeast (Keng et al 1992), pea and *Arabidopsis* (A. G. Smith, personal communication). The substantial similarity between the sequences, about 20% identity, as shown in Fig. 4, suggests that these deaminases have closely related tertiary structures and operate by a similar mechanism. Most of the invariant residues are located in the catalytic cleft (Louie et al 1992). In *E. coli* (Jordan et al 1987, Sasarman et al 1987) and *B. subtilis* (Hansson et al 1991) the *hemC* gene, which encodes porphobilinogen deaminase, overlaps with the *hemD* gene, which encodes uroporphyrinogen III synthase.

Determination of the crystal structure of porphobilinogen deaminase

The availability of milligram amounts of *E. coli* porphobilinogen deaminase from recombinant strains (Thomas & Jordan 1986) has led to the most exciting recent developments in the study of the deaminase—the crystallization of the enzyme (Jordan et al 1988b) and the determination of the structure at 1.9 Å resolution by X-ray crystallography (Louie et al 1992). The structure comprises three domains, each of about 100 amino acids, with the large catalytic cleft lying between domains 1 and 2. The dipyrromethane cofactor is linked to Cys-242 on a loop of domain 3 and projects into the catalytic cleft. The acetate and propionate side chains of the cofactor make numerous interactions through salt

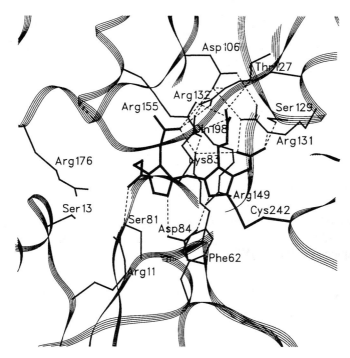

FIG. 5. The catalytic cleft of porphobilinogen deaminase showing the extensive contacts between the cofactor and the protein. The C1 ring of the cofactor is linked to Cys-242. The C2 ring of the cofactor provides the free α-position for reaction with substrate.

links and hydrogen bonds with the side chains of conserved residues, Arg-131, Arg-132 and Arg-155, Lys-83, Ser-129 and Thr-127, to form the C site (Fig. 5). The substrate-binding site (S) includes Arg-11, Arg-149, Arg-155 and Ser-13 (Fig. 5). There are three pyrrole-binding sites, one for the substrate and two for the cofactor. The catalytic Asp-84 is located within hydrogen-bonding distance of the pyrrole NH groups of the cofactor and substrate (see below). If the dipyrromethane cofactor is oxidized it adopts a more planar conformation with the C2 ring of the cofactor interacting with amino acid side chains at the substrate-binding site (Louie et al 1992). However, in the native dipyrromethane form, the C2 ring cannot interact with the substrate-binding site, which may explain why the cofactor is not subject to catalytic turnover and is a permanent component of the deaminase. The precise location of amino acids 47–57 was not revealed by the crystal structure and it is possible that another key residue from this sequence is involved in the mechanism. The crystal structure suggests there is considerable mobility between domain 1, which largely forms the substrate-binding site (S), and domain 2, which forms the cofactor-binding site (C) (Louie et al 1992). A large space found behind the cofactor could accommodate the growing tetrapyrrole chain. The crystal structure is discussed in detail elsewhere in this volume (Lambert et al 1994).

Site-directed mutagenesis of porphobilinogen deaminase

Invariant arginine residues

Site-directed mutagenesis of the conserved arginine residues in the *E. coli* enzyme at positions 11, 131, 132, 149, 155, 176 and 232 to histidine (Jordan & Woodcock 1991) or leucine (Lander et al 1991) has provided vital clues about the process of dipyrromethane cofactor assembly and tetrapyrrole chain elongation. It is interesting that replacement of arginine by histidine or leucine yields mutants with remarkably similar properties. With either His or Leu in place of Arg-131 or Arg-132, mutants are unable to assemble the dipyrromethane cofactor and inactive apodeaminases are formed. All the other arginine mutants assemble the cofactor, but their ability to form the tetrapyrrole is affected at different stages, as shown in Table 1. Substitution of either Arg-11 or Arg-155 inhibits the formation of the ES complex, indicating that these residues are particularly important for substrate binding. The effects of the mutations are consistent with the key positions of the conserved arginines in the catalytic cleft (Louie et al 1992).

The catalytic Asp-84

The importance of this invariant residue was highlighted by the crystal structure (Louie et al 1992). Changing this residue by site-directed mutagenesis to

TABLE 1 Effect of substitution of histidine residues for conserved arginine residues on the ability of *E. coli* porphobilinogen deaminase to assemble the dipyrromethane cofactor and catalyse chain elongation

Mutation	Specific activity[a]	K_m (µM)	Presence of cofactor[b]	Stage affected[c] 10 µM PBG	200µM PBG
Wild-type	43	17	+	None	None
R11H	0.1	nd	+	E→ES	E→ES
R131H	nd	nd	−	Cofactor assembly	Cofactor assembly
R132H	nd	nd	−	Cofactor assembly	Cofactor assembly
R149H	11.1	200	+	ES→ES$_2$	All
R155H	0.5	nd	+	E→ES	ES$_3$→product
R176H	6.0	30	+	ES→ES$_2$	ES$_2$→ES$_3$

[a]Specific enzyme activity is expressed as µmol uroporphyrinogen formed/h/mg protein.
[b]Determined by reaction with Ehrlich's reagent.
[c]Determined by FPLC.
nd, not determinable.
PBG, porphobilinogen.
See Jordan & Woodcock (1991) for further details.

glutamate, alanine or asparagine has provided fundamental information about the nature of catalysis. The enzyme from mutant D84E has a k_{cat} two orders of magnitude lower than that of the wild-type enzyme but an unchanged K_m, and forms exceptionally stable enzyme–intermediate complexes (Woodcock & Jordan 1994) that are amenable to crystallization for X-ray analysis. Deaminases from mutants D84A and D84N are completely inactive, yet they react with Ehrlich's reagent and appear to contain the dipyrromethane cofactor (Woodcock & Jordan 1994). The presence of the cofactor in enzymes in which the catalytic group is absent (D84A and D84N), or the substrate-binding site is altered (R11H/L and R155H/L), suggests that deamination of the porphobilinogen units destined for the cofactor may occur by an enzyme-facilitated reaction at the cofactor-binding site (see below). The complete absence of a cofactor in Arg-131 and 132 mutants supports this view.

Reactions catalysed by porphobilinogen deaminase

Deamination and amination reactions

Porphobilinogen forms preuroporphyrinogen in a step-wise assembly mechanism by the deamination of one porphobilinogen unit at a time, starting with ring A. The growing tetrapyrrole chain is assembled while anchored to the dipyrromethane cofactor. On incubation of the deaminase with porphobilinogen and high concentrations of bases such as ammonia (NH_2R) there is preferential release of the enzyme-bound monopyrrole (ES) as $NHRCH_2AP$ or the enzyme-bound bilane (ES_4) as $NHRCH_2APAPAPAP$. The intermediates ES_2 and ES_3 are less susceptible to release (Davies & Neuberger 1973), suggesting that the ES and ES_4 complexes have common properties. The deaminase can thus catalyse both deamination and amination reactions.

Dehydration and hydration reactions

Hydroxyporphobilinogen (hydroxyPBG), porphobilinogen with the amino group replaced by a hydroxyl group, is a good substrate for the enzyme (Battersby et al 1979c). The enzyme is thus able to catalyse dehydration as well as deamination. The release of the product 1-hydroxymethylbilane, preuroporphyrinogen, from ES_4 is, conversely, a hydrolysis reaction. Furthermore, in the absence of substrate, the intermediate complexes ES, ES_2 and ES_3 release the terminal pyrrole unit at the α-free end of the chain as hydroxyporphobilinogen which is then accepted as a substrate, resulting in the production of preuroporphyrinogen (Warren & Jordan 1988). The enzyme can thus catalyse both dehydration and hydrolysis reactions.

The above observations, together with the information provided by the crystal structure, suggest that a single catalytic site can catalyse the deamination of

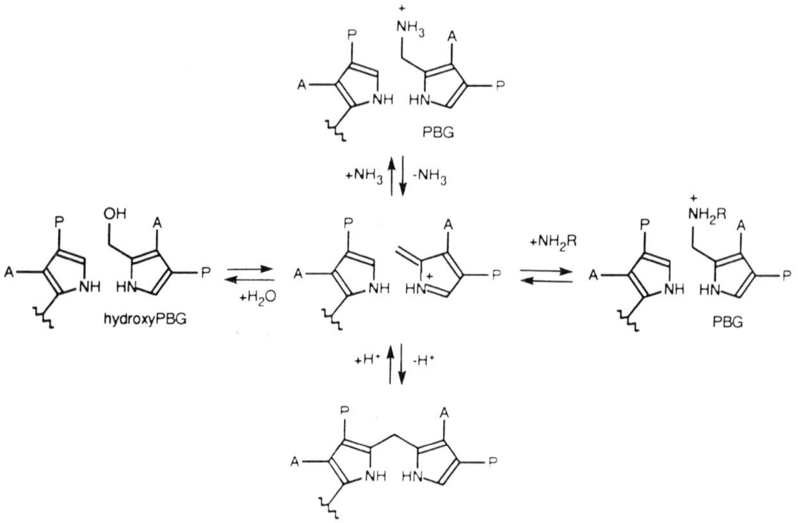

FIG. 6.   Reactions catalysed by porphobilinogen (PBG) deaminase.

porphobilinogen or the dehydration of hydroxyporphobilinogen, as well as the reverse reactions which involve the addition of ammonia (or a base) or water to the enzyme–intermediate complexes (Warren & Jordan 1988). These reactions are summarized in Fig. 6. These observations implicate a reactive azafulvene as a common species in all these reactions (Battersby 1978).

Mechanism of the tetrapolymerization reaction

The step-wise addition of the four molecules of porphobilinogen required for the assembly of preuroporphyrinogen is outlined in Fig. 2. The first substrate (ring A) binds to the vacant S site (Fig. 5) and is deaminated to the azafulvene; reaction with the α-position of the cofactor C2 ring follows. Aspartate 84 may facilitate the deamination and stabilize the positively charged nitrogens of the pyrrole rings. Deprotonation of the α-position would result in the formation of intermediate ES. The loss of the highly acidic hydrogen atom from the α-position and the ensuing change from sp^3 to sp^2 hybridization could provide the necessary driving force for the translocation process. Evidence from a study of the properties of enzyme–intermediate complexes (Warren & Jordan 1988) suggests that after each reaction the S site is vacated. In the absence of a further molecule of substrate the terminal pyrrole residue may slowly regain accessibility to the S site through protonation of the α-position in the reverse reaction and may be released by reaction of the azafulvene with water to yield hydroxyporphobilinogen (Warren & Jordan 1988) or with ammonia to yield

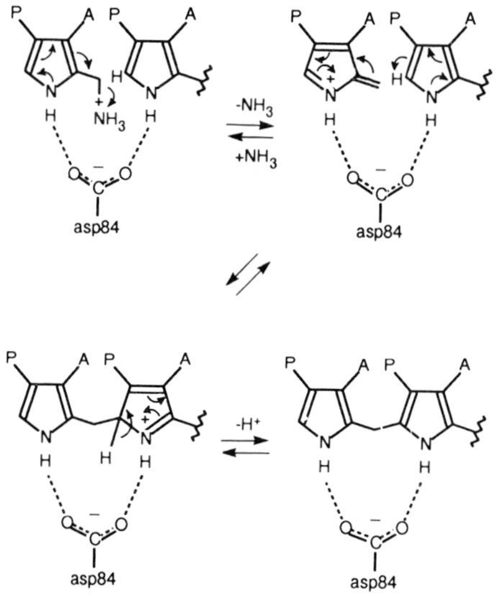

FIG. 7. The reaction catalysed by porphobilinogen (PBG) deaminase. Aspartate 84 facilitates the deamination of ammonia and stabilizes the positively charged pyrrole nitrogens during successive C–C bond formations. A reaction analogous to the release of PBG or hydroxyPBG occurs at the tetrapyrrole stage to liberate either the 1-aminomethylbilane (in the presence of ammonia) or the product preuroporphyrinogen. R = H, OH or OCH_3.

porphobilinogen. Normally, the tetrapolymerization reaction continues with the binding and reaction of further substrates (rings B, C and D) leading to ES_2, ES_3 and finally ES_4. The mechanism of addition of substrate is shown in Fig. 7.

Evidence that the conformation of porphobilinogen deaminase changes progressively during the tetrapolymerization reaction came from the observation (Warren & Jordan 1988) that the enzyme, which is essentially unreactive to the thiol reagent N-ethylmaleimide (NEM), becomes increasingly susceptible to inactivation by the reagent as substrate molecules progressively bind. Thus ES is partially inactivated, ES_2 is substantially inactivated, but ES_3 is almost completely inactivated by the reagent (Warren & Jordan 1988). The susceptible cysteine is located between domains 2 and 3 at position 134 (Louie et al 1992) and, although buried in the native state, must become exposed as the catalytic cycle proceeds so that it can act as a reporter group for the progressive conformational change.

If the same catalytic machinery promotes the release of preuroporphyrinogen by the reverse of the forward mechanism, the ES_4 intermediate would need to adopt a position in which ring A occupies the S site. It is well established that a 1-aminomethylbilane can bind in this way because the enzyme is able to deaminate the synthetic aminomethylbilane $NH_2CH_2APAPAPAP$ to preuroporphyrinogen (Burton et al 1979). Although the hydrolytic activity is normally confined to preuroporphyrinogen release, any α-free terminal ring of the enzyme–intermediate complexes can be cleaved slowly by such a hydrolysis mechanism (Warren & Jordan 1988).

Assembly of the dipyrromethane cofactor and the role of positively charged groups

The ability of mutant deaminases to assemble the cofactor even in the absence of an intact substrate-binding site (R11H/L and R155H/L mutants) or the catalytic aspartate (D84A/N mutants) implies that there could be a mechanism for the assembly of the cofactor which does not require these groups. Binding of the porphobilinogen unit (C1) to the positively charged residues at the C site of the apoenzyme could be sufficient to facilitate deamination and the resulting reactive azafulvene would be in a position to alkylate Cys-242 (Fig. 8, upper reaction). The binding of the second porphobilinogen unit (C2) at the C site could also result in deamination and coupling to complete the cofactor assembly process. Such an assembly mechanism could also explain why the cofactor is not subject to catalytic turnover (Jordan & Warren 1987, Warren & Jordan 1988). The cofactor assembly process must involve considerable conformational change because the protein becomes much more compact and extraordinarily stable to heat (Scott et al 1989). The association of domains 1 and 2 around the cofactor would bring the substrate-binding site (S), largely on domain 1,

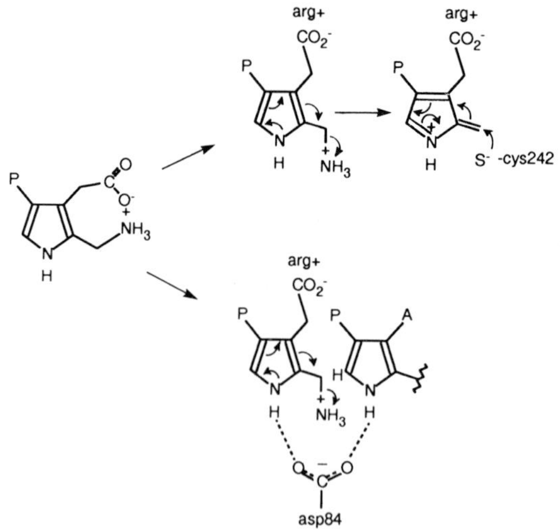

FIG. 8.   Unpairing of the stabilizing ion pair between the negatively charged acetate and protonated side chains of porphobilinogen to facilitate alkylation of Cys-242 or C-alkylation of a pyrrole α-position.

into the vicinity of the C site, largely on domain 2, to create the complete catalytic ensemble.

Whatever the precise mechanism of cofactor assembly and substrate polymerization (Fig. 8), the considerations above lead us to conclude that the major role of the enzyme is to unfetter the intrinsic chemical reactivity of the porphobilinogen by breaking the ion pair between the negatively charged acetate side chain and the protonated aminomethyl side chain. This would destabilize the $-\overset{+}{N}H_3$ leaving group by allowing it to adopt a favourable conformation perpendicular to the plane of the pyrrole ring for deamination. The acid-catalysed non-enzymic polymerization of porphobilinogen may also be facilitated by this mechanism, although in this case the ion pair would be broken by protonation of the acetate side chain.

Stereochemical studies

The steric course of preuroporphyrinogen synthesis has been investigated by incubating [$11RS$-$^3H_2$;$2,11$-$^{14}C_2$] porphobilinogen with enzymes of the haem biosynthesis pathway from avian erythrocytes (Jones et al 1984). These experiments established that 50% of the ^3H label was incorporated into protoporphyrin. In further experiments with [$11S$-^3H; $2,11$-$^{14}C_2$] porphobilinogen a single ^3H label was incorporated into position 10 of protoporphyrin IX, suggesting that three labelled hydrogens had been lost, from positions 5, 15 and 20, during the oxidation of protoporphyrinogen IX (Jones et al 1984). Similar conclusions were reached in studies with [$11R$-^2H] porphobilinogen (Jackson et al 1987). These experiments established that during the porphobilinogen deaminase- and uroporphyrinogen III synthase-catalysed reactions, all four methylene groups of preuroporphyrinogen (and of uroporphyrinogen III) are handled by totally stereospecific processes (Jones et al 1984). Most importantly, these experiments proved that the hydroxyl group in preuroporphyrinogen must have been added by an enzymic process at the active site of the deaminase and not by the non-enzymic trapping of any released methylenepyrrolenine-type intermediate. In addition to providing information about the deaminase/synthase reactions, these observations have an important bearing on the mechanism of action of protoporphyrinogen oxidase (EC 1.3.3.4), dealt with elsewhere in this volume (Akhtar 1994).

The steric course of the formation of the 1-aminomethylbilane NH_2CH_2-APAPAPAP at the 1-aminomethyl group has been established by incubating [$11R$-^2H] porphobilinogen and [$11S$-^2H] porphobilinogen with the deaminase in the presence of ammonia. Analysis of the 1-aminomethylbilane by derivatization and NMR spectroscopy established that the overall steric course is one of retention (Neidhart et al 1985). A similar conclusion was reached when the preuroporphyrinogen (hydroxymethylbilane) was enzymically synthesized from [$11R$-^3H] porphobilinogen and [$11S$-^3H] porphobilinogen. In this case, chirality of the 1-hydroxymethyl group was determined by chemical degradation and enzymic analysis (Schauder et al 1987).

The overall stereochemistry of the reaction catalysed by porphobilinogen deaminase is not yet elucidated, but the active site conformation suggested by the crystal structure of the enzyme may be more consistent with an inversion mechanism than retention.

Uroporphyrinogen III synthase

The second stage in the transformation of porphobilinogen into uroporphyrinogen III is catalysed by uroporphyrinogen III synthase. In this reaction, ring D of the 1-hydroxymethylbilane, preuroporphyrinogen, is inverted and cyclization to uroporphyrinogen III macrocycle is achieved (Fig. 1).

Uroporphyrinogen synthases have been isolated in homogeneous form from a variety of sources (see Jordan 1991 for a review). The synthases are all monomeric, of 28 000–33 000 M_r, with isoelectric points around pH 5. The K_m values for preuroporphyrinogen vary but are in the low μmolar range. Uroporphyrinogen synthases have turnover numbers of at least $200\,s^{-1}$, much higher than the deaminases, which ensures that preuroporphyrinogen never accumulates appreciably *in vivo*, so minimizing the formation of uroporphyrinogen I.

The *hemD* gene, encoding uroporphyrinogen III synthase, from *E. coli* (Sasarman et al 1987, Jordan et al 1987) and from *B. subtilis* (Hansson et al 1991) has been sequenced. In both cases the *hemD* gene immediately follows the *hemC* gene in a *hem* operon. The primary structure of the human uroporphyrinogen III synthase derived from the cDNA sequence (Tsai et al 1988) shows minimal similarity to the bacterial enzymes, which is surprising in view of the similarity between the deaminases. Overexpression of the *E. coli* enzyme (Alwan et al 1989) and mutagenesis has shown that the conserved Arg-138 residue is essential (P. M. Jordan & A. E. Elseed, unpublished work).

Studies with several isomeric 1-hydroxymethylbilanes, synthesized chemically, have highlighted the importance of the A and B rings for recognition by the enzyme, because changes to the A and P constituents of these rings prevent the reaction from proceeding normally (Battersby et al 1981). For the natural substrate, preuroporphyrinogen, binding in the 'uroporphyrinogen III conformation' must be optimal so that the ensuing chemistry is directed towards uroporphyrinogen III as the sole product. A review of the experiments on the deaminase and synthase carried out in Sir Alan Battersby's laboratory has been published recently (Battersby & Leeper 1990). In this review possible mechanisms for the rearrangement have been discussed. The mechanism of the synthase (cosynthase) reaction is also covered in detail elsewhere in this volume (Leeper 1994).

Acknowledgements

Funding from the SERC (MRI) and AFRC (PMB) is gratefully acknowledged. Thanks are due to Professor M. Akhtar (Southampton) and Dr M. J. Warren (QWM) for discussions about this manuscript. The mutants were generated by Dr S. C. Woodcock. I am indebted to Drs G. V. Louie, P. D. Brownlie, R. Lambert, J. B. Cooper and S. P. Wood and Professor T. L. Blundell at the Crystallographic Department, Birkbeck College and especially to Dr G. V. Louie for help with the preparation of Fig. 4.

References

Akhtar M 1994 The modification of acetate and propionate side chains during the biosynthesis of haem and chlorophylls: mechanistic and stereochemical studies. In: The biosynthesis of the tetrapyrrole pigments. Wiley, Chichester (Ciba Found Symp 180) p 131–155

Alwan AF, Mgbeje BIA, Jordan PM 1989 Purification and properties of uroporphyrinogen III synthase (cosynthase) from an overproducing recombinant strain of *Escherichia coli* K-12. Biochem J 264:397–402

Anderson PM, Desnick RJ 1980 Purification and properties of uroporphyrinogen I synthase from human erythrocytes. J Biol Chem 255:1993–1999

Battersby AR 1978 The discovery of nature's biosynthetic pathways. Sep Exper 34:1–13

Battersby AR, Leeper FJ 1990 Biosynthesis of the pigments of life: mechanistic studies on the conversion of porphobilinogen to uroporphyrinogen III. Chem Rev 90:1261–1274

Battersby AR, Fookes CJR, Matcham GWJ, McDonald E 1978 Biosynthesis of natural porphyrins. Enzymic experiments on isomeric bilanes. J Chem Soc Chem Commun, p 1064–1066

Battersby AR, Fookes CJR, Gustafson-Potter KE, Matcham GWJ, McDonald E 1979a Proof by synthesis that unrearranged hydroxymethylbilane is the product from deaminase and the substrate for cosynthase in the biosynthesis of uro'gen III. J Chem Soc Chem Commun, p 1155–1158

Battersby AR, Fookes CJR, Matcham GWJ, McDonald E 1979b Order of assembly of the four pyrrole rings during biosynthesis of the natural porphyrins. J Chem Soc Chem Commun, p 539–541

Battersby AR, Fookes CJR, Matcham GWJ, McDonald E, Gustafson-Potter KE 1979c Biosynthesis of the natural porphyrins: experiments on the ring-closure steps and with the hydroxy-analogue of porphobilinogen. J Chem Soc Chem Commun, p 316–319

Battersby AR, Fookes CJ, Matcham GWJ, Pandey PS 1981 Biosynthesis of natural porphyrins: studies with isomeric hydroxymethylbilanes on the specificity and action of cosynthetase. Angew Chem Int Ed Engl 20:293–295

Beaumont C, Porcher C, Picat C, Nordmann Y, Grandchamp B 1989 The mouse porphobilinogen deaminase gene. J Biol Chem 264:14829–14834

Beifuss U, Hart GJ, Miller AD, Battersby AR 1988 ^{13}C NMR studies on the pyrromethane cofactor of hydroxymethylbilane synthase. Tetrahedron Lett 29: 2591–2594

Berry A, Jordan PM, Seehra JS 1981 The isolation and characterization of catalytically competent porphobilinogen deaminase–intermediate complexes. FEBS (Fed Eur Biochem Soc) Lett 129:220–224

Burton G, Fagerness PE, Hosozawa S, Jordan PM, Scott AI 1979 ^{13}C NMR evidence for a new intermediate, preuroporphyrinogen, in the enzymic transformation of porphobilinogen into uroporphyrinogens I and III. J Chem Soc Chem Commun, p 202–204

Davies RC, Neuberger A 1973 Polypyrroles formed from porphobilinogen and amines by uroporphyrinogen synthetase of *Rhodopseudomonas spheroides*. Biochem J 133:471–492

Hansson M, Rutberg L, Schroder I, Hederstedt L 1991 The *Bacillus subtilis hemAXCDBL* gene cluster, which encodes enzymes of the biosynthetic pathway from glutamate to uroporphyrinogen III. J Bacteriol 173:2590–2599

Hart GJ, Leeper FJ, Battersby AR 1984 Modification of hydroxymethylbilane synthase (porphobilinogen deaminase) by pyridoxal 5′ phosphate: demonstration of an essential lysine residue. Biochem J 222:93–102

Hart GJ, Miller AD, Leeper FJ, Battersby AR 1987 Biosynthesis of the natural porphyrins: proof that hydroxymethylbilane synthase (porphobilinogen deaminase) uses a novel binding group in its catalytic action. J Chem Soc Chem Commun, p 1762–1765

Jackson AH, Lertwanawatana W, Procter G, Smith SG 1987 A synthesis of stereospecifically labelled porphobilinogen and its incorporation into protoporphyrin IX. Experientia 43:892–894

Jones C, Jordan PM, Akhtar MA 1984 Mechanism and stereochemistry of the porphobilinogen deaminase and protoporphyrinogen IX oxidase reactions: stereospecific manipulation of hydrogen atoms at the four methylene bridges during the biosynthesis of haem. J Chem Soc Perkin Trans I, p 2625–2633

Jordan PM 1991 The biosynthesis of 5-aminolaevulinic acid and its transformation into uroporphyrinogen III. In: Jordan PM (ed) Biosynthesis of tetrapyrroles. Elsevier, Amsterdam (New Compr Biochem 19), p 1–66

Jordan PM, Berry A 1980 Preuroporphyrinogen: a universal intermediate in the biosynthesis of uroporphyrinogen III. FEBS (Fed Eur Biochem Soc) Lett 112:86–88

Jordan PM, Seehra JS 1979 Biosynthesis of uroporphyrinogen III: order of addition of the four porphobilinogen rings in the formation of the tetrapyrrole ring. FEBS (Fed Eur Biochem Soc) Lett 104:364–366

Jordan PM, Warren MJ 1987 Evidence for a dipyrromethane cofactor at the catalytic site of *Escherichia coli* porphobilinogen deaminase. FEBS (Fed Eur Biochem Soc) Lett 225:87–92

Jordan PM, Woodcock SC 1991 Mutagenesis of arginine residues in the catalytic cleft of *Escherichia coli* porphobilinogen deaminase that affects dipyrromethane cofactor assembly and tetrapyrrole chain initiation and elongation. Biochem J 280:445–449

Jordan PM, Burton G, Nordlov H, Schneider M, Pryde L, Scott AI 1979 Preuroporphyrinogen, a substrate for uroporphyrinogen III cosynthetase. J Chem Soc Chem Commun, p 204–205

Jordan PM, Mgbeje IAB, Alwan AF, Thomas SD 1987 Nucleotide sequence of *hemD*, the second gene in the *hem* operon of *Escherichia coli K-12*. Nucleic Acids Res 15:10583

Jordan PM, Warren MJ, Williams HJ et al 1988a Identification of a cysteine residue as the binding site for the dipyrromethane cofactor at the active site of *Escherichia coli* porphobilinogen deaminase by [13]C n.m.r. FEBS (Feb Eur Biochem Soc) Lett 235:189–193

Jordan PM, Thomas SD, Warren MJ 1988b Purification, crystallization and properties of porphobilinogen deaminase from a recombinant strain of *Escherichia coli* K12. Biochem J 254:427–435

Keng T, Richard C, Larocque R 1992 Structure and regulation of yeast *HEM3* the gene for porphobilinogen deaminase. Mol & Gen Genet 234:233–243

Lambert R, Brownlie PD, Woodcock SC et al 1994 Structural studies on porphobilinogen deaminase. In: The biosynthesis of the tetrapyrrole pigments. Wiley, Chichester (Ciba Found Symp 180) p 97–110

Lander M, Pitt AR, Alefounder PR, Bardy D, Abell C, Battersby AR 1991 Studies on the mechanism of hydroxymethylbilane synthase concerning the role of arginine residues in substrate binding. Biochem J 275:447–452

Leeper FJ 1994 The evidence for a spirocyclic intermediate in the formation of uroporphyrinogen III by cosynthase. In: The biosynthesis of the tetrapyrrole pigments. Wiley, Chichester (Ciba Found Symp 180) p 111–130

Louie GV, Brownlie PD, Lambert R et al 1992 Structure of porphobilinogen deaminase reveals a flexible multidomain polymerase with a single catalytic site. Nature 359:33–39

Neidhart W, Anderson PC, Hart GJ, Battersby AR 1985 Synthesis of (*11S*)- and (*11R*)-[11-^{2}H$_1$]porphobilinogen: stereochemical studies on hydroxymethylbilane synthase (porphobilinogen deaminase). J Chem Soc Chem Commun, p 924–927

Petricek M, Rutberg L, Schroder I, Hederstedt L 1990 Cloning and characterization of the *hemA* region of the *Bacillus subtilis* chromosome. J Bacteriol 172:2250–2258

Pluscec J, Bogorad L 1970 A dipyrrolmethane intermediate in the enzymic synthesis of uroporphyrinogen. Biochemistry 9:4736–4743

Raich N, Romeo P-H, Dubart A, Beaupain D, Cohen-Sohal M, Goossens M 1986 Molecular cloning and complete primary sequence of human erythrocyte porphobilinogen deaminase. Nucleic Acids Res 14:5955–5968

Sasarman A, Nepveu A, Echelard Y, Dymetryszyn J, Drolet M, Goyer C 1987 Molecular cloning and sequencing of the *hemD* gene of *Escherichia coli* and preliminary data on the *uro* operon. J Bacteriol 169:4257–4262

Schauder JR, Jendrezejewski S, Abell A, Hart GJ, Battersby AR 1987 Stereochemistry of formation of the hydroxymethyl group of hydroxymethylbilane, the precursor for uroporphyrinogen III. J Chem Soc Chem Commun, p 436–439

Scott AI, Clemens KR, Stolowich NJ, Santander PJ, Gonzalez MD, Roessner CA 1989 Reconstitution of apo-porphobilinogen deaminase: structural changes induced by cofactor binding. FEBS (Fed Eur Biochem Soc) Lett 242:319–324

Sharif AL, Smith AG, Abell C 1989 Isolation and characterisation of a cDNA clone for a chlorophyll synthesis enzyme from *Euglena gracilis*. Eur J Biochem 184:353–359

Stubnicer AC, Picat C, Grandchamp B 1988 Rat porphobilinogen deaminase cDNA: nucleotide sequence of erythropoietic form. Nucleic Acids Res 16:3102

Thomas SD, Jordan PM 1986 Nucleotide sequence of the *hemC* locus encoding porphobilinogen deaminase of *Escherichia coli K-12*. Nucleic Acids Res 14:6215–6226

Tsai SF, Bishop DF, Desnick RJ 1988 Human uroporphyrinogen III synthase. Molecular cloning, nucleotide sequence and expression of full length cDNA. Proc Natl Acad Sci USA 85:7044–7053

Warren MJ, Jordan PM 1988 Investigation into the nature of substrate binding to the dipyrromethane cofactor of *Escherichia coli* porphobilinogen deaminase. Biochemistry 27:9020–9030

Woodcock SC, Jordan PM 1994 Evidence for the participation of aspartate-84 as a catalytic group at the active site of porphobilinogen deaminase from site-directed mutagenesis of the *hemC* gene of *Escherichia coli*. Biochemistry vol 33, in press.

DISCUSSION

K. Smith: What causes the pink coloration that you described?

Jordan: In acid the dipyrromethane is protonated to give a pink chromophore. Oxygen can also react with the free α-position of the dipyrromethane to give a dipyrromethanone. The dipyrromethane can also be oxidized to a dipyrromethene.

K. Smith: It is fairly difficult to oxidize a dipyrromethane to a dipyrromethene; the oxidation doesn't take place normally with only oxygen.

Jordan: I didn't say that. The slow addition of oxygen to the free α-position of the cofactor will result, we believe, in a dipyrromethanone. In acid, the dipyrromethane cofactor is not bound as tightly as it is normally because the acetate and propionate groups are protonated. If you heat the enzyme to 60–70 °C at pH 8 the dipyrromethane cofactor is not affected. If you heated a free dipyrromethane to that temperature it would react. The enzyme is therefore stabilizing the dipyrromethane. When you reduce the pH to 3 you change the structure of the enzyme so that it no longer binds the cofactor tightly, and the cofactor can be oxidized more easily.

Scott: But when it becomes oxidized it gets a new carbonyl at the α-free end, to become a dipyrromethanone, doesn't it?

Jordan: Not necessarily.

Hädener: Perhaps at low pH the dipyrromethane cofactor is cleaved from the enzyme by hydrolysis, so the first reaction will be dimerization to uroporphyrinogen I.

Battersby: The colour and absorption spectrum of the oxidized form of the cofactor do not match those of a porphyrin.

Jordan: The change is an oxidation. If you add borohydride the colour disappears.

Scott: Does that restore activity?

Jordan: We didn't dare try that at pH 3. When we did those experiments we didn't realize that the enzyme retained activity after treatment with acid.

Beale: As I recall, some earlier work by Umanoff et al (1988) was puzzling until your work was published. They had a mutant of *E. coli* which lacked 5-aminolaevulinic acid (ALA) dehydratase and therefore couldn't make porphobilinogen. The mutant also showed no porphobilinogen deaminase activity *in vitro*. I am still puzzled by that. Why when you add porphobilinogen to apoporphobilinogen deaminase don't you immediately regenerate the cofactor to get an active enzyme?

Jordan: The apoenzyme is quite unstable. Formation of the dipyrromethane cofactor stabilizes the enzyme dramatically. By the time you have grown the organism, much of the apoenzyme would have been degraded by proteases or denatured. I don't know what conditions Umanoff et al (1988) used, but I would have thought some deaminase activity would have been seen.

Scott: It takes some time to reconstitute, probably longer than their incubation time, so activity would not have been observed.

Jordan: Cofactor assembly is a much slower reaction than the actual enzymic reaction.

Beale: Does your original reason, that the apoenzyme is unstable, still hold?

Jordan: Yes.

Akhtar: I was under the impression that when the deaminase is biosynthesized as the apoenzyme, without the cofactor, the enzyme won't be folded properly and would be partially degraded. You seem to be saying that you can remove the dipyrromethane cofactor and still obtain a correctly folded protein.

Scott: That's not a problem. The apoenzyme is easily prepared from a genetically engineered strain that is *hemA*-deficient (ALA synthase-deficient).

Battersby: The apoenzyme can be produced by controlled acidic cleavage of the dipyrromethane cofactor from the holoenzyme (Hart et al 1988). There is some denaturation but the protein can be renatured and the inactive apoenzyme can be converted into active holoenzyme by incubation with porphobilinogen. The apoprotein has a built-in ability to attach and assemble its own cofactor.

Jordan: The enzyme assembles its own cofactor, but that reaction is a lot slower than the reaction generating preuroporphyrinogen.

Arigoni: Something similar is known to happen with a set of FMN- or FAD-dependent enzymes in which the cofactor is covalently bound to the surface through a methyl group. Mixing the enzyme with the oxidized form of the cofactor is sufficient to generate the covalently bound form of the reduced coenzyme.

Rebeiz: What is known about the conformational changes accompanying the combination of the dipyrromethane with the apoenzyme?

Jordan: Cofactor assembly is accompanied by a great increase in stability. I don't know if anyone has used circular dichroism to see what is actually happening.

Scott: We did some preliminary work which I would like to go back to (Scott et al 1989). We reconstituted the unfolded apoprotein, which gave fairly sharp lines in the NMR spectrum, then incubated with porphobilinogen for 10–15 min to get the active enzyme, and, as you would expect, saw line broadening.

Warren: The apoenzyme is heat labile, and will precipitate out at 45 °C, whereas the holoenzyme is stable to 65 °C. Likewise, the apoenzyme will be digested quickly by a small amount of protease, whereas the holoenzyme is degraded by proteases in a regimented, step-wise fashion.

Eschenmoser: With respect to the mechanism and the role of the carboxylate ion of Asp-84, isn't it also necessary that the enzyme must, under all circumstances, fix the conformation of the leaving ammonium group in a position perpendicular to the plane of the pyrrole ring? Can such a role for the carboxylate ion be inferred from the structure of the enzyme?

Jordan: I would agree that the leaving group has to be in the correct orientation.

Eschenmoser: Because no reactivity at all is expected when the leaving group is in the plane of the pyrrole ring, the conformational control of that leaving group might well be the most important part of the enzyme's function.

Jordan: Porphobilinogen is unusually stable. This stability is due to an ion pair between the negatively charged acetate group and the positively charged amino group of the side chains (Fig. 8).

Arigoni: In such an ion pair the hydrogen bond would be in the plane of the pyrrole ring and that, in turn, would keep the ammonium group in a stereoelectronically unfavourable conformation.

There is an independent way of checking what goes on, and this has to do with the other question: why does one cleave the bond between the second and the third pyrrole rings, but not between the first and the second? The obvious answer would be that the first and the second pyrrole rings are linked together in a conformation that is stereoelectronically unfavourable for the cleavage reaction. That should be borne out by the crystal structure.

Jordan: The bound cofactor probably cannot get close enough to the catalytic machinery once it has been assembled, and therefore remains permanently bound to the enzyme.

Arigoni: If covalent assembly with the cofactor is spontaneous, one wonders why there is no microscopic reversibility. There must be some trick by which the cofactor is locked into position.

Jordan: The locking is achieved by binding to the enzyme.

Battersby: It would be quite inappropriate for me to give a second lecture about the work on this topic we did at Cambridge, but it may interest the group to hear about the experiment which led to the discovery of the dipyrromethane cofactor.

It had always been felt that the covalent binding between porphobilinogen and the deaminase was through oxygen or sulphur or possibly nitrogen. We tackled this problem in Cambridge by first overproducing the deaminase, then using the large amount of enzyme available to generate the ES_1, ES_2 and ES_3 complexes. These were then separated by HPLC, to give quantities sufficient for ^{13}C NMR spectroscopy. Having generated the ES_1 complex, from porphobilinogen specifically labelled with ^{13}C at C-11, we had a complex with one labelled porphobilinogen unit attached to the enzyme. The ^{13}C signal in the NMR spectrum of this ES_1 complex showed that the binding was not to nitrogen or oxygen, but it fitted perfectly something with which we were very familiar, namely pyrrole. That showed clearly that we were dealing with the connection of the first porphobilinogen unit onto a pyrrole unit. I shall never forget that day. By experiments involving multiple ^{13}C-labelling and ^{14}C-labelling followed by degradation, some similar to those Peter Jordan described, we established that the dipyrromethane was present. Also, we demonstrated direct attachment of the first porphobilinogen unit to the free α-position of the cofactor (Hart et al 1987, 1988, 1990, Beifuss et al 1988, Miller et al 1988, Alefounder et al 1989).

Jordan: The deaminase has been very exciting, giving us surprise after surprise. The first time we really realized there was a dipyrrole cofactor was in 1984 when we were trying to find the amino acid to which the substrate was linked. Dr Alan Berry labelled the enzyme with radioactive porphobilinogen to make ^{14}C-labelled ES$_2$. We degraded this with proteolytic enzymes, and isolated a pink dipyrrole chromophore linked to a peptide. However, the pink peptide, amazingly, contained no ^{14}C label. We thus knew that there was a dipyrrole system of some sort present already in the enzyme. When you realize there is a primer, it's obvious, because porphobilinogen deaminase is a polymerase.

Kannangara: There are some bacteria which produce tripyrrole pigments. Do they make these pigments in this way?

Arigoni: The pathway is totally different.

Rebeiz: Professor Jordan, you suggested a catalytic mechanism involving Asp-84. Do you have any evidence for this mechanism, or is it just hypothetical?

Jordan: Substitution of Glu for Asp-84 makes the enzyme reaction very slow. If there is no acidic group in that location the enzyme doesn't work at all—the Asp-84→Ala mutant, for example, is inactive. No other potential catalytic groups are present; it's not as though you have a choice from five or six possible groups.

Rebeiz: How far removed is the aspartate from the cavity of the enzyme?

Jordan: It is at the catalytic site, hydrogen-bonded to the dipyrromethane cofactor (Fig. 4).

Rebeiz: It would be interesting to do a three-dimensional simulation on that model.

Hädener: Some years ago you described heterogeneity of the deaminase (Jordan et al 1988), showing five FPLC peaks of similar activity, but more recently you have shown a single peak. Does this peak correspond to one of those five peaks? Have you found out more about the other active species?

Jordan: We don't get as many peaks as we used to get, which baffles me. I think this may be a problem with oxidation. There is an indication that two species are present, but I don't know the reason for that. We see a single M_r on electrospray mass spectrometry of the mixture. The other possibility is that the two peaks are different conformers.

Hädener: We always get, and have always got, a single peak (Hädener et al 1990). Were the crystals for X-ray crystallography grown from the mixture, or did you separate the mixture before crystallization?

Jordan: Whether we should isolate and crystallize a single species was one of my big concerns at the outset. We ended up crystallizing the mixture.

Wood: We screened materials of variable quality over a long period, and at one stage we got much better crystals from protein that was less pure. The purity of dissolved crystals was always much improved.

Warren: A similar phenomenon is seen with the expressed *E. coli* deaminase. The two major bands on a non-denaturing polyacrylamide gel both seem to be reasonably active. It is tempting to think that they might be different conformers, but it's difficult to prove.

Hädener: It's intriguing that we have never seen anything like that.

Pitt: The only time we see multiple bands is when we isolate the individual ES complexes and run these together on native gels. An extra band on a native gel could be an enzyme–substrate complex that has come through the purification procedure. The ES_1 complex is relatively stable, and could be carried forward through the purification if it's done quickly enough.

Scott: Our second band is never the ES complex.

Warren: This is much easier to visualize in non-denaturing gel electrophoresis whereby the different forms of the enzyme separate out. It is more difficult to separate them by anion-exchange FPLC because the elution profiles of the different forms are sometimes superimposed.

Pitt: We have never seen more than one peak in FPLC of the native enzyme.

Warren: If you take the first part of the deaminase peak that comes off the FPLC column, and compare that on non-denaturing gels with the last part of the peak, you might see a difference.

Pitt: The only time we have found two distinct peaks on FPLC is with some of the Arg→Leu mutants, and there we have two very different proteins.

Akhtar: Non-denaturing gel electrophoresis is an entirely different beast from FPLC.

Pitt: In Jordan et al (1988) the different species observed by native polyacrylamide gel electrophoresis were also separated by FPLC.

Battersby: It would be helpful at this point if Andrew Pitt could tell us something about the effects of mutation of conserved arginine residues.

Pitt: While Peter Jordan's group were substituting histidines for arginines in the *E. coli* deaminase, we at Cambridge had substituted leucines for the same arginines. Our findings were similar, despite the different substitutions chosen. The only major difference is that whereas Peter Jordan's group did not isolate an ES_4 complex with any mutant, with the Arg-155→Leu mutation we were able to isolate cleanly an ES_4 complex after incubation with a high concentration of porphobilinogen. We have done this with radioactively labelled porphobilinogen and isolated ES_1, ES_2, ES_3 and ES_4 complexes. The enzyme appears to be blocked at the stage of product release. Even when the ES_4 complex is isolated and reincubated with porphobilinogen, it does not release product.

Jordan: If I were to predict which mutants would make ES_4, I would go for substitutions at Arg-11 and Arg-155.

Pitt: We were careful to make sure that we really had the ES_4 complex by labelling, because we find that ES_4 runs on FPLC in exactly the same place as the ES_3 complex from the native enzyme.

Jordan: How long do you need to react the enzyme with porphobilinogen?

Pitt: We react either for a long time, overnight, with a low concentration, $250\,\mu M$, or for a short time, 10 min, with a high concentration, $>10\,mM$.

Jordan: How can you be certain that you are not making a tetrapyrrole (aminomethylbilane) non-enzymically which is then binding to the enzyme?

Pitt: That may be possible with the longer reaction times, but because we see progressive formation of the higher order ES complexes with increasing porphobilinogen concentration, and we can isolate the ES_4 complex, incubate with dilute porphobilinogen and re-isolate only ES_4, this seems unlikely.

Elder: Professor Jordan, one of the mutations you described, Arg-149→His, rendered the enzyme almost totally inactive at the alkaline pH you use for your assay. Is it more active at lower pH?

Jordan: Yes. One of the human mutant enzymes (equivalent to Arg-149→His) also has a lower pH optimum than the native enzyme, working better at pH 6 than at pH 8.4, at which we normally do the assays. I have no satisfactory explanation for this.

Arigoni: You have suggested that arginine is the counter ion used to neutralize the carboxyl group; this is of course eminently reasonable, but how many arginine residues are there at the periphery of the cavity?

Jordan: There are five, at positions 11, 131, 132, 149 and 155, and a further one at 176, at the back of the catalytic cleft.

Arigoni: You have a total of twelve carboxylate groups. What happens to the remaining ones?

Jordan: Perhaps they are not all bound to the enzyme. Perhaps it's not that important for the enzyme to bind each unit. If there were 12 positive counterions, the product might never be able to leave the enzyme. The enzyme must have enough counter-ions to bind the reacting pyrrole units but not enough to prevent mobility or product release.

Arigoni: Is it known whether binding through salt formation is specific for propionate versus acetate side chains?

Pitt: We have modified the structure of porphobilinogen, replacing the acetate or propionate side chains with either a methyl or an ethyl group individually. If you remove the acetate side chain, there is no binding to the enzyme. The compounds with substituted propionate side chains are good inhibitors of the enzyme, with μM K_i values.

Jordan: Are they incorporated and elongated?

Pitt: That is an interesting question. Treatment of the enzyme with inhibitor produces only the complex with one inhibitor bound. However, in the presence of substrate the inhibited enzyme is reactivated over about 30 min, and this appears to be a result of reaction with porphobilinogen rather than hydrolysis of the ES_1 complex, because in the absence of substrate it is not reactivated. The inhibitors may therefore be turned over, but very, very slowly.

Jordan: So you get a pseudoproduct.

Pitt: We think so. You can bind inhibitor to the ES_1, ES_2 and ES_3 complexes as well. If you treat the ES_4 complex with inhibitor you release a bilane.

References

Alefounder PR, Hart GJ, Miller AD et al 1989 Biosynthesis of the pigments of life: structure and mode of action of a novel enzymic cofactor. Bioorg Chem 17:121–129

Beifuss U, Hart GJ, Miller AD, Battersby AR 1988 ^{13}C NMR studies on the pyrromethane cofactor of hydroxymethylbilane synthase. Tetrahedron Lett 29:2591–2594

Hädener A, Alefounder PR, Hart GJ, Abell C, Battersby AR 1990 Investigation of putative active site lysine residues in hydroxymethylbilane synthase: preparation and characterization of mutants in which (a) Lys-55, (b) Lys-59 and (c) both Lys-55 and 59 have been replaced by glutamine. Biochem J 271:487–491

Hart GJ, Miller AD, Leeper FJ, Battersby AR 1987 Biosynthesis of the natural poprhyrins: proof that hydroxymethylbilane synthase (porphobilinogen deaminase) uses a novel binding group in its catalytic action. J Chem Soc Chem Commun, p 1762–1765

Hart GJ, Miller AD, Battersby AR 1988 Evidence that the pyrromethane cofactor of hydroxymethylbilane synthase (porphobilinogen deaminase) is bound through the sulphur atom of the cysteine residue. Biochem J 252:909–918

Hart GJ, Miller AD, Beifuss U, Leeper FJ, Battersby AR 1990 Biosythesis of porphyrins and related macrocycles. Part 35. Discovery of a novel dipyrrolic cofactor essential for the catalytic action of hydroxymethylbilane synthase (porphobilinogen deaminase). J Chem Soc Perkin Trans I, p 1979–1993

Jordan PM, Thomas SD, Warren MJ 1988 Purification, crystallization and properties of porphobilinogen deaminase from a recombinant strain of *Escherichia coli* K12. Biochem J 254:427–435

Miller AD, Hart GJ, Packman LC, Battersby AR 1988 Evidence that the pyrromethane cofactor of hydroxymethylbilane synthase (porphobilinogen deaminase) is bound to the protein through the sulphur atom of cysteine-242. Biochem J 254:915–918

Scott AI, Clemens KR, Stolowich NJ, Santander PJ, Gonzalez MD, Roessner CA 1989 Reconstitution of apo-porphobilinogen deaminase: structural changes induced by cofactor binding. FEBS (Fed Eur Biochem Soc) Lett 242:319–324

Umanoff H, Russell CS, Cosloy SD 1988 Availability of porphobilinogen controls appearance of porphobilinogen deaminase activity in *Escherichia coli* K-12. J Bacteriol 170:4969–4971

Structural studies on porphobilinogen deaminase

R. Lambert, P. D. Brownlie, S. C. Woodcock*, G. V. Louie, J. C. Cooper, M. J. Warren*, P. M. Jordan*, T. L. Blundell and S. P. Wood

*ICRF Structural Molecular Biology Unit, Department of Crystallography, Birkbeck College, Malet Street, London WC1E 7HX and *School of Biological Sciences, Queen Mary and Westfield College, Mile End Road, London E1 4NS, UK*

Abstract. The X-ray crystallographic analysis of porphobilinogen deaminase (hydroxymethylbilane synthase, EC 4.3.1.8) shows the polypeptide chain folded into three domains, (1) N-terminal, (2) central and (3) C-terminal, of approximately equal size. Domains 1 and 2 have a similar overall topology, a modified doubly wound parallel β-sheet. Domain 3 is an open-faced three-stranded antiparallel β-sheet, with one face covered by three α-helices. The active site is located between domains 1 and 2. The dipyrromethane cofactor linked to cysteine 242 protrudes from domain 3 into the mouth of the cleft. Flexible segments between domains 1 and 2 are thought to have a role in a hinge mechanism, facilitating conformational changes. The cleft is lined with positively charged, highly conserved, arginine residues which form ion pairs with the acidic side chains of the cofactor. Aspartic acid 84 has been identified as a critical catalytic residue both by its proximity to the cofactor pyrrole ring nitrogen and by structural and kinetic studies of the Asp-84→Glu mutant protein. The active site arginine residues have been altered by site-directed mutagenesis to histidine residues. The mutant proteins have been studied crystallographically in order to reconcile the functional changes in the polymerization reaction with structural changes in the enzyme.

1994 The biosynthesis of the tetrapyrrole pigments. Wiley, Chichester (Ciba Foundation Symposium 180) p 97–110

Our first attempts to crystallize porphobilinogen deaminase (PBGD; hydroxymethylbilane synthase, EC 4.3.1.8) for X-ray analysis started in 1986. Although we obtained crystals almost immediately, they were thin and of little use for diffraction experiments. The major advance occurred when we realized that the protein crystallization was sensitive to light. We were soon able to grow large ($1 \times 0.5 \times 0.5$ mm^3) straw-yellow crystals by precipitation of the protein with polyethylene glycol around pH 5.0–5.5. The crystals would not grow unless the solutions were protected from light. Furthermore, well-formed crystals became degraded on prolonged light exposure (Jordan et al 1992). We now understand some of these early observations, particularly with respect to the oxidation state of the dipyrromethane cofactor, a central component of the enzyme which at that time had yet to be fully characterized.

The PBGD crystals produced very strong diffraction patterns and reacted readily with a number of heavy atom reagents. Our only disappointment was their extreme sensitivity to thiol-directed reagents. The method of multiple isomorphous replacement was used to calculate a good electron density map at 3 Å resolution and the molecular model has subsequently been refined at 1.9 Å (Louie et al 1992) and 1.76 Å resolution (Louie et al 1994). We therefore have a detailed description of the enzyme which can be used to advance understanding of its mechanism. Here, we briefly review the structure and discuss the implications it has for current views on the mechanism. Recent results on the state of the cofactor (Hädener et al 1993) and the structure and properties of mutants (Jordan & Woodcock 1991, Lander et al 1991) will be discussed together with the likely role played by conformational changes in the catalytic cycle.

The structure

The polypeptide chain of PBGD from *Escherichia coli* comprises 313 amino acids, and the refined atomic model at 1.76 Å resolution describes residues 3–48, 58–307, the cofactor, 249 water molecules and one acetate ion with an estimated coordinate error of 0.2 Å. The molecule has approximate dimensions of $57 \times 43 \times 32$ Å^3 with the chain folded into three domains of roughly equivalent size (see Fig. 1). Domain 1 (the N-terminal domain) and domain 2 have a similar five-stranded β-sheet arrangement with α-helices flanking each of the β-sheets and running approximately parallel with them. The strand in each sheet that immediately follows a cross-over from the other domain is antiparallel. The approximate two-fold symmetry between domains is broken by two prolines in the $\beta3_1$–$\beta4_1$ loop, which corresponds to the α-helix $\alpha3_2$ of domain 2. This organization of the polypeptide chain is very similar to that in the transferrins and bacterial periplasmic binding proteins (Baker et al 1987, Spurlino et al 1991, Louie 1993). Domain 3 is an open-faced three-stranded sheet with three helices packed on one face. Each domain has a distinct hydrophobic core but the contacts between domains are dominated by polar interactions. There are few direct interactions between domains 1 and 2, but both form a complex array of ion pairs and hydrogen bonds with the dipyrromethane cofactor which lies in a large cleft between them. The cofactor is covalently attached to Cys-242 (the ring attached to the enzyme being termed C1, and the second pyrrole ring C2; see Jordan 1994, this volume), which resides on the $\alpha1_3$–$\beta1_3$ loop interconnecting a helix in domain 1 with a strand in domain 3. The acidic side groups of the cofactor make interactions with conserved basic side chains of the protein which surround the cofactor cavity, while the pyrrole nitrogens, whose protonation state defines the characteristic chemistry of the pyrrole nucleus,

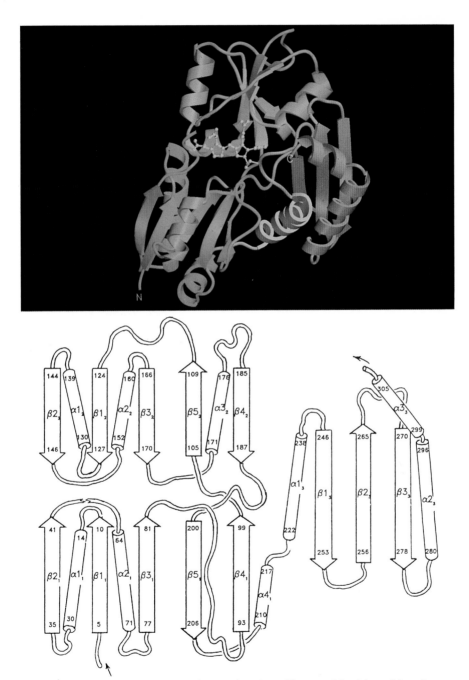

FIG. 1.   The structure of porphobilinogen deaminase illustrated (*top*) in a ribbon format and (*bottom*) as a topology diagram, showing the organization of the polypeptide chain in three domains and the nomenclature of secondary structure elements. (Domain 1, blue; domain 2, green; domain 3, violet.)

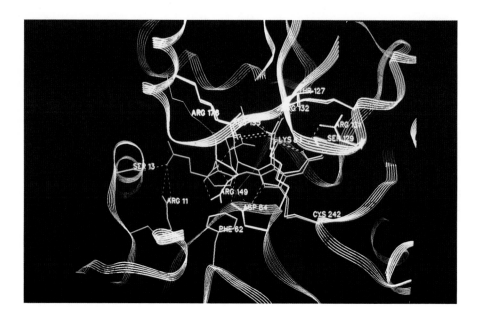

FIG. 2. Structure of the active site of porphobilinogen deaminase derived by X-ray crystallography, showing three pyrrole-binding sites. Cofactor ring C1 occupies essentially the same site in the oxidized (yellow) and reduced (brown) states whereas ring C2 occupies different positions.

are hydrogen-bonded to Asp-84. Thus the X-ray crystallographic structure has highlighted the active site region, a potential catalytic residue and general architectural features which imply that conformational flexibility at least as extensive as observed in the transferrins may be possible in porphobilinogen deaminase.

Detail at the active site

The floor and ceiling of the active site cleft are formed by the C-termini of the β-sheets, the N-termini of the α-helices and their connecting loops of domains 1 and 2. One face of helix $\alpha 3_2$ and the interdomain cross-over strands make up the rear wall defining a cavity approximately $15 \times 13 \times 12$ Å^3. In the original structural analysis (Louie et al 1992) no thiol reagent was used to protect the protein and in these conditions the electron density attributed to cofactor ring C2 was indicative of two different conformations. About 80% of the electron density occupied a position towards the front of the cleft such that the two pyrrole rings were only 11° from coplanarity. The remaining electron density was distributed towards the back of the cleft with ring planes 61° apart (see Fig. 2). The near coplanarity of the dominant cofactor conformer, the yellow crystal coloration and the extra lobe of electron density on the α-position of ring C2 suggested an enhanced susceptibility to oxidation in the crystallization conditions and that the cofactor was predominantly in a dipyrromethenone form. In this form it could undergo a light-induced isomerization, the most likely explanation for the deleterious effect of light during crystallization. Inclusion of β-mercaptoethanol in our crystallizations produced colourless crystals, but these were always very small. Good crystals of a selenium-substituted enzyme have been prepared with high concentrations of dithiothreitol (Hädener et al 1993). The efficacy of this recipe has been confirmed in our laboratory with native enzyme and with some mutant enzymes. Under these reducing conditions the crystals are colourless and the rear position of the cofactor is fully occupied. It is interesting to note that the isomorphous replacement experiments would have followed a different path in reducing conditions because the cofactor ring C2 occupies the binding site for the $K_2Pt(Cl_4)$ derivative and the reducing agent might well have modified the reactivity of other heavy atom reagents used.

X-ray analysis has thus revealed three binding sites for pyrrole rings; two are the native binding sites for the two cofactor rings (C1 and C2), and the third is available for substrate binding (site S). All these positions are stabilized by a number of interactions with the protein. Side groups on ring C1 form ion pairs with Arg-131, Arg-132, Lys-83 and Arg-155. This ring has the lowest thermal parameters and the best electron density. At the rear site, cofactor ring C2 shows interactions with Lys-83 and Arg-131. The position occupied by ring C2 in the oxidized structure (site S) shows side group ion pairs with Arg-11, Arg-149 and Arg-155 and a ring-stacking interaction with Phe-62. The electron density is less well defined than that for the C1 ring. Other hydrogen bonds are formed with Ser-13, Ser-81, Thr-127 and Ser-129. In view of the stability of each cofactor ring position it seems likely that each is relevant to enzyme function. Before discussing this further, however, we should briefly outline the general chemistry of the mechanism.

The mechanism

There is general agreement that the enzyme reaction begins with the binding of one molecule of porphobilinogen, which is deaminated to generate an azafulvene. This species is subject to nucleophilic attack by the α-position of the C2 cofactor ring or of the terminal ring of enzyme–substrate intermediates. A hydrogen must be removed from this position. This series of reactions must be repeated four times and be followed by a reversed reaction whereby a water molecule attacks the tetrapyrrolic azafulvene to yield 1-hydroxymethylbilane (preuroporphyrinogen) (Fig. 3; see Pichon et al 1992 for a recent discussion and Jordan 1994, this volume). Aspartic acid 84 would appear to be the only residue suitably positioned to donate/accept a proton and to stabilize positive charges throughout the reaction cycle. The environment of Asp-84 is modified when the C2 or S sites are occupied. A pyrrole ring occupying the S site forms a stacking interaction with Phe-62 and screens Asp-84 from solvent. The Phe-62 side chain is less defined when site S is empty. This provides a route to modulation of the pK_a of Asp-84. There is no evidence of replication of catalytic groups elsewhere in the cleft.

   It seems most likely that the rear position of the C2 cofactor ring existing in reducing conditions represents the resting cofactor conformer. The front site (site S) is a strong candidate for a substrate-binding site. However, the substrate molecule would need to dock in a reversed orientation so as to position the azafulvene's exocyclic methide appropriately relative to the free α-carbon for reaction. Aspartic acid 84 could hydrogen bond with the substrate pyrrole nitrogen in either orientation. It is also possible that one of the conserved residues in the 48–58 region, invisible in the crystal structure owing to disorder, is involved. Alternatively, the azafulvene might be reoriented after deamination. These possibilities can be tested only by structural analysis of enzyme–intermediate complexes (currently underway). Presumably, when the coupling reaction is complete the newly added ring must vacate the substrate-binding site, and it seems reasonable that it should move into the site formerly occupied by cofactor ring C2. This scheme is broadly consistent with the properties of enzymes with substitutions at Arg-11, Arg-149 and Arg-155, all of which supply interactions to site S—the putative substrate-binding site; such enzymes have an impaired ability to form the first enzyme–substrate complex, ES. Once moved to the back site the terminal ring is likely to be less susceptible to reactions that would lead to its release. The C2 cofactor ring is stable in this site, and the

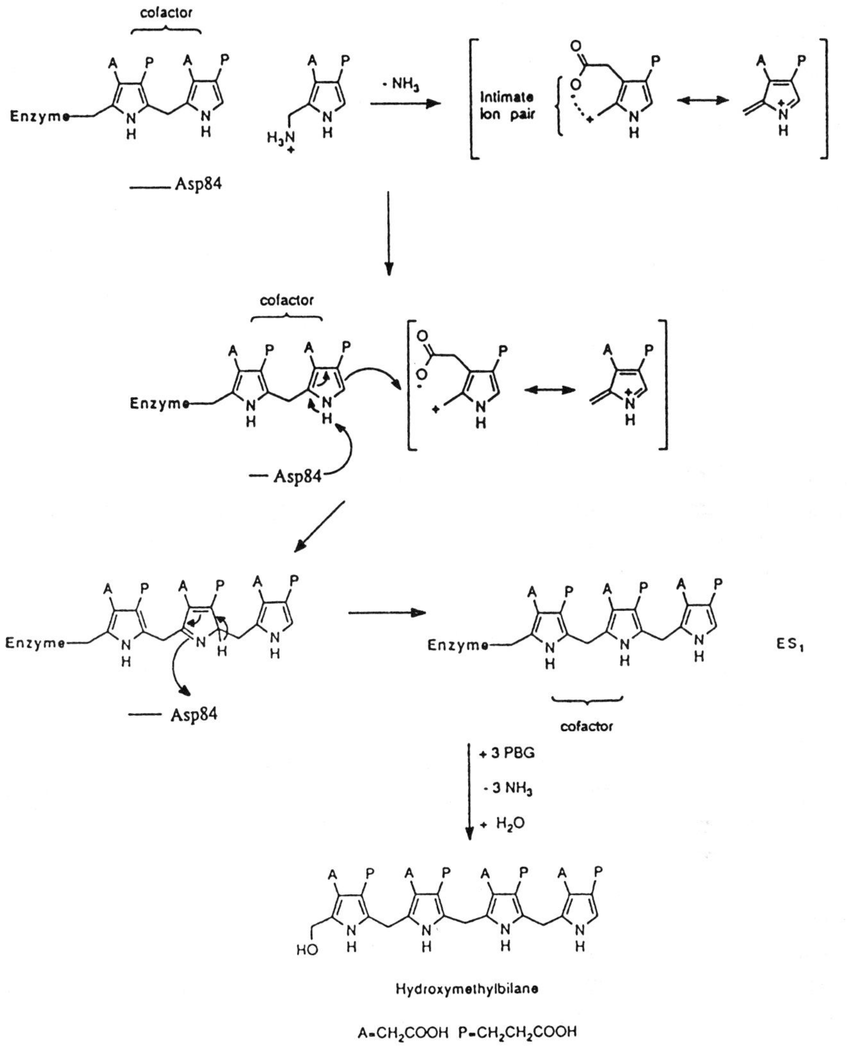

FIG. 3. The putative reaction mechanism of porphobilinogen deaminase (after Pichon 1992).

known lability of the terminal rings of intermediates must relate to their occupation of the front site (Warren & Jordan 1988).

These speculations take us to the heart of the problem of this intriguing mechanism. How does the enzyme accommodate the labile growing polypyrrole chain of escalating negative charge, employing perhaps only a single catalytic residue, and sense chain completion before hydrolytic release?

A number of general mechanisms can be envisaged, each of which is consistent with some experimental data. In the simplest, the reacting groups might move in procession past the active centre, with some return mechanism to catalyse the release reaction. Such a mechanism would involve conformational changes. Gross conformational changes are known to occur on ligand binding in the structurally analogous transferrins and periplasmic binding proteins. The self-contained nature of the deaminase domains and their polar interactions are the central features suggestive of a propensity for conformational change. Several features interrupting the two-fold symmetry of domains 1 and 2 also imply conformational change is likely. The absence of the expected third helix of domain 1 has already been noted (see Fig. 1) and the substituted $\beta3_1$–$\beta4_1$ loop carries Asp-84 at its N-terminus. Domain 1 forms interactions mainly with the substrate-binding site, whereas domain 2 forms most interactions with both cofactor ring sites. The two cross-over strands between these domains provide a hinge region for movement of the domains relative to each other and it may be significant that Val-196 in this region occupies two distinct orientations. The loop in domain 3 to which the cofactor is attached forms few contacts with domains 1 and 2 and could accommodate or initiate cofactor movements. Cysteine 134, which resides at the interface of domains 2 and 3, shows greatly increased susceptibility to chemical modification in the presence of substrate. The substitution of His for Arg-232, which is distant from the active site but involved in an interdomain ion pair, causes a severe loss of activity. All these factors point to a substantial structural change occurring during chain elongation.

An alternative mechanism would be for the growing chain to be folded into the cavity. The volume available corresponds approximately with 3½ pyrrole rings and is derived partly from the large number of small conserved side chains that line the cleft. This volume limitation explains why the elongation does not proceed beyond a tetrapyrrolic product. Furthermore, mutation of Arg-176, which makes a water-mediated hydrogen bond with C2, perturbs a late stage of the extension reaction.

The mechanism most probably therefore combines features from both of these schemes. As the first ring addition is completed, a concerted movement of the terminal ring into the C2 site, with binding of a further substrate molecule in site S, must coincide with a repositioning of the cofactor rings either by convolution in the cleft or by displacement. Cofactor ring C1 could be replaced by ring C2 as it is pulled through by a conformational adjustment of the orientation of domain 3 with respect to domains 1 and 2.

Structural study of enzyme–intermediate complexes is clearly required to gain further understanding. The mutants described elsewhere in this volume (Jordan 1994) provide a feasible route, as different patterns of intermediates accumulate in different mutants. As a starting point we have prepared isomorphous crystals of PBGD from some of these mutant strains and have collected and processed

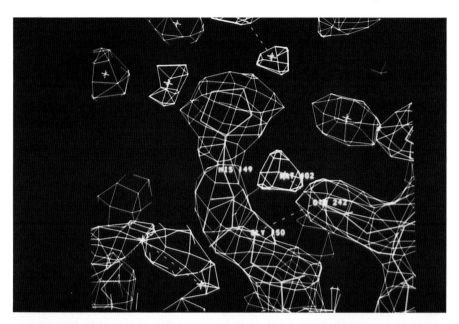

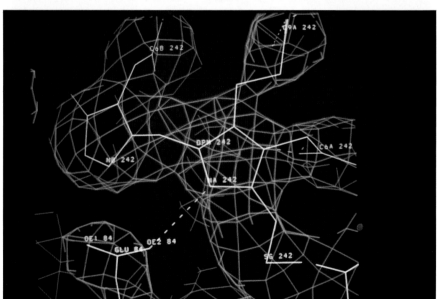

FIG. 4. (*top*) The conformation of the histidine 149 residue in the Arg-149→His mutant PBGD. (*bottom*) The difference density of the glutamate 84 residue in the Asp-84→Glu mutant protein.

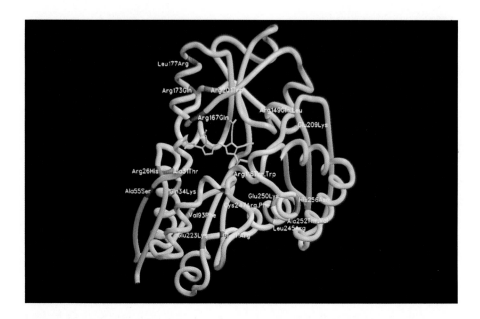

FIG. 5. The positions of known amino acid substitutions resulting from mutations in the PBGD gene in patients with acute intermittent porphyria are illustrated as their equivalents in the *E. coli* fold.

X-ray data for those with Arg-176→His, Arg-149→His and Asp-84→Glu mutations. The first two mutant protein crystals were prepared in oxidizing conditions and the third under the reducing conditions. In the oxidized native structure, Arg-149 interacts with the acetic acid side chain of cofactor ring C2 whereas in the reduced state it does not. Difference electron density maps for Arg-149→His confirm the sequence change and show no evidence for major conformational changes. The side chain of the substituted His-149 fails to interact with the cofactor and swings away to interact with solvent (Fig. 4). This mutant enzyme has an increased K_m, consistent with the loss of an interaction at the putative substrate-binding site. Arginine 176 makes no direct interactions with the cofactor but does form a water-mediated hydrogen bond to the propionate side chain of ring C2 in the reduced state. The electron density shows that His-176 cannot make this interaction and instead stacks against an adjacent tryptophan residue. The amino acid substitution interferes with the formation of ES_2 and ES_3. This is consistent with the mechanism discussed above in which the intermediates fold into the cleft. The Asp-84→Glu mutant has only 1% of the activity of the native enzyme, confirming the central importance of this residue that was implied by structural observations. Difference electron density maps of this mutant enzyme show that the most likely explanation for this effect is the loss of the hydrogen bond to the pyrrole nitrogen of cofactor ring C2 (see Fig. 4b). The hydrogen bond to the site S position is also unlikely to form. Thus, there may be value in examining both reduced and oxidized states of mutant enzymes because this may provide a view approximating the ES complex.

Some of the mutations generated in the *E. coli* enzyme have counterparts associated with acute intermittent porphyria in humans (Kappas et al 1989). More than 20 human PBGD mutants are known. We have modelled the human enzyme on the basis of the strong conservation of structurally important residues, and tried to correlate the behaviour of the mutant enzymes (see Fig. 5) with our structural understanding of the protein. In some cases we can explain the aberrant behaviour of the mutant protein (Brownlie et al 1994). For example, the Leu-177→Arg mutation (position 159 in *E. coli*) places a charged residue in the hydrophobic core of the protein which would be deleterious to its folding (Mgone et al 1992). Replacement of Arg-149 by Gln (human) would be expected to perturb addition of cofactor to the apoenzyme in a similar way to the equivalent Arg-131→His mutation in *E. coli* (Delfau et al 1990). In some cases, however, the deleterious nature of the mutation is not understood, and we need to study these cases carefully to understand the full mechanistic implications.

Acknowledgements

We would like to acknowledge the funding from SERC (Molecular Recognition Initiative) and AFRC (P. M. Jordan). G. V. Louie is the recipient of an MRC Canada Postdoctoral fellowship. We gratefully acknowledge Dr F. J. Leeper, University Chemical

Laboratory, University of Cambridge, UK, Dr A. Hädener, Institut für Organische Chemie der Universität Basel, Switzerland and Dr D. Rice, Krebs Institute, University of Sheffield, UK.

References

Baker EN, Rumball SV, Anderson BF 1987 Transferrins: insights into structure and function from studies on lactoferrin. Trends Biochem Sci 12:350–353

Brownlie PD, Wood SP, Louie GV et al 1994 Towards understanding the molecular structural basis of acute intermittent porphyria. Science, submitted

Delfau H, Picat C, de Rooji FWM et al 1990 Two different point G to A mutations in exon 10 of the porphobilinogen deaminase gene are responsible for acute intermittent porphyria. J Clin Invest 89:1511–1516

Hädener A, Matzinger PK, Malashkevich VN et al 1993 Purification, characterization, crystallization and X-ray analysis of selenomethionine-labeled hydroxymethylbilane synthase from *Escherichia coli*. Eur J Biochem 211:615–624

Jordan PM 1994 Porphobilinogen deaminase: mechanism of action and role in the biosynthesis of uroporphyrinogen III. In: The biosynthesis of the tetrapyrrole pigments. Wiley, Chichester (Ciba Found Symp 180), p 70–96

Jordan PM, Woodcock SC 1991 Mutagenesis of arginine residues in the catalytic cleft of *Escherichia coli* porphobilinogen deaminase that affects dipyrromethane cofactor assembly and tetrapyrrole chain initiation and elongation. Biochem J 280:445–449

Jordan PM, Warren MJ, Mgbeje BIA et al 1992 Crystallisation and preliminary X-ray investigation of *Escherichia coli* porphobilinogen deaminase. J Mol Biol 224:269–271

Kappas A, Sassa S, Galbraith RA, Nordmann Y 1989 The porphyrias. In: Scriver CR, Beaudet A, Sly WS, Valle D (eds) The metabolic basis of inherited disease, 6th edn. McGraw-Hill, New York, p 1305–1365

Lander M, Pitt AR, Alefounder PR, Bardy D, Abell C, Battersby A 1991 Studies on the mechanism of hydroxymethylbilane synthase concerning the role of arginine residues in substrate binding. Biochem J 275:447–452

Louie GV 1993 Porphobilinogen deaminase and its structural similarity to the binding proteins. Curr Opin Struct Biol 3:401–408

Louie GV, Brownlie PD, Lambert R et al 1992 Structure of porphobilinogen deaminase reveals a flexible multidomain polymerase with a single catalytic site. Nature 359:33–39

Louie GV, Brownlie PD, Lambert R et al 1994 The three dimensional structure of *Escherichia coli* porphobilinogen deaminase at 1.76-Å resolution. Protein Sci, submitted

Mgone CS, Lanyon WG, Moore MR, Connor JM 1992 Detection of seven point mutations in the porphobilinogen deaminase gene in patients with acute intermittent porphyria, by direct sequencing of in vitro amplified cDNA. Hum Genet 90:12–16

Pichon C, Clemens KR, Jacobson AR, Scott AI 1992 On the mechanism of porphobilinogen deaminase: design, synthesis, and enzymic reactions of novel porphobilinogen analogues. Tetrahedron 23:4687–4712

Spurlino J, Lu G-Y, Quiocho FA 1991 The 2.3-Å resolution structure of the maltose- or maltodextrin-binding protein, a primary receptor of bacterial active transport and chemotaxis. J Biol Chem 266:5202–5219

Warren MJ, Jordan PM 1988 Investigation into the nature of substrate binding to the dipyrromethane cofactor of *Escherichia coli* porphobilinogen deaminase. Biochemistry 27:9020–9030

DISCUSSION

Rebeiz: Do all the mutations shown in Fig. 5 result in altered enzyme activity?

Wood: Yes. All these mutations lead to substantially reduced enzyme activity. They have been described in heterozygotes for acute intermittent porphyria who show half the expected level of enzyme activity.

Rebeiz: Is there a relationship between the influence of an amino acid substitution on enzymic activity and its distance from the reaction centre?

Wood: The inactivating mutations shown are widely distributed throughout the structure. We can explain the effects of substitution of active site residues. Some substitutions far from the cleft are likely to be deleterious due to effects on the overall structure. Others are less easy to understand, such as His-256→Asn. The usual side chain forms hydrogen bonds to cap the C-terminus of $\alpha 1_3$ and with Asn-311 of the adjacent helix $\alpha 2_3$. Helix $\alpha 1_3$ precedes the cofactor-carrying loop and the expected disturbance to the hydrogen bonding in the mutant must somehow transmit to the active site.

Hädener: I gather that Andy Pitt has suggested that in the site which binds the substrate the substrate ring should be rotated 180°. That seems reasonable.

Pitt: There's something else to consider here. If you turn the detached ring corresponding to C2 of the cofactor in the oxidized form through 180° (Fig. 2, yellow), and assume it is the substrate ring, it effectively sits almost face to face with the C2 pyrrole ring of the reduced cofactor which is pointing in towards the back of the active site cleft. You can then start to do the chemistry with the two pyrrole rings effectively face to face, but as you form the methane bridge between the two pyrrole rings, you must change the orientation of the two rings that are sitting face to face, because you must end up with something that has a 109° angle at the methane bridge.

Arigoni: This corresponds to what Albert Eschenmoser alluded to in an earlier discussion (p 91). The angle between the two rings is the one required by the stereoelectronic theory.

Pitt: That's right to begin with, but as the reaction proceeds, the angle between the two rings must change; perhaps this drives the translocation process, through a pushing force from the chemical reaction at this end of the growing chain rather than a pulling force from the cysteine end resulting from conformational change.

Jordan: During the loss of the α-hydrogen atom a change from sp^3 to sp^2 hybridization would change the conformation and could provide the driving force for the translocation.

Wood: PBGD isn't a highly active enzyme, providing only a modest rate enhancement for the reaction. Electrostatic effects involving the basic groups of the cleft are probably important for substrate binding but the energetics of the extension reaction might also be met in part by the displacement of ordered water molecules from the cleft.

Arigoni: Can you visualize the water molecules?

Wood: Yes. The water molecules in the cleft are quite well defined. There are about 17.

Arigoni: Are they random?

Wood: No; they're all hydrogen-bonded to the protein or the cofactor. The degree of saturation of the hydrogen-bonding potential of the whole structure is high, around 75%.

Hädener: Isn't there also a geometrical problem? For the growing chain to fit into the cavity, it has to be bent, and this can be done only if the rings flip into a position more horizontal than the one shown are in Fig. 2.

Wood: That's correct.

Pitt: Everyone feels conformational change may well take place, and there is quite a lot of circumstantial evidence for that. It is our feeling that the major conformational change takes place between the ES_2 and ES_3 complexes.

Wood: The selection of small conserved side chains around the back of the cleft suggests that space is required for some part of the extension reaction involving folding of intermediates into the cleft, but there is not sufficient room for this to be the whole story.

Pitt: If the major conformational change is at the ES_2 to ES_3 stage, the cavity may be large enough to accommodate the first two rings but may start to open up for the final stages of the process.

Scott: When you do simple molecular mechanics modelling on tetrapyrrolic or hexapyrrolic ensembles you begin to get helical structures with hydrogen bonds from the acetate side chain carboxylates back into the pyrrole NHs. I had a naïve idea that the ES_4 complex came out of the cleft like a miniature helix, with the dipyrromethane concealed at the bottom. Have you tried any molecular mechanics modelling?

Wood: No, though we have talked about such a helical model. More experimental results are needed before useful modelling can be done with longer intermediates. So far we have attempted only to rotate the pyrrole ring in the putative substrate-binding site by 180°. This rearrangement still allows formation of a hydrogen bond between the pyrrole nitrogen and Asp-84.

Beale: If there's one active site, and the cofactor is also assembled at that active site, the joining of the two initial pyrrole rings of the cofactor must precede the attachment of the first pyrrole to the cysteine, because once it's attached to the cysteine, the dipyrrole bridge region can no longer reach the active site.

Arigoni: Why do you say that?

Beale: Because the active site is where rings 2 and 3 come together. The active site is where the distal ring of the dipyrrole cofactor can meet up with the incoming third pyrrole ring, two pyrrole rings away from the cysteine. When the first two were put together, there could not have been sufficient reach between the cysteine and the active site. That suggests that an enzyme in which the cysteine is substituted by a homologous but non-thiol-containing amino acid might still be able to act as a dipyrrole synthase. Does it?

Pitt: No. The mechanism by which the cofactor is inserted may not be exactly the same as that by which the later pyrrole rings are put on.

Scott: We certainly failed to reactivate the enzyme with synthetic APAP-aminomethyldipyrromethane (1).

1

Jordan: The apoenzyme doesn't have the benefit of the dipyrromethane cofactor to hold domains 1 and 2 together. If there is more space between them, as is possible in the apoenzyme, Cys-242, in domain 3, could actually get nearer to the catalytic machinery than it can when the cofactor is assembled. That would be a sophisticated evolutionary accomplishment. The fact that mutants with substitutions at Arg-131 and 132 cannot assemble the cofactor does not prove this is so, but could mean that binding at that particular site is important in the process of cofactor assembly.

Rebeiz: Could I play devil's advocate, and inject a note of caution? How closely related is an X-ray crystallographic structure to the structure of the native enzyme *in situ*?

Wood: One should always remember that question. It is certainly possible that crystal-packing forces influence the degree of cleft opening or closure; but the overall conformation of the polypeptide chain within each domain should not differ significantly between solution and the crystal.

Scott: An NMR solution structure should be available fairly soon. Ad Bax (NIH) has ^{13}C- and ^{15}N-enriched PBGD and is going to attempt three- and four-dimensional NMR experiments. This structure won't necessarily be different from the crystal structure, but it will be good to have a solution conformation.

Arigoni: But there is something missing here, as in most crystal structures. The stage is set. We know what kind of drama is being played out from the overall result, but we do not know how many acts there have been, or who has played the major role. The dynamics are missing. That is a challenge to the organic chemists.

Scott: Another approach would be to crystallize the ES$_2$ and ES$_3$ complexes.

Eschenmoser: Dr Wood, would it have been easy or difficult to discover the presence of the two cofactor pyrrole rings in the crystal structure of the enzyme if you hadn't known that they were there?

Wood: Ring C1 has good electron density and was clearly visible at an early stage in the analysis of the oxidized crystal form. Ring C2 was less well defined and distributed between the two sites, but nevertheless clearly a pyrrole ring in refinement. Had the structural analysis proceeded from reduced crystals, however, it is not clear that we would have quickly located the putative substrate site (Fig. 2).

Scott: What resolution factor would be necessary for the dipyrromethane to stick out like a sore thumb?

Wood: One could begin to see it at 3Å resolution.

Kräutler: The assembly of a tripyrrole unit from a dipyrrole unit has three essential steps—the removal of the amino function, the formation of the C–C bond, and the deprotonation of the adduct. Is it known which of these steps is rate-determining in the enzymic process?

Pitt: No. We are not even sure whether the mechanism involves the azafulvene, although that's the mechanism that's always drawn. You could just as easily draw a mechanism using the α-free position of the pyrrole as a nucleophile to displace the ammonium directly.

Akhtar: Surely that wouldn't happen, on chemical grounds; even if there was nothing else wrong with that mechanism, one would predict an inversion of stereochemistry.

Arigoni: What's wrong with that?

Akhtar: When the stereochemistry is determined, there will prove to be a retention!

Arigoni: The only stereochemical information available stems from the work of Alan Battersby and his colleagues (Schauder et al 1987); there is overall retention in the transformation of the aminomethyl compound to the linear hydroxymethyl tetrapyrrole. There could be two retentions, or two inversions. That raises an intriguing question. If one had not known what the experimental result was, could it have been predicted from the crystal structure?

Wood: We do not yet have enough structural evidence about the organization of the reacting groups to make such predictions.

Arigoni: The fact that you can substitute water with other nucleophiles such as hydroxylamine or ammonia tells you that the direction in which the nucleophile approaches the azafulvene must be from the exposed face of the molecule, not from the inner face of the cavity where there is no space for hydroxylamine.

Kräutler: The first step can then be eliminated from my earlier considerations: the removal of the amino group may or may not be synchronized with the second step, but you cannot make the two later steps occur at the same time. The C–C bond formation cannot be synchronous with the deprotonation. To find out about the importance of that last step, kinetically, perhaps one should use the corresponding compound deuterated at the α-position. The hybridization of that position seems to be important for the geometry of the active complex which the enzyme should stabilize.

Arigoni: Whether the expected intrinsic isotope effect can be detected or not depends on the degree of commitment of that step in the entire reaction profile. It's worth a try though.

Jordan: With porphobilinogen deuterated at the α-position there is no effect on the rate. In all the ES complexes the α-hydrogen is retained on the terminal ring. This was shown with α-tritiated porphobilinogen.

Hädener: Perhaps I should add a few comments about our selenomethionine-labelled PBGD. We produced the selenoprotein with the intention of doing multiple wavelength anomalous dispersion (MAD) measurements. Essentially, the producing strain was made methionine dependent then grown in the presence of selenomethionine rather than methionine (Hädener et al 1993). A methionine-dependent organism grown in this way will produce all its proteins, including overexpressed ones, in selenomethionine-labelled form. This does not seem to affect the growth of these mutants greatly, a fact that continues to amaze me. All vital functions which require a derivative of methionine in a protein or cofactor remain essentially undisturbed when methionine is fully replaced by selenomethionine.

Akhtar: Is the rate of growth the same as that of the ordinary organism?

Hädener: It's a little slower, but of the same order of magnitude. This was originally described by Cowie & Cohen in 1957. It is interesting that the same cannot be done with the cysteine residues.

Arigoni: Are you implying that these organisms have a selenium analogue of *S*-adenosylmethionine which works in exactly the same fashion?

Hädener: Yes.

Arigoni: Has that been checked independently?

Hädener: It follows from circumstantial evidence. Mudd & Cantoni (1957) showed that selenomethionine is an even better substrate for the methionine-activating enzyme from yeast than the natural sulphur analogue. In addition, the abilities of *Se*-adenosylselenomethionine and *S*-adenosylmethionine to serve as methyl donors in the formation of creatine were similar.

Arigoni: There may still be traces of the endogenous methionine.

Hädener: You can keep these cultures for more than 100 generations without supplying methionine.

Arigoni: Is there any source of sulphur in your medium?

Hädener: There is sulphate, but the mutant is not able to make methionine from it, although it does make cysteine. We don't see any trace of residual ordinary methionine when we analyse the amino acid composition of proteins produced in this way. I assume the same is true for *S*-adenosylmethionine. Also, deaminase produced in this way shows kinetic parameters exactly the same as those of the wild-type enzyme.

If you calculate a difference map with the data sets of the native protein and the selenomethionine variant, you see electron density at all the sites where there are methionine sulphur atoms in the native protein's structure. That is what you would expect, but confirms that the methionines really have been replaced.

We found electron density at the N-terminus in the difference map (Hädener et al 1993). Has this now been assigned to a residue?

Louie: There are six methionines in *E. coli* PBGD, and five of the six have difference density in an $F_{Se-Met} - F_{S-Met}$ difference map. The remaining methionine is the N-terminal methionine. An electron-density peak in the difference map near the N-terminus is unaccounted for, but it is unclear whether or not this peak represents the sulphur-to-selenium substitution of Met-1. In our current atomic model for PBGD, which begins at residue 3, it does not appear that residues 1 and 2 can be built onto residue 3 such that the selenium/sulphur of Met-1 is positioned in the difference peak. However, the terminal residues of proteins can be subject to misinterpretation because they are poorly ordered.

Akhtar: I can accept that a protein can have selenomethionine instead of methionine, but how can selenomethionine participate in *S*-adenosylmethionine-dependent reactions? Does *E. coli* require a methylation at any stage?

Arigoni: It has to make spermine and spermidine, and their biosynthesis requires *S*-adenosylmethionine.

Leeper: It also has to make sirohaem.

Beale: You can really be fooled with sirohaem. If a mutant that cannot make sirohaem contains any ampicillin-selected plasmid, the mutant will grow, because it can use the ampicillin as a source of reduced sulphur and bypass the need for sirohaem.

Arigoni: What happens to lipoic acid, or to biotin?

Scott: There is a natural arsenic analogue related to *S*-adenosylmethionine (Cullen & Reiner 1989).

Akhtar: Surely that doesn't transfer the methyl group, does it?

Scott: I don't think that's been tested.

Battersby: It is amazing that no one has synthesized seleno-*S*-adenosyl-methionine. Someone should.

Arigoni: Selenobiotin has been made, and it is biologically active (Piffeteau et al 1976).

References

Cowie DB, Cohen GN 1957 Biosynthesis by *Escherichia coli* of active altered proteins containing selenium instead of sulfur. Biochim Biophys Acta 26:252–261

Cullen WR, Reiner KJ 1989 Arsenic speciation in the environment. Chem Rev 89:713–764

Hädener A, Matzinger PK, Malashkevich VN et al 1993 Purification, characterization, crystallization and X-ray analysis of selenomethionine-labeled hydroxymethylbilane synthase from *Escherichia coli*. Eur J Biochem 211:615–624

Mudd SH, Cantoni GL 1957 Selenomethionine in enzymatic transmethylations. Nature 180:1052

Piffeteau A, Gaudry M, Marquet A 1976 Biological properties of selenobiotin. Biochem Biophys Res Commun 73:773–778

Schauder JR, Jendrezejewski S, Abell A, Hart GJ, Battersby AR 1987 Stereochemistry of formation of the hydroxymethyl group of hydroxymethylbilane, the precursor of uro'gen III. J Chem Soc Chem Commun, p 436–438

The evidence for a spirocyclic intermediate in the formation of uroporphyrinogen III by cosynthase

Finian J. Leeper

University Chemical Laboratory, Lensfield Road, Cambridge CB2 1EW, UK

Abstract. In the course of the cyclization of the linear tetrapyrrole hydroxymethylbilane to uroporphyrinogen III, catalysed by uroporphyrinogen III synthase (cosynthase), ring D of the bilane becomes inverted. Many different mechanisms have been proposed for this transformation but the most economical is one involving a spirocyclic pyrrolenine. Synthesis of a spirolactam, and other compounds closely related to the spirocyclic pyrrolenine, has shown that such compounds are not impossibly strained. The spirolactam is a powerful inhibitor of the enzyme, which suggests it does resemble an intermediate in the enzymic process. In the synthetic procedure to make an ester of the spirolactam the two products obtained were initially thought to be conformational isomers. However, molecular mechanics calculations on a model of the spirolactam predicted that several low energy conformations should exist and that the energy barriers for their interconversion are all lower than 32 kJ/mol. Reinvestigation revealed that one of the two products is in fact a macrocyclic dimer with a 28-membered ring. On the basis of the predicted preferred conformations of the spirolactam and of uroporphyrinogen III, a detailed three-dimensional mechanism is proposed, along with a rationalization of how the rearrangement of ring D may be directed by the enzyme.

1994 The biosynthesis of the tetrapyrrole pigments. Wiley, Chichester (Ciba Foundation Symposium 180) p 111–130

The key step in the formation of the tetrapyrrole macrocycle is catalysed by uroporphyrinogen (uro'gen) III synthase (EC 4.2.1.75), also known as cosynthase. This enzyme catalyses the cyclization of the linear head-to-tail tetrapyrrole, hydroxymethylbilane (**1**), with a concomitant inversion of ring D, to give the unsymmetrical uroporphyrinogen III (**6**). Non-enzymic cyclization of **1**, which is very rapid in acid and has a half-life of only four minutes even at pH 7, gives only the unrearranged symmetrical uro'gen I (**2**). The inversion of the substituents on ring D is the hallmark of all the naturally occurring tetrapyrroles; in fact, uro'gen III is the last precursor that is common to the biosynthesis of all tetrapyrroles, with the pathway leading to vitamin B_{12},

111

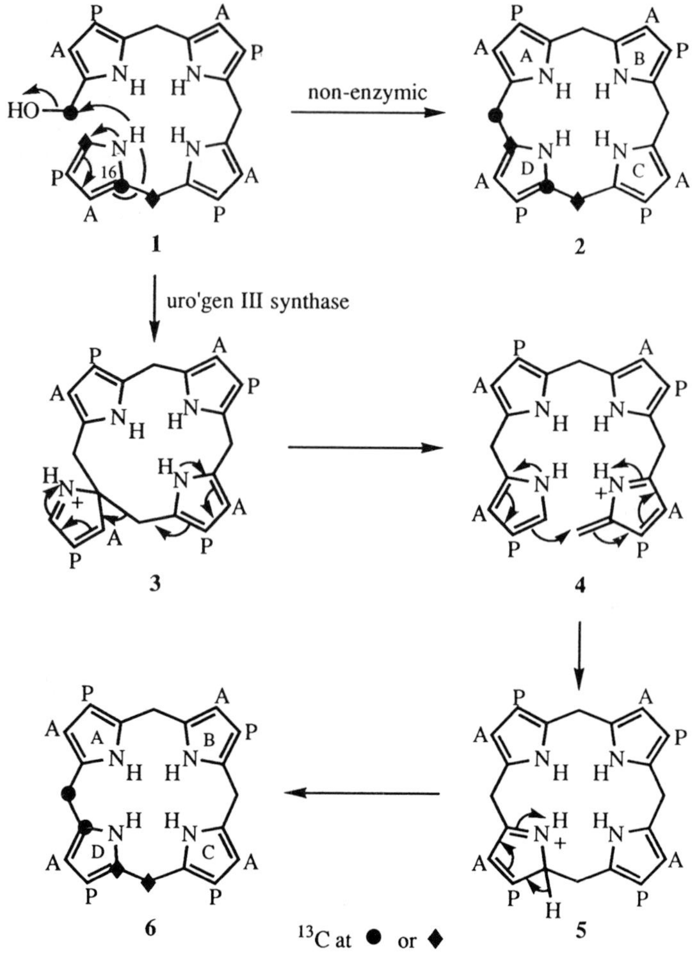

^{13}C at ● or ◆

A = CH_2CO_2H, P = $CH_2CH_2CO_2H$

FIG. 1. The spiro mechanism proposed for the reaction catalysed by cosynthase, uroporphyrinogen III synthase (EC 4.2.1.75). ^{13}C labels were at positions marked ● in one experiment and at ◆ in another.

sirohydrochlorin and factor F_{430} diverging from that to protohaem, chlorophylls and bilins at this stage.

Labelling studies by Battersby et al (1978), of the type indicated in Fig. 1, showed that it is ring D of the bilane that is inverted, that the whole pyrrole ring along with its substituents is turned around, and that the rearrangement is intramolecular. Several different mechanisms have been proposed to account for this process. In essence, they all involve addition of an electrophile to C-16 to facilitate cleavage of the C-15 to C-16 bond. The simplest possible electrophile is a proton, but addition of a proton would involve ring D being detached as a free monopyrrole during the course of the mechanism and the enzyme has not been observed to utilize this monopyrrole. Furthermore, a subsequent enzyme-bound intermediate would probably be the hydroxymethylbilane in which ring D is inverted. This bilane has been synthesized (Battersby et al 1981) and, although it is a substrate for the enzyme, it is largely rearranged back to uro'gen I and not simply cyclized to uro'gen III; it cannot therefore be a normal intermediate in the formation of uro'gen III. Another electrophile which might be used is a methylene iminium ion such as might be provided by methylene tetrahydrofolate. This would avoid the problems associated with H^+, but it has been shown that cosynthases from various sources are still highly active when all folate derivatives are absent (Smythe & Williams 1988).

The most economical mechanism, in terms of the number of individual steps required, uses the hydroxymethyl group of the substrate as the electrophile (as shown in Fig. 1), because the bond so formed (between the two carbons marked ●) has to be formed at some stage during the mechanism anyway. Formation of this bond would generate a spirocyclic pyrrolenine (3), which could then fragment in the alternative direction to give the azafulvene (4); this would then cyclize normally, via 5, to give uro'gen III (6). Essentially this mechanism was first suggested by Mathewson & Corwin (1961) but they thought that the spiro intermediate (3) would be too strained and therefore suggested that the pyrrole rings would have to be protonated at their α-positions to offer more flexibility.

Synthesis of the spirolactam

To investigate whether the spiro intermediate (3) would be too strained, we set out to synthesize compounds containing the same macrocyclic ring. Our first target was the dicyano macrocycle (7, Fig. 2) and the synthesis of this compound was successful (Stark et al 1985). It is a stable crystalline compound which enabled us to obtain its crystal structure (Fig. 2); this structure showed that the macrocycle has a highly puckered conformation with the central pyrrole ring pointing down and the other two pyrrole rings pointing up. We also synthesized a spirocyclic lactam (8). The successful syntheses of these compounds proved that the macrocycle of the proposed spiro intermediate (3) is not too strained to exist and that there is no necessity for protonation at the pyrrolic α-positions to increase flexibility.

We then turned to the synthesis of a spirolactam having all the natural acetate and propionate side chains. This was made possible by the discovery of a remarkable reaction of iodopyrroles such as 9: when these react with acetoxymethyl pyrroles such as 10 under Lewis acid catalysis, coupling occurs with attack from the non-iodinated α-position of the iodopyrrole *even though*

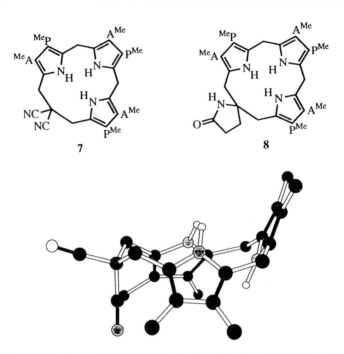

FIG. 2. Structure of the dicyano macrocycle (7) derived by X-ray crystallography. The side chains have been truncated to one carbon atom and the hydrogen atoms have been omitted for the sake of clarity. Structure 8 is a spirocyclic lactam also synthesized.

this position is already substituted (Fig. 3). Thus 9 and 10 give the bis(pyrrolylmethyl)lactam (11) (Stark et al 1986), initially as the iodopyrrolenine which is then hydrolysed. The similarity between the mechanism of this coupling reaction and the one proposed for formation of the spiro intermediate (3), (Fig. 1) is obvious. From 11, the same chemistry that was used in the synthesis of 7 and 8, involving attachment of a third pyrrolic ring and then acid-catalysed cyclization of the alcohol (12), allowed the synthesis of the spirolactam (13).

The cyclization reactions producing 8 and 13 both yielded not one but two products, both of which gave NMR and high resolution mass spectrometry data consistent with the desired spirolactams. This had not been observed in the synthesis of the dicyano macrocycle (7). On the basis of the crystal structure of 7, two likely conformations of the spirolactam (13) are 13a and 13b (Fig. 3). The strained nature of the macrocyclic ring suggested that the pyrrole rings of 13 might not be able to rotate and thus that these conformations would be unable to interconvert; inspection of a space-filling Corey–Pauling–Kolten model of 13 supported this view. The two compounds obtained from the cyclization reaction were therefore thought to be the conformational isomers (or atropisomers) 13a and 13b. In the case of the dicyano macrocycle (7), the two

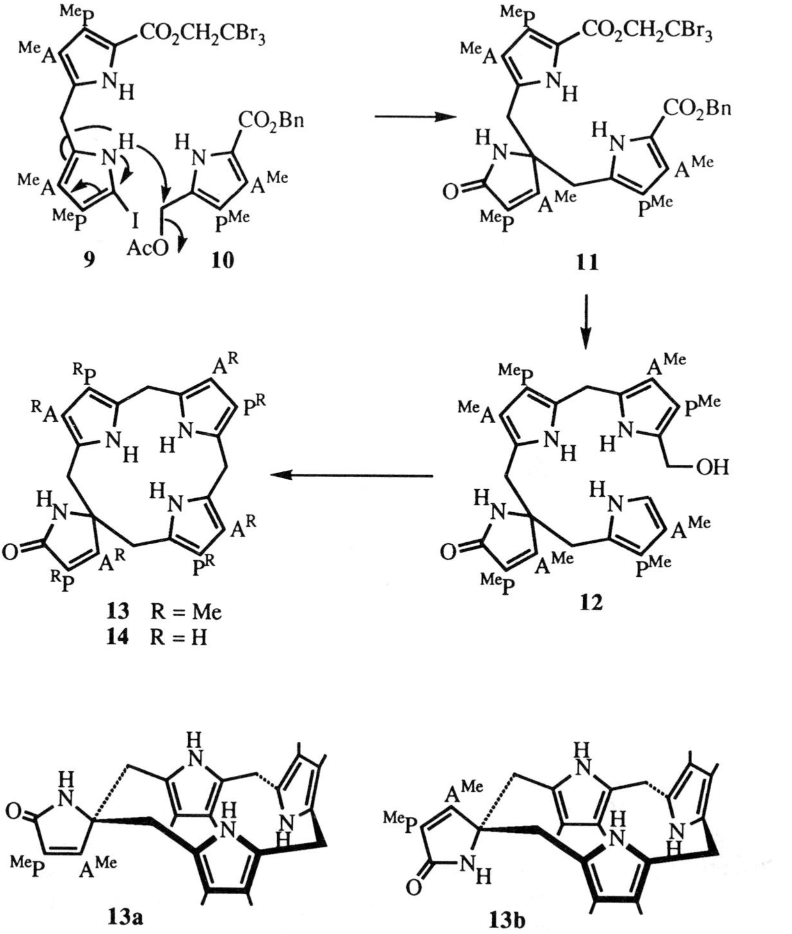

FIG. 3. Synthesis and suggested structures of the spirolactam (13).

conformations would not be different (but merely enantiomeric) because of the equivalence of the two cyano groups, explaining why there was only one product from this cyclization.

Inhibition of cosynthase

The spirolactam (13), after hydrolysis to the free acid (14), differs from the proposed spiro intermediate (3) only by having an amide in place of the imine; all the side chains and other rings are identical. If 3 is a true intermediate, therefore, cosynthase would be expected to bind 14 strongly, and so it proved. The octaacid (14) derived from the major form of 13 showed an inhibition constant (K_i) of $1-2\,\mu$M, 10-fold lower than the K_m value for the substrate (1). On the other hand, the free acids from the minor form of 13 and from the simpler macrocycles 7 and 8 did not inhibit cosynthase at comparable concentrations. This shows that the presence of the three correctly substituted pyrrole rings is not sufficient for tight binding: the presence and correct orientation of the five-membered spirocyclic ring with its side chains is also required. More recently, we have made the free acid form of the spirolactam (14) enantiomerically pure by incorporating a resolution procedure in the synthesis. One enantiomer again inhibited cosynthase strongly, with a K_i value of about $1\,\mu$M, whereas the other had a K_i value more than 20 times higher (Cassidy et al 1991). This is further evidence of the specificity of the interaction of cosynthase with the spirolactam. Studies are in progress to determine the absolute configuration of the inhibitory enantiomer.

Chemical studies of the rearrangement reaction

Having got so close, we obviously hoped to be able to convert the spirolactam (13) into the spiro intermediate (3) itself. The chemistry required was worked out on model systems. Treatment of the bis(pyrrolylmethyl)lactam (15, Scheme 1) with Lawesson's reagent gave the thiolactam (16) and reduction with nickel boride gave the desired pyrrolenine (17) (Hawker et al 1987). Unfortunately, the same conditions did not succeed with the spirolactam (13): none of the corresponding thiolactam could be isolated and decomposition occurred, with uroporphyrin methyl esters being the only products identified. It seems that spiro compounds such as this are prone to fragmentation–recombination reactions of the very type proposed for the mechanism of the cosynthase reaction.

One interesting result to have come out of this study, however, was the discovery that the fragmentation of these compounds does not take place equally on the two sides of the pyrrolenine. For example, 17, when heated or treated with acid, gave fragmentation–recombination products 18 and 19, as well as cross-over products; the ratio of the products indicated that fragmentation occurred three times more frequently on the side equivalent to the ring C–D cleavage observed for cosynthase, making 18 the major product. This tendency can be explained by the greater electron-withdrawing effect of the acetate side chains over the propionate ones. This effect would make the cation produced by fragmentation less stable on ring A, where the charge would be partially localized on the carbon bearing the acetate side chain, than on ring C.

Molecular mechanics studies

Because one of the two forms of the spirolactam inhibited cosynthase and the other did not, we decided to use molecular mechanics to try to predict the

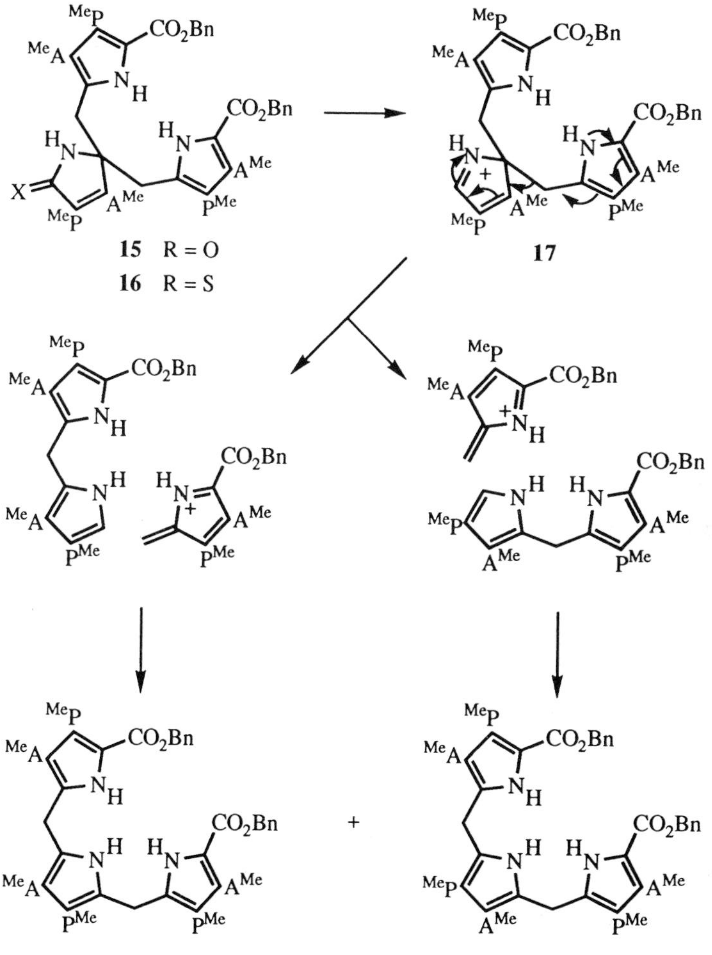

18 + **crossover products** **19**

Scheme 1

structures of the two atropisomers and which is the more stable. The molecular mechanics calculations were carried out with the MacroModel program (Mohamadi et al 1990). In all the calculations the side chains of the pyrroles were reduced to just methyl groups, i.e., spirolactam **20** (Scheme 2), to avoid ending up with multiple minimum-energy conformations differing only in the configuration of the side chains. It is unlikely that this change affected the results to any great extent, because in all the conformations found the acetate and propionate side chains, if present, would point away from the centre of the molecule, avoiding any steric interactions.

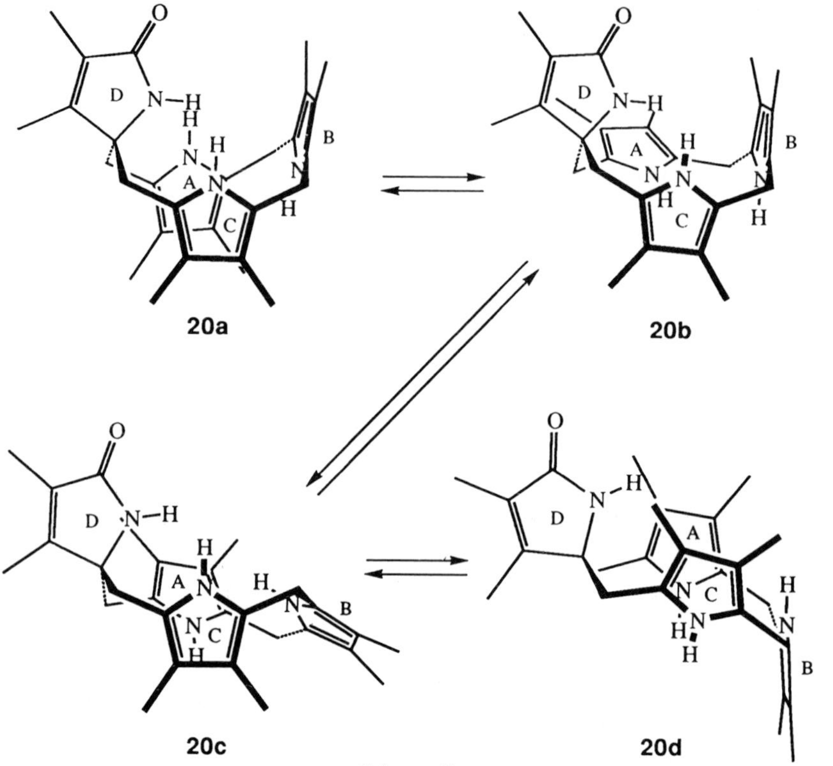

20a   20b

20c   20d

Scheme 2

After minimizing a large number of starting conformations, we found the global energy minimum to be as shown in structure **20a**, which corresponded to one of the predicted atropisomers. Another energy minimum, **20d**, 12.6 kJ/mol higher in energy, corresponded to the other atropisomer. However, two other conformations, **20b** and **20c**, came at about the same energy as **20d**. Furthermore, it seemed that **20b** and **20c** could be intermediates on a pathway for the interconversion of **20a** and **20d**. Twisting first of ring A would convert **20a** into **20b**, then of ring B would give **20c** and finally of ring C would produce **20d**. We had not observed four atropisomers and so some of these conformations clearly must be interconverting—but which? To try to answer this question, we investigated the energy barriers of each of these ring inversions using molecular mechanics.

In the method used to calculate the energy barriers for each ring inversion, the dihedral angle about a chosen bond was constrained to values intermediate between those for the starting and finishing conformations. For example, in the interconversion of **20a** and **20b** the dihedral angle N_A—C4—C5—C6 was varied from $-60°$ to $+55°$ in small incremental steps. At each chosen angle the

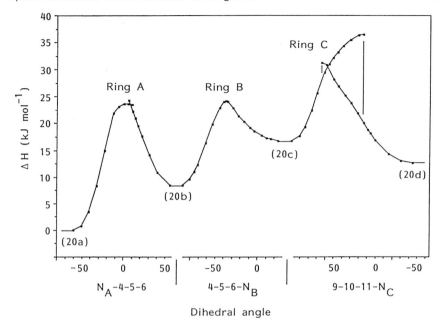

FIG. 4.   The energy profile calculated by molecular mechanics for the inversion of each of the three pyrrolic rings of spirolactam **20** (see Scheme 2) in turn. For the interconversion of **20c** and **20d**, two stable conformations were found at intermediate dihedral angles, one accessed from **20c** and the other from **20d**. Hence, the curves cross each other rather than meeting in a common transition state. Each conformation becomes unstable beyond a certain angle and at that point minimizes to the corresponding position on the other curve. The lower of the two energies at which this occurs has been taken as the upper limit on the energy of the true transition state for this interconversion.

structure was minimized and then the angle constraint was removed and the energy recalculated. This procedure was also followed in the reverse direction, i.e., starting with **20b** and varying the torsion angle to get back to **20a**. The results are plotted in Fig. 4; the largest of the three barriers, that between **20c** and **20d**, was found to be no more than 32 kJ/mol above the global minimum. This is less than the barrier between chair and twist-boat conformations of cyclohexane and would imply an interconversion rate of at least $10^7 \, \mathrm{s}^{-1}$ at room temperature. As a check, the same procedure was used to calculate the barrier to flipping of cyclohexane; the result, 44.0 kJ/mol, compared extremely well with the experimental value of 44.8 kJ/mol (Anet & Anet 1975).

Confirmation of the dimeric structure

With the indication from molecular mechanics that the two compounds obtained in the synthesis of the spirolactam could not be atropisomers, alternative explanations were sought. The possibility that one was the dimer (21) had not been considered previously because field desorption mass spectrometry (which generally gives molecular ions with little, if any, fragmentation) had given the same mass for both compounds ($\frac{m}{z}$ 964.4). On reinvestigation, however, it was found that the higher R_F minor product also gave a peak at $\frac{m}{z}$ 1928.8. Furthermore, the presence of a small isotope peak 0.5 mass units higher than the peak at 964.4 indicated that the latter is due, at least in part, to the M^{2+} ion (J. D. Lewis, unpublished work 1991).

21

It is unusual to find a dimer at a higher R_F value than the corresponding monomer, so this result was confirmed by measuring the yields of the two compounds at different dilutions of the reaction mixture (P. M. Petersen, unpublished work 1991). As can be seen from Table 1, the yields of the monomer 13 rose while those of the dimer 21 fell with increasing dilution. This is as expected for a monomer and a dimer but not for two different monomeric atropisomers. The yield of the monomer at the highest dilution was a significant improvement over previous results.

The conclusion that the minor, higher R_F, compound is in fact the dimer (21) does not affect in any way our previous conclusion about the mechanism of uroporphyrinogen III synthase because it is the major, lower R_F, product (13) which, after hydrolysis, inhibits the enzyme. Thus, we believe the evidence that the mechanism proceeds through the spirocyclic intermediate (3) is as strong as ever.

A proposal for the three-dimensional course of the cosynthase-catalysed reaction

The molecular mechanics calculations suggest that the lowest energy conformation of the spirolactam (13) is the one represented by 20a. This is supported by NOE (nuclear Overhauser effect) studies, which showed strong enhancements between the lactam NH and the adjacent pyrrolic NHs. We have also calculated energies for various different conformations of porphyrinogens, which indicate that the

TABLE 1   Yields of the monomeric (13) and the dimeric (21) form of the spirolactam from acid-catalysed cyclization of alcohol (12) at different dilutions

	Yield (%)	
Concentration	Monomer (13)	Dimer (21)
5.4 mM	31	26
1.8 mM	32	17
0.6 mM	59	12

highly puckered conformation in which the rings point alternately up and down is the most stable by a large margin. If we assume that the enzyme-bound spiro intermediate (3) and the enzyme-bound product, uro'gen III (6), adopt these most stable conformations, we can begin to build up a three-dimensional picture of the mechanism of cosynthase.

One of the most difficult facts to explain is that cosynthase does not specifically catalyse the formation of uro'gen III; it will catalyse substantial inversion of ring D of its substrate regardless of the arrangement of the acetate and propionate side chains. To explain this, I have assumed that the negatively charged side chains of ring D interact with positively charged groups on the enzyme in such a way that there is a twisting force on this pyrrole ring that prevents cyclization from C-19 of the bilane to the hydroxymethyl carbon. This is illustrated in Scheme 3(a), in which the propionate side chain on ring D of hydroxymethylbilane is bent out of the plane of the paper while the acetate side chain is bent in the other direction; each is bound to a complementary residue on the enzyme. Such an arrangement of the side chains would prevent the cyclization without rearrangement. However, the flexibility of the side chains would permit the formation of the spiro intermediate (3), Scheme 3(c), and its fragmentation to the alternative ring C azafulvenium ion (4), Scheme 3(d).

In principle, the pyrrolic ring D of 4 could attack either the front (*re*) face or the back (*si*) face of the ring C azafulvenium ion. However, in the model that I have constructed ring D would be unable to rotate into a position from which it could attack the back face. Instead, ring C would have to rotate until it was more or less horizontal, Scheme 3(e), presenting what was the front face to attack from above by ring D. Cyclization to (5) could then occur and a final deprotonation would then give uro'gen III (6).

Although it is possible to draw a plausible scheme on paper such as that in Scheme 3, it is important to test whether it is feasible in three dimensions also. Molecular mechanics was used again to investigate this point. Starting from the minimum energy conformation (20a), acetate and propionate side chains were added to rings B and D. Ammonium ions were then included in a position optimum for them to bind to each of the four carboxylates, thus creating a model of the situation in Scheme 3(c). A methane molecule (not

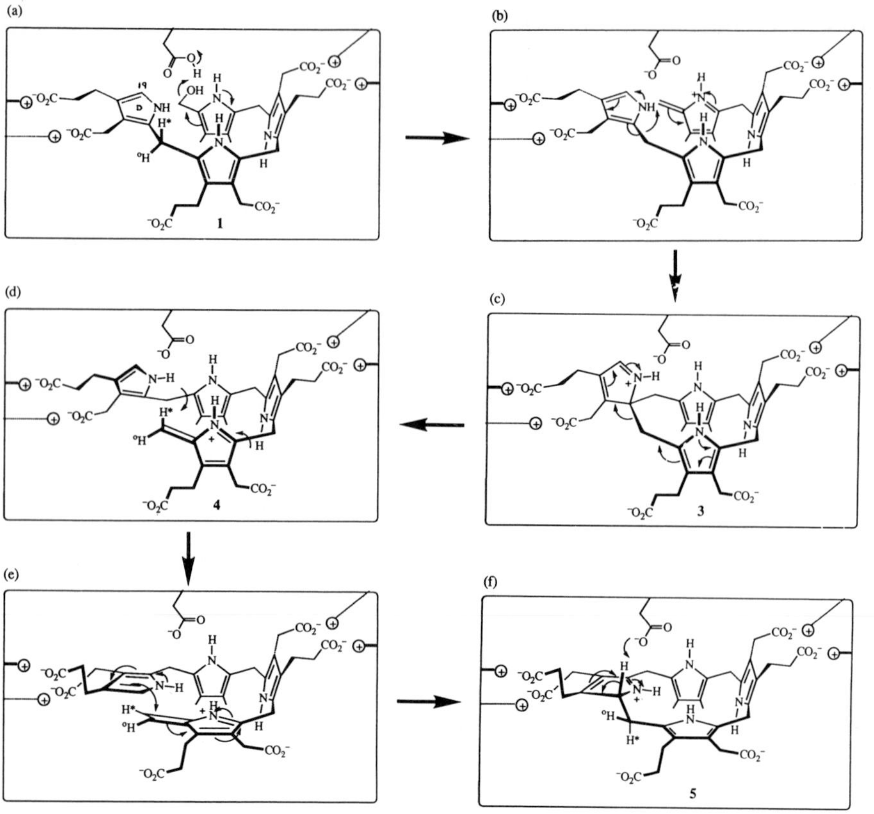

Scheme 3

shown) was also placed beyond ring A so as to model the back wall of the active site and prevent the tetrapyrrole from unwrapping. The four ammonium ions and the methane molecule make up a minimal model of the enzyme active site and from here on their positions were kept fixed. The model of the spiro intermediate (3) was then modified to give, in turn, models of hydroxy-methylbilane (1) and all the proposed intermediates. These molecules were then allowed to relax into minimum energy conformations within this 'active site'. In each case, minimum energy conformations corresponding to those shown in Schemes 3(a)–(f) were found and the carboxylates all remained within reasonable hydrogen-bonding distance of their respective counter-ions.

These molecular mechanics results demonstrate that the mechanism proposed in Scheme 3 is indeed feasible; we cannot, however, take this as a justification for excluding other mechanisms which might be equally possible. Our model suggests that the stereochemistry at the C-15 methylene group (shown by hydrogen atoms H° and H* in Scheme 3) may be inverted in the course of the reaction.

This stereochemistry has not yet been determined but we are now developing methods that will enable us to undertake the appropriate experiments. Similarly, although the model depicts one particular absolute stereochemistry of the spiro intermediate (3), the opposite enantiomer is equally possible. We should be able to tell which enantiomer is used by determining the absolute configuration of the highly inhibitory enantiomer of spirolactam (14).

It is hoped that the model of the mechanism and the active site of cosynthase presented here will allow more detailed questions to be formulated and further predictions to be tested, and thus lead to a deeper understanding of this remarkable enzyme.

Acknowledgements

I should like to thank Professor Sir Alan Battersby for his help and guidance and the many research workers in his group, named in the references, who contributed to this work. I should also like to thank Professor W. Clark Still for supplying the MacroModel program.

References

Anet FAL, Anet R 1975 Conformational processes in rings. In: Jackman LM, Cotton FA (eds) Dynamic NMR spectroscopy. Academic Press, London, p 543–619
Battersby AR, Fookes CJR, McDonald E, Meegan MJ 1978 Biosynthesis of type-III porphyrins: proof of intact enzymic conversion of the head-to-tail bilane into uro'gen-III by intramolecular rearrangement. J Chem Soc Chem Commun, p 185–186
Battersby AR, Fookes CJR, Matcham GWJ, Pandey PS 1981 Biosynthesis of natural porphyrins: studies with isomeric hydroxymethylbilanes on the specificity and action of cosynthetase. Angew Chem Int Ed Engl 20:293–295
Cassidy MA, Crockett N, Leeper FJ, Battersby AR 1991 Synthetic studies on the proposed spiro intermediate for biosynthesis of the natural porphyrins: the stereochemical probe. J Chem Soc Chem Commun, p 384–386
Hawker CA, Stark WM, Battersby AR 1987 Selectivity in the rearrangement of a di(pyrrolylmethyl)-2*H*-pyrrole. J Chem Soc Chem Commun, p 1313–1315
Mathewson JH, Corwin AH 1961 Biosynthesis of pyrrole pigments: a mechanism for porphobilinogen polymerization. J Am Chem Soc 83:135–137
Mohamadi F, Richards NGJ, Guida WC et al 1990 MacroModel—an integrated software system for modeling organic and bioorganic molecules using molecular mechanics. J Comp Chem 11:440–467
Smythe E, Williams DC 1988 Rat liver uroporphyrinogen III synthase has similar properties to the enzyme from *Euglena gracilis*, including absence of a requirement for a reversibly bound cofactor for activity. Biochem J 253:275–279
Stark WM, Baker MG, Raithby PR, Leeper FJ, Battersby AR 1985 The spiro intermediate proposed for biosynthesis of the natural porphyrins: synthesis and properties of its macrocycle. J Chem Soc Chem Commun, p 1294–1296
Stark WM, Hart GJ, Battersby AR 1986 Synthetic studies on the proposed spiro intermediate for biosynthesis of the natural porphyrins: inhibition of cosynthetase. J Chem Soc Chem Commun, p 465–467

DISCUSSION

K. Smith: We have X-ray crystallographic structures of half a dozen different porphyrinogens (though they're not in the series you discussed) and their conformations can vary dramatically. There are four possible isomers (22–25) of octaethyltetraphenylporphobilinogen. The most common conformation has two opposite pyrroles with the NH groups in the same plane, pointing inwards, and two opposite pyrroles with two NHs up or one NH up and one NH down. A variety of different stable conformations of porphyrinogens clearly exist. It is the regiochemistry of the placing of the phenyls on the *meso* positions which determines whether the pyrroles are up, down or in the plane.

Is there any temperature dependence in the NMR spectrum of the inhibitory spirolactam (13)? It is rather saddle-shaped, and you might expect a dynamic process in which the saddle could invert with pyrroles flapping up and down.

Leeper: We didn't see any temperature dependence for the monomer. The dimer gave broad signals and there was large temperature dependence. That's probably because of the intramolecular hydrogen bonding that is possible in the dimer because it is more flexible.

Scott: The construction of your lactam goes through the iodo intermediate. Would it be possible to make the 19-iodohydroxymethylbilane, which in principle could form the lactam with cosynthase?

Leeper: Iodine is probably too big, though we haven't tried.

Hädener: You described the conversion of a tripyrrolic model lactam (15) into a pyrrolenine (17). Did you try the analogous conversion with the macrocyclic spiro compound?

Leeper: I thought someone would ask that! We tried, of course, and failed— we couldn't even make the thiolactam. All we isolated from that reaction was uroporphyrinogen. The compound is simply too sensitive, and undergoes the same reaction as we expected in cosynthase but non-enzymically.

Arigoni: Which uro'gen was isolated?

Battersby: The products were examined as the aromatized materials, as uroporphyrin I and uroporphyrin III esters.

Scott: Can you recall what the ratio was?

Battersby: It certainly wasn't pure type III, nor would you expect it to be. My recollection is that on NMR analysis we could see the relevant signals from the *meso* positions of uroporphyrin I and III octamethyl esters. We did not make a quantitative determination in this one experiment because we wanted to press ahead towards the main objective. We shall return to this topic. We have looked at the acid-catalysed rearrangement of a related system (26) which yields, after aromatization of the initial products, a 2:1 ratio of uroporphyrin III to uroporphyrin I esters (C. J. Hawker & A. R. Battersby, unpublished work 1988).

26

Eschenmoser: You probably have a nice rationalization of that not unimportant observation.

Leeper: In all our syntheses, we find that the acetate side chain is more electron-withdrawing than the propionate, as you would expect because the carbonyl is closer to the pyrrole ring. In the protonated azafulvene required for the formation of **18**, the positive charge cannot be delocalized to the carbon bearing the acetate side chain, but is delocalized to the one bearing the propionate; the other fragmentation, leading to **19**, involves an azafulvene in which the positive charge can be delocalized onto the carbon bearing the electron-withdrawing acetate.

Scott: In all of our rationalizations of the mechanisms of deaminase and cosynthase, we embrace the attractive idea of azafulvenes, for which there is no direct evidence. At least in the case of deaminase, one can trap the azafulvene (if it's there) with water, ammonia or hydroxylamine. If the mechanism of cosynthase involves fragmentation through two different azafulvenes, why has it been so difficult to trap either of them with external nucleophiles such as cyanoborohydride or hydroxylamine?

Battersby: We have made an extensive study of the trapping by ammonia and hydroxylamine of reactive intermediates, with the properties expected of azafulvenes, generated by the action of deaminase (Battersby et al 1983a). The trapped products were characterized spectroscopically. It was also shown that hydroxylamine does not affect the activity of cosynthase (Hart & Battersby 1985). This negative result does *not* mean that azafulvenes are not generated during the rearrangement process catalysed by the enzyme; it could be simply that the intermediates in the active site are not available to external reagents in the way they are for deaminase. Because the product from deaminase, hydroxymethylbilane, is in our view formed by water trapping an azafulvene, it is perhaps not so surprising that other nucleophiles can do so as well.

Leeper: Also, the cosynthase reaction is rapid, much faster than the deaminase reaction.

Scott: You can slow it down by using a very poor substrate, the amino-methylbilane.

Leeper: But that doesn't necessarily increase the lifetime of the azafulvene. Once the azafulvene is made, it is short-lived.

Scott: Could you remind us of the experiments done in Cambridge in which the ring D acetates and propionates were transformed into alkyl groups (Battersby et al 1983b). As I remember, the acetate in ring D is essential for the mechanism, but the propionate is not. How does that tie in with both carboxyls being bound in your model?

Battersby: The hydroxymethylbilane with methyl in place of acetate on ring D was barely a substrate for cosynthase but the bilane with ethyl in place of propionate on ring D was the best modified substrate we have synthesized in the hydroxymethylbilane series. The latter modified substrate was ring-closed at about one quarter of the rate of the natural bilane and about 65% of the enzymically formed product (i.e., after allowing for non-enzymic cyclization) had undergone inversion of ring D. So, of the two groups on ring D, the acetate is more important that the propionate (Battersby et al 1983b).

Scott: We've done an experiment recently (unpublished) which is interesting in connection with Dr Leeper's idea of ionization or binding of the carboxyls in ring D. We made a specifically ^{13}C-labelled acetate carbonyl (at C-17) on the hydroxymethylbilane which was substituted with bromine on the α-free position. This acts as an inhibitor and won't undergo cyclization. The ^{13}C NMR spectrum of the inhibitor bound to cosynthase revealed two forms of the acetate, one partially ionized and the other completely ionized. We think one is bound and the other is not. We are now going to do the same experiments with the propionate and the other carboxyls labelled with ^{13}C to see which are involved in binding to cosynthase. Although this effect is not large, we've determined that there is certainly something different from the control experiment where the carboxyls are not sitting in the enzyme's active site; all of them are ionized in this case at pH 7.

Arigoni: What would happen if you prepared the linear tetrapyrrole corresponding to **4** as the hydroxybilane? Would it be accepted by the enzyme?

Leeper: We have synthesized the bilane in which the order, from the hydroxymethyl end, is PAPAPAAP (inverted in rings A, B and C, but not D). It is not a substrate for the enzyme, as far as we can detect; non-enzymic cyclization is so rapid that you cannot detect a small amount of enzymic turnover (Battersby et al 1981).

Arigoni: Another case you mentioned was the linear tetrapyrrole with an inverted ring D. This is a substrate for the enzyme, and is converted, to a large extent, into uro'gen I, by flipping of the ring, but you said it is also converted to uro'gen III by a direct condensation.

Leeper: As far as we could determine, the enzymic component of the reaction, as opposed to the concomitant non-enzymic component, proceeded with about 45:55 flipping and not flipping (Battersby et al 1981).

Arigoni: Are you saying that the non-rearranged one is the product of the non-enzymic reaction?

Leeper: That does happen, but after you have taken that amount of non-enzymic cyclization away from the product, what you are left with is still half flipped and half not flipped.

Arigoni: That raises an interesting question. What is the driving force by which the normal reaction is channelled one way? Showing, as you have, that with the modified substrates both flipped and non-flipped products are made opens the way for a similar duality with the natural substrate. This, however, is not being observed.

Leeper: In Scheme 3, if the side chains on ring D are not the right length, if one is too short and the other is too long, they might not bind at all to their respective counter-ions; or, if the molecule does bind in this sort of conformation (Scheme 3a) it won't be strongly bound, so there's little to stop it cyclizing without rearrangement.

Arigoni: But when you interchange the two chains, it still works and goes partly with ring flipping and partly without.

Leeper: Yes, but it doesn't work well, only slowly.

Jordan: When the enzyme binds the real substrate it must bind it essentially as uroporpyrinogen III with ring D positioned so it is *already* in the correct conformation to generate uroporphyrinogen III.

Arigoni: This does not seem to be on the way to the spiro compound.

Jordan: I do not really favour the spiro compound. If you bind the substrate with ring D at one angle, it is more like uroporphyrinogen III; if you bind it at another angle it is more like uroporphyrinogen I. I suspect the enzyme is specifically binding ring D in a flipped conformation, so that when the bond-breaking and bond-forming events occur, one product, uroporphyrinogen III, results.

Arigoni: I'm afraid we are simply verbalizing the problem.

Jordan: Dr Leeper, in your model, you positioned some positively charged, fixed points. Cosynthases seem to be unstable, and are probably non-rigid structures. Why did you actually position fixed points? Was it because this was your first bash, and have you now tried to move these points around during the modelling?

Leeper: That was our first bash, and we obviously don't know where the best point to place them is.

Jordan: The other problem is that there are few conserved groups in the cosynthase structure that you can use. The primary structures are not well conserved between species. There is only one arginine (Arg-138 in the *E. coli* sequence) which is highly conserved, and mutation of this to histidine completely inactivates the *E. coli* enzyme. If you make a double substitution, Arg-138→His:Glu-139→Asp, the enzyme will, albeit slowly, make type I, although this needs to be confirmed.

Leeper: Modification of arginine and lysine residues destroys enzymic activity (Hart & Battersby 1985, Crockett et al 1991). Nothing else has been identified as far as I know.

Battersby: Another important experiment, which we shall be doing shortly, is to look at the stabilizing effect on cosynthase of binding the spirolactam (14) which is closely related to the putative spiro intermediate.

Arigoni: Is there now a general consensus for the existence of the spiro intermediate, or does anybody dare to challenge this pathway?

Scott: I do. There is still no direct evidence for it, although it's an attractive process which could, in spite of the fragmentation models, still involve [1,5]-sigmatropic shifts.

Arigoni: It would then be difficult to understand why the Cambridge compounds act as well as they do.

Battersby: The mechanism involving the spiro intermediate is the only one for which there is any evidence at all. The inhibition of binding, together with the results gained from the resolved lactams, provide really strong evidence for the spiro mechanism.

Eschenmoser: Professor Scott, would the transition state of your alternative pathway, involving the [1,5]-sigmatropic rearrangement, be related structurally to the one which leads to the spiro intermediate?

Scott: The pathway would still go through the spiro intermediate: only the way of arriving at the product would differ.

Eschenmoser: Should we then speak of different mechanisms?

Scott: I doubt if calculations have been done on the sigmatropic shift mechanism.

Arigoni: What is implied is that all the open intermediates with azafulvene structure are bypassed if a sigmatropic rearrangement is involved. They are transition states, not intermediates.

Scott: That's right. We wouldn't be able to trap them, but one reason why we haven't been able to trap azafulvenes might be their low concentration.

Battersby: All the rearrangements of synthetic pyrrolic materials that we have studied have occurred largely or entirely by fragmentation and the groups have not cleanly undergone [1,5]-shifts (Battersby et al 1987, Hawker et al 1987).

Leeper: If [1,5]-sigmatropic rearrangements were occurring, the stereo-chemistry at the migrating methylene group (C-15) would be retained.

Scott: That is a good probe, of course.

Eschenmoser: This question is reminiscent of the notoriously difficult problem of whether a rearrangement occurs via a classical or a non-classical cation or a concerted or non-concerted pathway. It can become extremely difficult to decide the answers to such questions, and I wonder whether there are not other more significant, more urgent, questions that could be answered more easily.

Arigoni: The question could be answered by double-labelling, using ^{13}C-labelled substrate labelled with deuterium.

Scott: Berson-type experiments (Berson 1968) could be used.

Jordan: Jack Baldwin has suggested a Diels–Alder mechanism for this reaction, which is intriguing.

Arigoni: How would an enzyme manage to catalyse an electrocyclic process efficiently?

Jordan: I don't know, but it does work, on paper at least.

Scott: But it would still involve a fragmentation–recombination eventually.

References

Battersby AR, Fookes CJR, Matcham GWJ, Pandey PS 1981 Biosynthesis of natural porphyrins: studies with isomeric hydroxymethylbilanes on the specificity and action of cosynthetase. Angew Chem Int Ed Engl 20:293–295

Battersby AR, Fookes CJR, Matcham GWJ, McDonald E, Hollenstein R 1983a Biosynthesis of porphyrins and related macrocycles. Part 20. Purification of deaminase and studies of its mode of action. J Chem Soc Perkin Trans I, p 3031–3040

Battersby AR, Fookes CJR, Pandey PS 1983b Linear tetrapyrrolic intermediates for biosynthesis of the natural porphyrins. Experiments with modified substrates. Tetrahedron 39:1919–1926

Battersby AR, Baker MG, Broadbent HA, Fookes CJR, Leeper FJ 1987 Biosynthesis of porphyrins and related macrocycles. Part 29. Synthesis and chemistry of 2,2-disubstituted 2*H*-pyrroles (pyrrolenines). J Chem Soc Perkin Trans I, p 2027–2048

Berson JA 1968 The sterochemistry of sigmatropic rearrangement. Acc Chem Res 1:152–160

Crockett N, Alefounder PR, Battersby AR, Abell C 1991 Uroporphyrinogen III synthase: studies on its mechanism of action, molecular biology and biochemistry. Tetrahedron 47:6003–6014

Hart GJ, Battersby AR 1985 Purification and properties of uroporphyrinogen III synthase (cosynthetase) from *Euglena gracilis*. Biochem J 232:151–160

Hawker CJ, Stark WM, Battersby AR 1987 Selectivity in the rearrangement of a di(pyrrolylmethyl)-2*H*-pyrrole. J Chem Soc Chem Commun, p 1313–1315

The modification of acetate and propionate side chains during the biosynthesis of haem and chlorophylls: mechanistic and stereochemical studies

M. Akhtar

Department of Biochemistry, School of Biological Sciences, University of Southampton, Bassett Crescent East, Southampton SO9 3TU, UK

Abstract. In the conversion of uroporphyrinogen III into protoporphyrin IX and thence into chlorophylls, all eight carboxylic side chains, as well as the four *meso* positions, are modified, and four enzymes are involved. In the uroporphyrinogen decarboxylase-catalysed reaction all four acetate side chains are converted into methyl groups by the same mechanism, to produce coproporphyrinogen III. Both methylene hydrogen atoms remain undisturbed and the reaction occurs with the retention of stereochemistry. Several questions regarding the enzymology of the decarboxylase are posed. Do all the decarboxylations occur at the same active site and, if so, are the four acetate chains handled in a particular sequence? Is the decarboxylation reaction aided by the transient formation of an electron-withdrawing functionality in the pyrrole ring? Coproporphyrinogen oxidase converts the two propionate side chains of rings A and B into vinyl groups, with an overall anti-periplanar removal of the carboxyl group and the H_{si} from the neighbouring position. Evidence is examined to evaluate whether a hydroxylated compound acts as an intermediate in the oxidative decarboxylation reaction. Protoporphyrinogen oxidase then converts the methylene-interrupted macrocycle of protoporphyrinogen IX into a conjugated system. The conversion has been suggested to involve three consecutive dehydrogenation reactions followed by an isomerization step. The face of the macrocycle from which the three *meso* hydrogen atoms are removed in the dehydrogenation reaction is thought to be opposite to that from which the fourth *meso* hydrogen is lost during the prototropic rearrangement. In an investigation of the *in vivo* mechanism for the esterification of the ring D propionic acid group with a C_{20} isoprenyl group 5-aminolaevulinic acid was labelled with ^{13}C and ^{18}O at C-1 and incorporated into bacteriochlorophyll *a*. The ^{18}O-induced shift of the ^{13}C resonance in the NMR spectrum showed that both oxygen atoms of the carboxyl group are retained in the ester bond. This and other results suggest that the reaction occurs by the nucleophilic attack of the ring D carboxylate anion on the activated form of an isoprenyl alcohol.

1994 The biosynthesis of the tetrapyrrole pigments. Wiley, Chichester (Ciba Foundation Symposium 180) p 131–155

The experimental findings from which the mechanistic and stereochemical conclusions discussed here are drawn are well known (Akhtar 1991). Therefore, my main emphasis here is to point to future challenges. I hope that it will become obvious that further advances in the field are dependent not on the availability of any single technique but rather on progress across several disciplines ranging from synthetic organic chemistry, enzymology and genetic engineering through to X-ray crystallography.

Uroporphyrinogen decarboxylase

Uroporphyrinogen decarboxylase (EC 4.1.1.37) has been purified from several sources and shown to have a subunit M_r of around 40 000 (for reviews see Akhtar 1991, Dailey 1990a). Of all the enzymes involved in tetrapyrrole biosynthesis, uroporphyrinogen decarboxylase shows the most promiscuous behaviour. Its physiological substrate is uroporphyrinogen III (1, Scheme 1; P, propionate) which is converted into coproporphyrinogen III (2) by the decarboxylation of the acetate side chains at positions 2, 7, 12 and 18 (Neve et al 1956). There are theoretically 14 intermediates between uroporphyrinogen III (1) and coproporphyrinogen III (2), which the late A. Jackson, in a Herculean effort, synthesized non-enzymically and showed to be the substrates for the decarboxylase (Jackson et al 1980). The enzyme also acts on uroporphyrinogen I, but the II and IV isomers are poor substrates (Neve et al 1956, Jackson et al 1980, Mauzerall & Granick 1958, Smith & Francis 1979). Despite its relatively broad substrate profile, the enzyme displays a subtle specificity which was highlighted by the meticulous analysis of porphyrins excreted in the urine of patients with porphyria or of animals treated with the drug hexachlorobenzene (see Akhtar 1991, Jackson et al 1976). These studies, initiated in the 1960s by Batlle and Grinstein (del C Batlle et al 1964, Tomio et al 1970), were brought to a climax

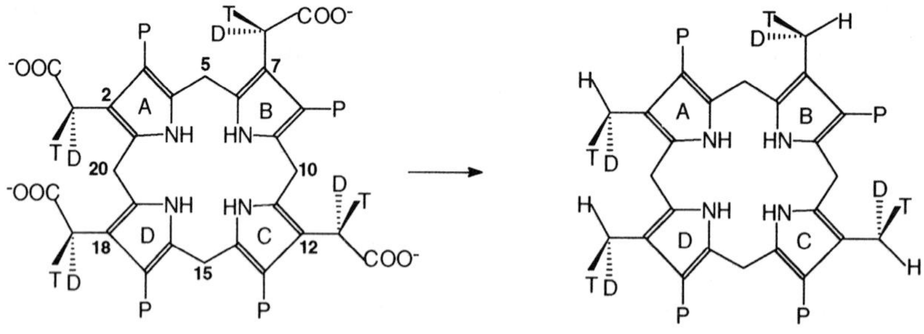

Uroporphyrinogen III (1, when D = H, T = H)

3

Coproporphyrinogen III (2, when D = H, T = H)

4

Scheme 1

by the comprehensive work of Jackson and Alder and their colleagues, who showed that unique isomers of hepta-, hexa- and pentacarboxylic acid porphyrins were found in the urine of patients suffering from porphyria (Jackson et al 1976). The structures of the intermediates led Jackson and colleagues to propose that in the conversion of uroporphyrinogen III into coproporphyrinogen III the acetate (A) side chains on rings D, A, B and C are decarboxylated in a clockwise fashion (Scheme 2). The contemporaneous isolation at Cambridge of a heptacarboxylic acid porphyrin decarboxylated at ring D (6) from an enzymic incubation (Battersby et al 1974) was consistent with Jackson's hypothesis. Luo & Lim (1993) confirmed Jackson's findings and showed that unique isomers of hepta-, hexa- and pentacarboxylic porphyrins (6, 7 and 8) were indeed excreted in the urine of patients suffering from porphyria cutanea tarda (acquired porphyria) whereas the urine of normal individuals contained all the isomers of variously decarboxylated porphyrin intermediates (Luo & Lim 1993, Lash 1991). Differing profiles of intermediates were also demonstrated in enzymic incubations. An erythrocyte lysate given uroporphyrinogen III as a substrate produced all four isomers each of the hepta-, hexa- and pentacarboxylic acid porphyrins, suggestive of a random decarboxylation sequence. Interestingly, however, with porphobilinogen as the substrate, a clockwise decarboxylation sequence was observed, with the resulting compounds mainly having decarboxylated acetate side chains on ring D, rings A and D and rings A, B and D (as in Scheme 2) (Luo & Lim 1993).

The contrasting profiles are illuminating and indicate that the substrate-binding site has a flexible architecture which enables the decarboxylase to combine specificity with promiscuity. It could be argued that when uroporphyrinogen III is generated slowly and is present in limiting amounts, as in the coupled system *in vitro* and in porphyria, the Jackson clockwise sequence is followed because binding interactions in the Michaelis complexes involved in the ordered pathway would be of high affinity, whereas at higher substrate concentrations lower affinity complexes would be formed through a random binding mode. The dissociation constants of the enzyme–substrate complexes need to be altered by only an order of magnitude to convert a highly ordered process into a non-specific decarboxylation.

The interaction between uroporphyrinogen decarboxylase and its substrates

From a physiological viewpoint, there is no necessity for a strict decarboxylation sequence, because with uroporphyrinogen III (1) as the substrate, all the possible decarboxylation sequences, of which there are 24, involving 14 different intermediates, will eventually generate the same product, coproporphyrinogen III (2). None the less, the formation of unique intermediates under stringent conditions emphasizes that the architecture of the decarboxylase allows preferential interactions. These interactions must be subtle, and in the absence of structural information can be considered only in general terms. For example, there must be a site on the enzyme containing a cluster of residues involved in decarboxylation of the acetate side chain and binding of the neighbouring propionate side chain. This site is designated the catalytic site (Scheme 2). In close proximity to this one can envisage a second site for the recognition of the four pyrrolic NH bonds (this site is not shown in Scheme 2). Structure 5 in Scheme 2 shows the binding of uroporphyrinogen III to the decarboxylase, with its ring D locked into the catalytic groove. With this mode of binding it is conceivable, indeed likely, that there are also distal interactions between the remaining side chains of uroporphyrinogen III for which six complementary sites on the decarboxylase would be required. These sites, in a clockwise arrangement, are symbolized as S_1, S_2, S_3, S_4, S_5 and S_6. When these interactions are fully operational, the ring D acetate (A) side chain is converted into a methyl group. For Jackson's clockwise sequence to be followed, the subsequent round of decarboxylations require the dissociation of the product and the binding of the porphyrinogen ring to the enzyme (6) from the face opposite to that used in the complex (5). This amounts to a flipping of the macrocyclic ring around the C-10–C-20 axis by 180°. The flipping allows the acetate (A) and propionate (P) side chains of the remaining three rings to be positioned correctly at the catalytic site. The next step is the decarboxylation of the ring A acetate side chain, followed by ring B and finally ring C. In all four complexes (5–8) identical interactions operate at the catalytic and –NH-binding sites. However, no general pattern of distal interactions at the S_1–S_6 sites is evident, and the only feature common to all the complexes is a carboxyl group belonging either to an acetate or to a propionate side chain in the S_6 position. For varying combinations of side chains of the (more than 14) different compounds which are substrates for the decarboxylase to be accommodated, the overall contribution to binding energy made by these distal interactions must be rather small.

X-ray crystallography is commonly used to elucidate the nature of specific interactions between a protein and its preferred ligand. Uroporphyrinogen decarboxylase offers an unusual challenge—to define the structural features which allow it to accommodate a large number of related compounds to produce a spectrum of complexes with free energies of formation differing by no more than 1 kcal.

The mechanism

Stereochemical approaches developed in my laboratory (Barnard & Akhtar 1979, and references therein) shed light on the mechanism of decarboxylation. We established that in the reaction, the two methylene hydrogen atoms of the acetate side chain remain undisturbed and the C–carboxylate is replaced by a new C–H bond with retention of stereochemistry (3→4, Scheme 1).

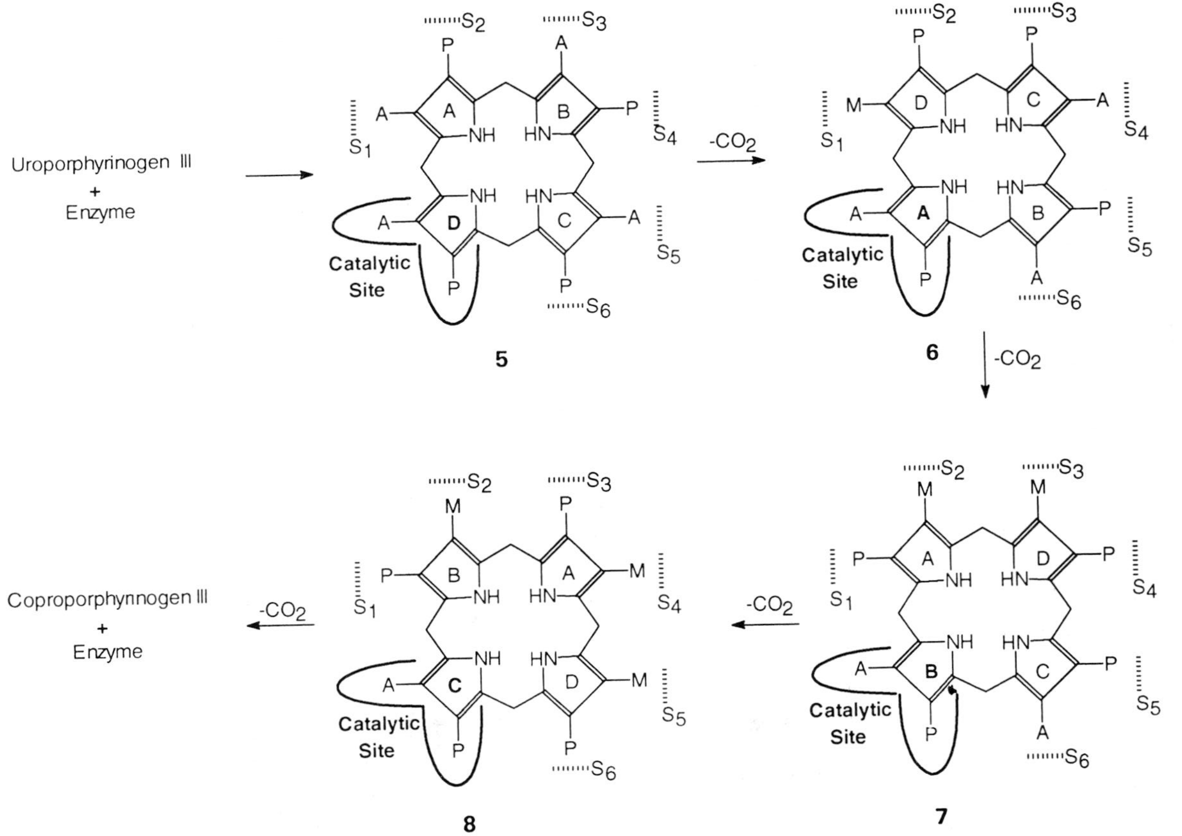

Scheme 2

The question of whether in the conversion of uroporphyrinogen III (**1**) into coproporphyrinogen III (**2**) the decarboxylation of the four acetate side chains is catalysed at a single active site or with the involvement of several distinct active centres has been debated (Akhtar 1991). The stereochemical course of the reaction suggests that whatever the answer, the stereochemical mechanism is the same for all four decarboxylations (**3**→**4**, Scheme 1). Stereochemistry is also retained in the decarboxylation of the acetate side chains in bacteriochlorophyll *a* biosynthesis as well as in the analogous conversion of the C-12 acetate side chain during corrin biosynthesis (Battersby et al 1983 and references therein).

The surprising fact that no coenzyme or metal requirement has been demonstrated for the enzyme prompted the proposal of a realistic alternative mechanism (Scheme 3). We suggested that the porphyrinogen nucleus of the substrate can be used to generate a conjugated electron sink if a tautomeric form of the pyrrole is produced through protonation of one of its α-positions (**9**→**10**). The scissile bond is now located in an electronic environment (i.e., $-\overset{+}{N}H=C-$) known in many enzymic reactions, particularly those involving pyridoxal 5′-phosphate, to promote decarboxylation. The overall reaction then occurs through the sequence **9**→**12** shown in Scheme 3. To explain the retention of configuration we proposed that the catalytic group involved either in removal of the hydrogen from the O–H bond of the carboxyl group or in the binding of a carboxylate anion also participates in the protonation of the intermediate (**11**). Our mechanism, which uses a single base to carry out two chemical steps, stems from the reasonable, but unproven, assumption that the catalytic sites of enzymes have evolved with due consideration for economy.

For the sake of illustrative convenience in the reaction **9**→**10** (Scheme 3), we have shown the proton adding onto the α-position of the pyrrole ring that is *syn* to the acetate side chain, but this is not mandatory mechanistically and the tautomeric form of **10**, that is, **13**, should be equally effective in promoting decarboxylation. Finally, we need to comment on the nature of the group $-X$: that interacts with the carboxylic group in the decarboxylation process. From the available X-ray crystallographic data from many protein–ligand complexes, the generalization which has begun to emerge is that a carboxylic group that does not undergo a chemical change is usually anchored to an enzyme by an arginine residue. What is the micro-environment around the carboxylic group which promotes its decarboxylation? All that can be said at this stage is that because at physiological pH the carboxylic group of the substrate will already be in an ionized form, a conventional base is not required for its deprotonation. Implicit in the mechanism of Scheme 3 is the use of the conjugate acid of $-X$: for engaging the carboxylic group in a hydrogen-bonded structure. The conversion of **10** to **11** then may be viewed as being driven by a multiplicity of favourable outcomes—the collapse of the hydrogen-bonded structure, release of CO_2 and the neutralization of the positive charge. X-ray crystallographic

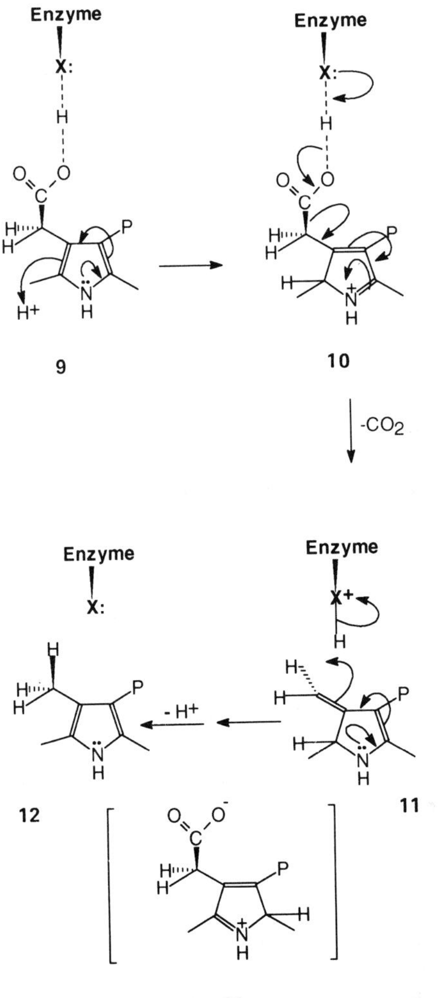

Scheme 3

studies on the decarboxylase could shed light on such mechanistic issues and also define the chemistry of the binding interactions which enable the enzyme to accommodate several related substrates with varying degrees of efficacy.

Coproporphyrinogen-III oxidase

The conversion of coproporphyrinogen III (2) into protoporphyrinogen IX (14, Scheme 4) involves the oxidative decarboxylation of the two propionate side chains in rings A and B (positions 3 and 8 in 2, Scheme 4) to vinyl groups. The

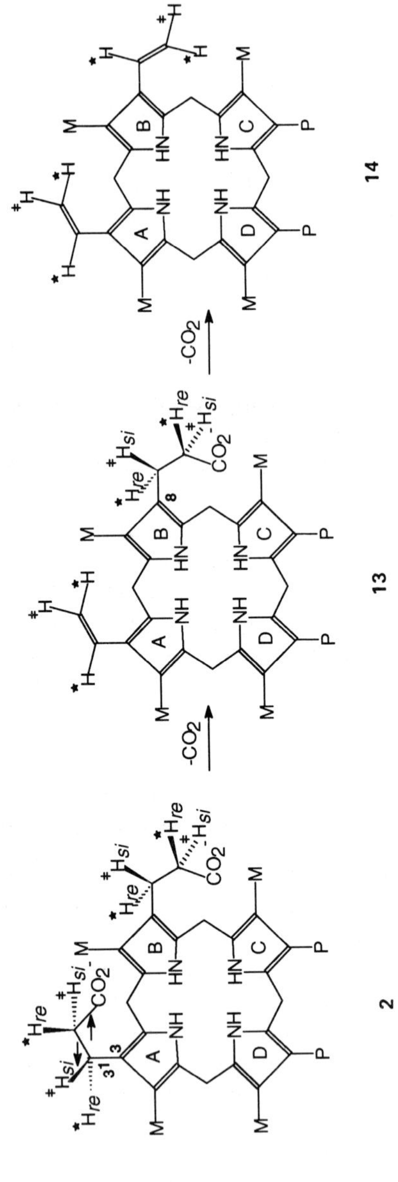

Scheme 4

elaboration of both propionate (P) groups is catalysed by a single enzyme, coproporphyrinogen-III oxidase (EC 1.3.3.3) (Sano & Granick 1961). The enzyme will not accept coproporphyrinogen I as a substrate, explaining why this compound accumulates in inherited diseases such as congenital erythropoietic porphyria.

The purification of coproporphyrinogen-III oxidase initiated in the laboratories of Granick (Sano & Granick 1961) and Rimington (del C Batlle et al 1965) during the 1960s was finally achieved by Yoshinaga & Sano (1980a,b) who reported the purification of homogeneity of bovine liver coproporphyrinogen III oxidase. The enzyme was a monomer* of M_r 71 600 which required O_2 for catalysis but no reducing agent. The coproporphyrinogen-III oxidase purified from yeast has an M_r of 75 000, but DNA sequencing of the gene for the enzyme, *HEM13*, showed it to encode a protein of 328 amino acids with a deduced M_r of 37 673 (Camadro et al 1986). Unlike the mammalian enzyme, the yeast enzyme is a dimer, but, like the former, it requires O_2 for its activity but no reducing agent. The O_2 required by eukaryotic coproporphyrinogen-III oxidases must be replaced by other oxidants for the corresponding enzyme in anaerobic bacteria and facultative anaerobes. Since the original report by Tait (1972) that extracts of photosynthetically grown *Rhodobacter sphaeroides* can anaerobically oxidize coproporphyrinogen III in the presence of NADP, ATP and L-methionine, no substantial progress in the field has been made (for a review see Dailey 1990b). Although Tait's basic observation was confirmed by Seehra et al (1983) and by Poulson & Polglase (1974) with partially purified yeast coproporphyrinogen oxidase, more recent attempts to demonstrate the anaerobic activity with homogeneous yeast enzyme have been unsuccessful (Camadro et al 1986).

The order in which the propionate side chains are removed and mechanistic studies on coproporphyrinogen-III oxidase

There are two main considerations in the conversion of coproporphyrinogen III into protoporphyrinogen IX—the order of the two decarboxylations and the mechanistic route followed.

The isolation of harderoporphyrinogen (Kennedy et al 1970, Cavaleiro et al 1974), which resembles coproporphyrinogen except that the 3-position bears a vinyl group, was seminal in establishing the order in which the two propionate side chains are handled and led to the proposal of a preferred biosynthetic sequence ($2 \rightarrow 13 \rightarrow 14$, Scheme 4) in which the decarboxylation of the ring A occurs before that of ring B.

The mechanism through which coproporphyrinogen III is oxidatively decarboxylated into protoporphyrinogen IX has been investigated by studying

---

*The enzyme has now been found to be a dimer of 37 000 M_r subunits (Kohno et al 1993).

the conversion of deuterium- or tritium-labelled propionate side chains into vinyl groups as shown in **a**. Porphobilinogen was synthesized non-enzymically

$$-\overset{\beta}{C}H_2-\overset{\alpha}{C}H_2-COOH \longrightarrow -CH=CH_2 + CO_2 \qquad\qquad \textbf{a}$$

with deuterium in the side chain (Battersby et al 1972) or enzymically labelled with ^{3}H (Zaman et al 1972, citations in Akhtar 1991, Akhtar & Jones 1986) and converted into haem with a haemolysed avian erythrocyte preparation. These studies showed that the *only* hydrogen atom removed from the propionate side chain was that from the β-position, occupying the H_{si} orientation (conversion **2→13**) (Battersby et al 1972, Zaman et al 1972, Akhtar & Jones 1986, Zaman & Akhtar 1976, Battersby 1978) with the overall process occurring through an anti-periplanar elimination, as indicated by the horizontal arrows in Scheme 4, **2** (Battersby 1978). That only one hydrogen atom is lost during the formation of the vinyl group means that acrylic or β-oxo acid intermediates are ruled out, but the results cannot distinguish between the equally plausible mechanisms shown in Scheme 5 (which assume that the acceptor in aerobic systems is O_2).

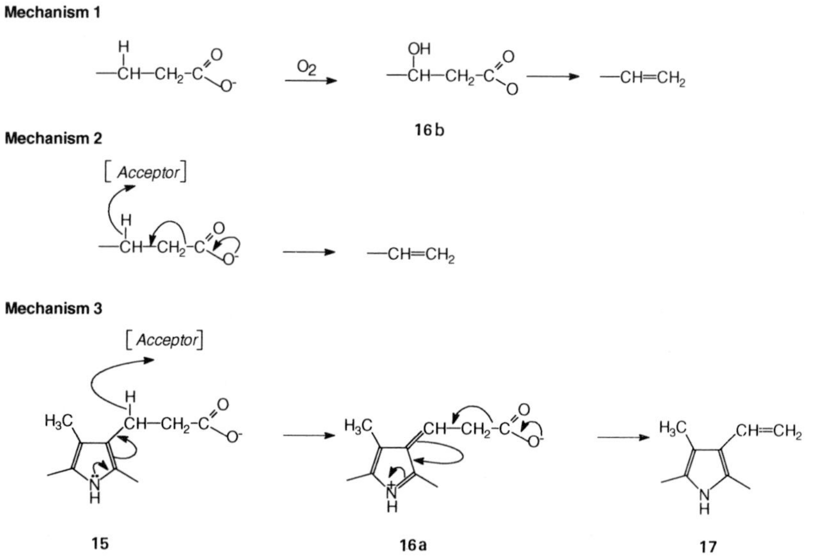

Scheme 5

Mechanism 1 assumes that reaction proceeds via an oxygen-dependent hydroxylation in which the first step, the hydroxylation, provides the energy for the decarboxylation–elimination process giving rise to the vinyl group. The support for this mechanism stems from the observation that synthetic

3-(β-hydroxypropionate)-8-propionate deuteroporphyrinogen IX (**2** hydroxylated at C-3') was efficiently converted into protoporphyrinogen IX by an avian erythrocyte preparation and by homogeneous coproporphyrinogen-III oxidase purified from bovine liver (Yoshinago & Sano 1980a,b). With the purified enzyme, the conversion of the propionate side chain into the vinyl group was dependent on O_2, whereas the corresponding conversion from the β-hydroxypropionate derivative was not. This behaviour is broadly consistent with Mechanism 1 (Scheme 5) for the formation of the two vinyl groups. The main problem, however, is that if the equation is to be stoichiometrically balanced, the oxygen-dependent hydroxylation process *must* include an hydride equivalent, but such a requirement has not been found for the reaction catalysed by coproporphyrinogen-III oxidase purified to homogeneity from rat liver, bovine liver or yeast.

Mechanism 2, originally proposed by Sano & Granick (1961), involves the removal of a hydride ion from the β-position with simultaneous decarboxylation resulting in the formation of the double bond. In Mechanism 3, a variant of Mechanism 2, the transformation occurs by a two-step sequence. The crucial C–H bond cleavage in the initial desaturation reaction (**15→16**) is facilitated by electron release from the vinylogous nitrogen, and the decarboxylation in the second stage (**16→**product) is driven by the electron sink provided by the positively charged ring nitrogen, $-C{=}\overset{+}{N}H-$. I should stress that both these reactions are mechanistically based on known enzymological precedents.

The stereochemistry and mechanism of the anaerobic coproporphyrinogen-III oxidase

I have already given the general impression that our current knowledge of the process by which coproporphyrinogen III is converted into protoporphyrinogen IX under anaerobic conditions is even scantier than that of the corresponding aerobic reaction. Notwithstanding this, the stereo–radiochemical approach described for the anaerobic reaction has been extended to two types of anaerobic system, the cell-free system from *Rhodobacter sphaeroides* originally described by Tait (1972), in which the anaerobic coproporphyrinogen oxidase activity is expressed with the addition of NADP, NADH, ATP and methionine, and *in vivo* systems for the biosynthesis of bacteriochlorophyl a_{phytyl} by *Rhodobacter sphaeroides* and bacteriocholorophyll $a_{geranylgeranyl}$ by *Rhodospirillum rubrum*. In all these studies only one hydrogen atom, H_{si}, from the β-position, was lost (Ajaz 1982), showing that the stereo- and regio-specificity of hydrogen elimination during the conversion of the propionate into the vinyl side chain is the same in the aerobic and anaerobic systems.

We have a preference for Mechanism 3. This mechanism can also be applied to anaerobic coproporphyrinogen oxidase reactions if another electron acceptor is substituted for O_2. This does not mean that the same protein is involved, but simply emphasizes that once a catalytic mechanism has emerged through natural selection it evolves further with subtle rather than dramatic changes.

Protoporphyrinogen-IX oxidase

Protoporphyrinogen-IX oxidase (EC 1.3.3.4) catalyses the conversion of protoporphyrinogen IX (14) to protoporphyrin IX (18) (Scheme 6). After several unsuccessful attempts, the enzyme has recently been purified to apparent homogeneity from several sources (Siepker et al 1987; for a review see Daily 1990b). Siepker et al (1987) were the first to draw attention to the tightly bound flavin adenine dinucleotide in the bovine liver enzyme. Bound flavin was also found in mouse liver enzyme by Proulx & Dailey (see Dailey 1990b), who characterized it as flavin mononucleotide. The bovine and mouse liver enzymes are monomers with M_r around 65 000, somewhat larger than the yeast enzyme (M_r 56 000–59 000) (see Labbe–Bois & Labbe 1990). Although the prosthetic group from the yeast enzyme has not been identified, the obligatory electron acceptor for all the three enzymes is O_2, as suggested earlier by Sano & Granick (1961). This requirement for O_2 raises an interesting question about haem and chlorophyll biosynthesis in anaerobic bacteria or facultative anaerobes. Protoporphyrin oxidases from such organisms should be able to use electron acceptors other than O_2, as suggested by several studies.

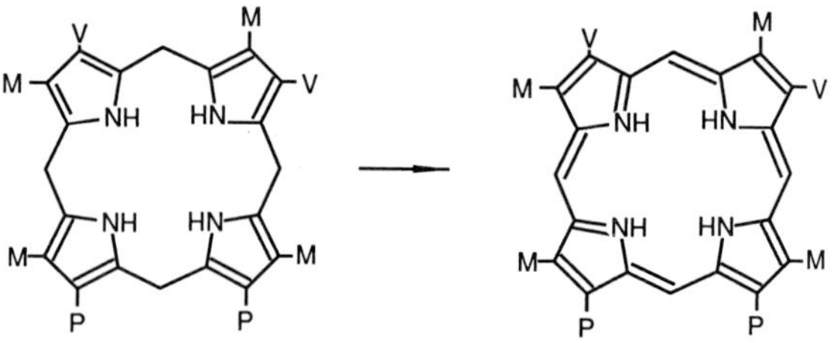

A= $CH_2.CO_2^-$

P= $CH_2.CH_2.CO_2^-$

V= $CH=CH_2$

Scheme 6

Stereochemical and mechanistic studies of protoporphyrinogen IX oxidase

The finding that $[5,10,15,20\text{-}^3H_8]$coproporphyrinogen III is converted into protoporphyrin IX with retention of 50% of the ^3H (Jackson et al 1974, Battersby et al 1976) provided independent evidence for the participation of an enzyme in the conversion of a porphyrinogen into a porphyrin nucleus, as did studies in our laboratory in which the *meso* positions of the precursor were labelled from $[11RS\text{-}^3H_2]$porphobilinogen (Jones et al 1984). Our approach paved the way for the evaluation of the stereochemical course of the oxidase-catalysed reactions.

The aminomethyl carbon of [11 S-^{3}H] porphobilinogen (19) is stereospecifically incorporated into the four methylene positions of hydroxymethylbilane and thence into the corresponding *meso* position of uroporphyrinogen III (Jordan 1994, this volume). Although the precise stereochemistry of the *meso* positions when uroporphyrinogen III is synthesized from porphobilinogen stereospecifically labelled at C-11 is not known, we have argued that the stereochemistry at C-5 and C-10 (the α and β positions) must be identical and intuitively believe that the C-15 and C-20 (γ and δ positions) must also have this same stereochemistry (see **20**). When four molecules of [11 S-^{3}H] porphobilinogen were incorporated into protoporphyrin IX, three of the ^{3}H atoms were lost and the fourth was found at C-10 (Jones et al 1984). The predicted orientation of ^{3}H in uroporphyrinogen III is shown in structure **20** (Scheme 7); the same stereochemistry at the *meso* positions is expected in protoporphyrinogen IX, the substrate for the oxidase.

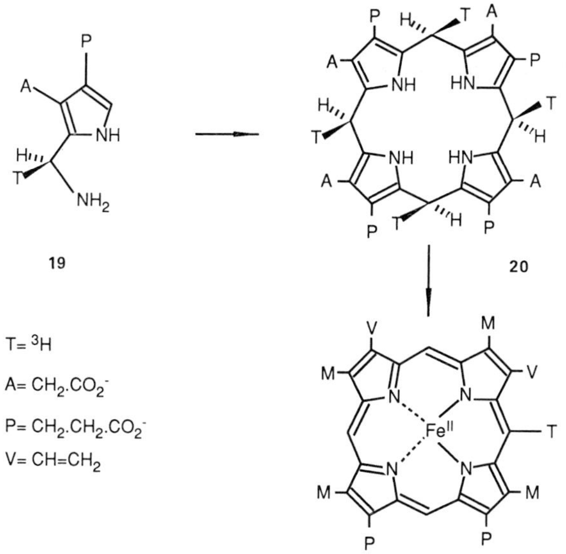

T = ^{3}H
A = $CH_2.CO_2^-$
P = $CH_2.CH_2.CO_2^-$
V = $CH=CH_2$

Scheme 7

The mechanism of the protoporphyrinogen-IX oxidase-catalysed reaction

Until the above results were obtained, it had been assumed that all four *meso* hydrogen atoms were removed from the same face of the macrocyclic ring during the protoporphyrinogen-IX oxidase-catalysed reaction, though no mechanism for this had been proposed. The stereochemical findings described above, embodied in Scheme 7, when taken in conjunction with the assertion that the ^{3}Hs at C-5 and C-10 are co-facial in protoporphyrinogen IX biosynthesized from [11 S-^{3}H] porphobilinogen (**19**), indicate that the C-5 and C-10 positions are treated differently in the overall reaction catalysed by protoporphyrinogen-IX oxidase. We have suggested, therefore, that one of these positions is the site of the oxidation reaction (loss of 'hydride') and the other of the tautomerization process (loss of H$^+$), and that the two processes use hydrogen atoms from opposite faces of the macrocycle (Scheme 8; H^a and H^b represent hydrogen atoms above and below the plane of the ring). The overall transformation so involves three step-wise desaturation reactions, each with the loss of a hydride from the *meso* position and a proton from the pyrrole N, giving, after three rounds, the intermediate **25**. Tautomerization to the porphyrin then occurs through the stereospecific loss of the fourth *meso* hydrogen as a proton (**25**→**26**). Because it is not yet possible to define the precise order in which the various positions of protoporphyrinogen IX are handled by the oxidase, our proposed mechanism is illustrated in Scheme 8 with an unsubstituted porphyrinogen nucleus.

The flavin prosthetic group associated with mouse and bovine liver protoporphyrinogen oxidases is mentioned above. The three desaturation

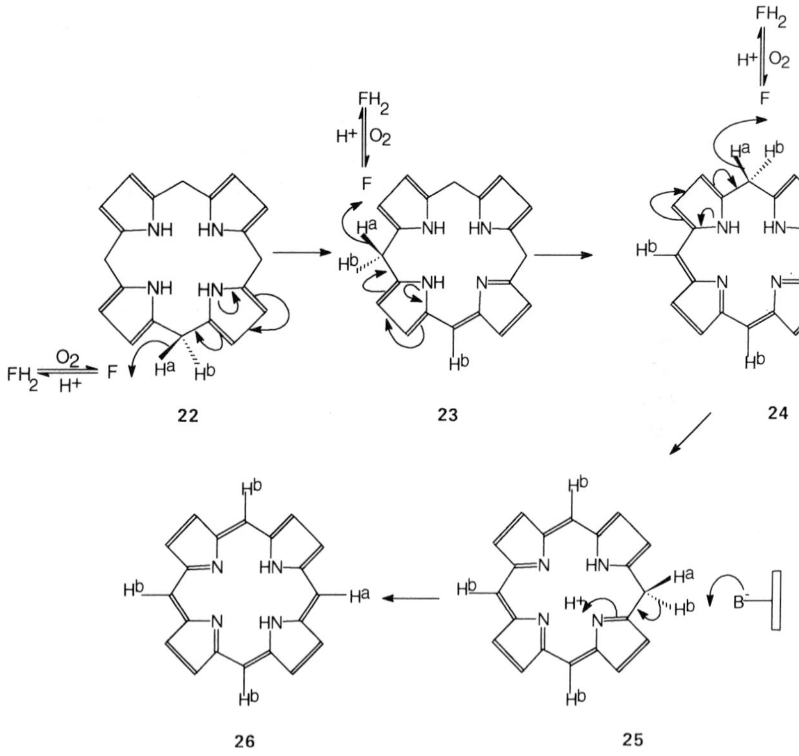

Scheme 8

reactions in Scheme 8 are mechanistically analogous to the insertion of unsaturated linkages between C–N and C–C bonds catalysed by D-amino acid oxidases and acyl-CoA dehydrogenases, respectively. These enzymes contain a flavin prosthetic group which acts as the acceptor of a 'hydride' equivalent. The reduced flavin thus produced is reoxidized by O_2 to regenerate the active enzyme. In Scheme 8, a similar scenario is envisaged for the desaturation reactions catalysed by eukaryotic protoporphyrinogen oxidases. It is not known whether the corresponding enzymes from anaerobic bacteria use flavin, but if they do, its reoxidation must be achieved by an alternative process, presumably by a coupling of the oxidation to the organism's respiratory chain.

The mechanism shown in Scheme 8 represents our efforts to rationalize the unexpected stereochemical results, but the same mechanism could have been deduced from first principles as the simplest way of achieving the enzymic conversion of the porphyrinogen into the porphyrin nucleus.

The elaboration of the ring D propionate side chain in chlorophyll biosynthesis: the chlorophyll synthase-catalysed reaction

A common feature of all known chlorophylls is the esterification of the ring D propionate carboxyl group at C-17[3] with a long-chain alcohol. The alcohol varies considerably, but has been shown to be phytol for bacteriochlorophyll *a* from *Rhodobacter sphaeroides* and higher plant chlorophylls. Investigation of this esterification can be traced back to Willstätter & Stoll (1913) who exploited the discovery made by Borodin (1882) that the formation of ethyl chlorophyllide from ethanol extracts of green leaves was due to the action of chlorophyllase, an enzyme present in leaves, on chlorophyll *a*. This enzyme, though clearly important in chlorophyll degeneration, was also assigned a role in its biosynthesis. This view prevailed until two complementary approaches, one taken by Rüdiger et al (1980) and the other developed by my group, pointed to the existence of another enzyme that catalysed the final esterification step in chlorophyll biosynthesis. Rüdiger et al (1980) showed incorporation of the pyrophosphorylated form of geranylgeraniol into chlorophyll *a* in maize seedling preparations and attributed this conversion to a new enzyme, chlorophyll synthase. We developed an ^{18}O-labelling method which provided quantitative information on the flux of chlorophyll biosynthesis through different mechanistic pathways. Theoretically, there are three distinct pathways through which an isoprenyl side chain can be elaborated, differentiated by the status of oxygen atoms during the conversion of a carboxyl group into the ester bond. In Scheme 9, esterification by Mechanism 1 is the reversal of the reaction catalysed by chlorophyllase; only one of the two carboxyl oxygen atoms is retained in the ester during the first cycle but both oxygen atoms will eventually be washed out owing to the facile reversal of the reaction. Mechanism 2 involves an active ester intermediate (where OX is a biological leaving group) and removal of only

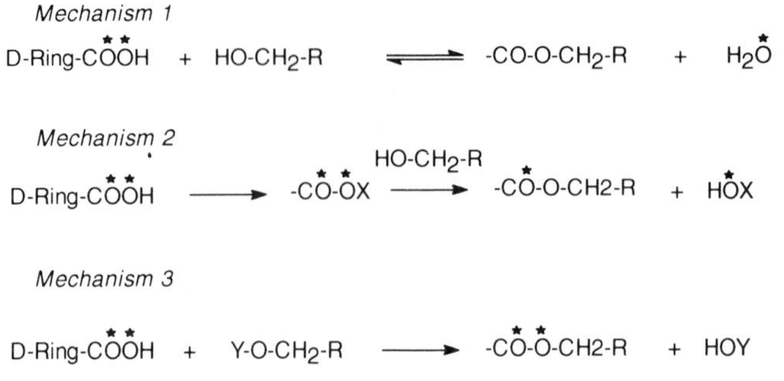

Mechanism 1

$$\text{D-Ring-CO\overset{*}{O}\overset{*}{H}} \quad + \quad \text{HO-CH}_2\text{-R} \quad \rightleftharpoons \quad \text{-CO-O-CH}_2\text{-R} \quad + \quad \text{H}_2\overset{*}{\text{O}}$$

Mechanism 2

$$\text{D-Ring-CO\overset{*}{O}\overset{*}{H}} \quad \longrightarrow \quad \text{-C\overset{*}{O}-O\overset{*}{X}} \quad \overset{\text{HO-CH}_2\text{-R}}{\longrightarrow} \quad \text{-C\overset{*}{O}-O-CH2-R} \quad + \quad \text{H\overset{*}{O}X}$$

Mechanism 3

$$\text{D-Ring-CO\overset{*}{O}\overset{*}{H}} \quad + \quad \text{Y-O-CH}_2\text{-R} \quad \longrightarrow \quad \text{-C\overset{*}{O}-\overset{*}{O}-CH2-R} \quad + \quad \text{HOY}$$

Scheme 9

one of the carboxyl oxygen atoms; Mechanism 3 is a carboxyl–alkyl transfer process requiring retention of both oxygen atoms (OY is a biological leaving group such as pyrophosphate, phosphate and adenylate).

To examine these alternatives, we fed non-enzymically synthesized 5-aminolaevulinic acid labelled at C-1 with ^{13}C and ^{18}O (27) to *Rhodobacter sphaeroides* under conditions which ensured that most of the bacteriochlorophyll *a* (29) biosynthesized, via the sequence in Scheme 10, originated from the added precursor rather than endogenous intermediates (Emery & Akhtar 1987). The high-resolution 100.53 MHz ^{13}C NMR spectrum of the bacteriochlorophyll *a* produced (Fig. 1) showed the C-17^3 resonance at 173.86 p.p.m. to consist of four components corresponding to the $-^{16}\text{O}-^{13}\text{C}=^{16}\text{O}$, $-^{18}\text{O}-^{13}\text{C}=^{16}\text{O}$, $-^{16}\text{O}-^{13}\text{C}=^{18}\text{O}$ and $-^{18}\text{O}-^{13}\text{C}=^{18}\text{O}$ species (Fig. 1B). The upfield isotope shifts for these species (1.5, 3.8 and 5.2 Hz) are consistent with those of phytyl acetate standards. The isotopic ratios demonstrated that retention of ^{18}O at *both* oxygen atoms was in excess of 90% (compare the peak intensities in the upper panel of Fig. 1 with those in B). As expected, the intensities of C-17^3 signals from $-^{18}\text{O}-^{13}\text{C}=^{16}\text{O}$ and $-^{16}\text{O}-^{13}\text{C}=^{18}\text{O}$ were equal and their sum (38%) is equivalent to the intensity of single-labelled oxygen species in the starting 5-aminolaevulinic acid (36%).

The results clearly show that both carboxyl oxygen atoms are retained in the formation of the ester bond at C-17^3, suggesting that the bond is elaborated through a carboxyl–alkyl transfer process (Mechanism 3, Scheme 9). The isotopic composition of the C-13^3 carbomethoxy group, which also originates from the C-1 of 5-aminolaevulinic acid, was the same as for the C-17^3 (see Fig. 1A). This is not surprising because the methyl group of the ester is derived from *S*-adenosylmethionine, most probably through a carboxyl–alkyl transfer mechanism.

This NMR approach with *Rhodobacter sphaeroides*, together with our earlier studies on *Rhodospirillum rubrum* using a mass spectrometric method (Akhtar et al 1984, Ajaz et al 1984) and the enzymological work of Rüdiger et al (1980) on the plant cell-free system, points to Mechanism 3 for the C-17^3 isoprenylation step. More importantly, however, the similarity of the ^{18}O

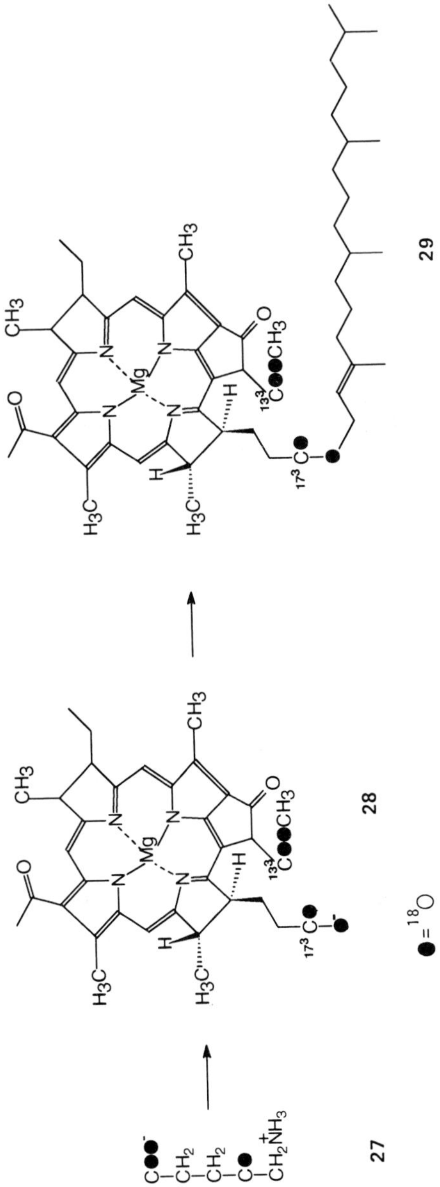

Scheme 10

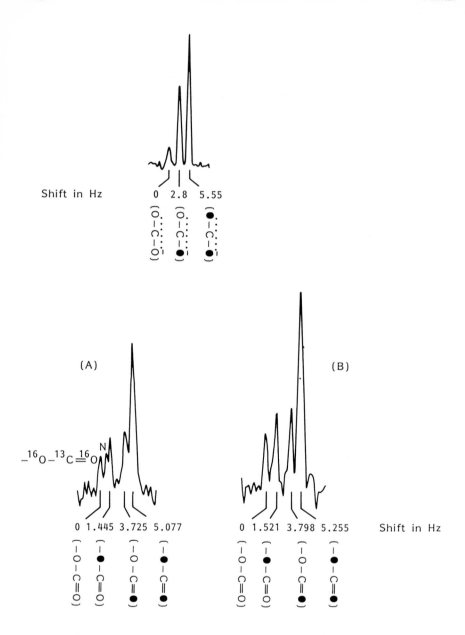

FIG. 1. ^{13}C NMR spectra showing the C-1 resonance of the precursor [1-^{13}C; 1,1,4-^{18}O$_3$] 5-aminolaevulinate (*upper panel*) and C-13^3 (A) and C-17^3 (B) resonances of bacteriochlorophyll a_{phytyl} biosynthesized in *Rhodobacter sphaeroides*. Filled oxygen atoms represent ^{18}O. The peak marked N in (A) is not identified.

content of the C-1 oxygen atoms of the precursor 5-aminolaevulinic acid and the two oxygen atoms of the isoprenyl ester revealed by the NMR studies gives quantitative information on the biosynthetic flux operating *in vivo* through Mechanism 3, which, for *Rhodobacter sphaeroides*, is estimated to be 90–95%.

Many reactions are now known in which an isoprenyl unit is transferred to an acceptor atom (Scheme 11). The classical examples are the reactions in which the acceptor nucleophile is an sp^2 carbon atom as in terpene (a) and haem *a* (e) biosynthesis (Scheme 11). The involvement of hetero atoms then emerged, for example the use of N in the isoprenylation of tRNAs, of O in the biosynthesis of chlorophylls (b), and the lipid component of archaebacteria (c); the most recent discovery is of a cysteine S being modified to produce *S*-isoprenylated proteins (d) involved in a variety of signalling pathways. From a chemical viewpoint, all these reactions conform to the mechanistic principle exemplified in Scheme 11. It would be interesting to explore whether there is structural resemblance between the proteins which catalyse these varied isoprenyl transfer reactions.

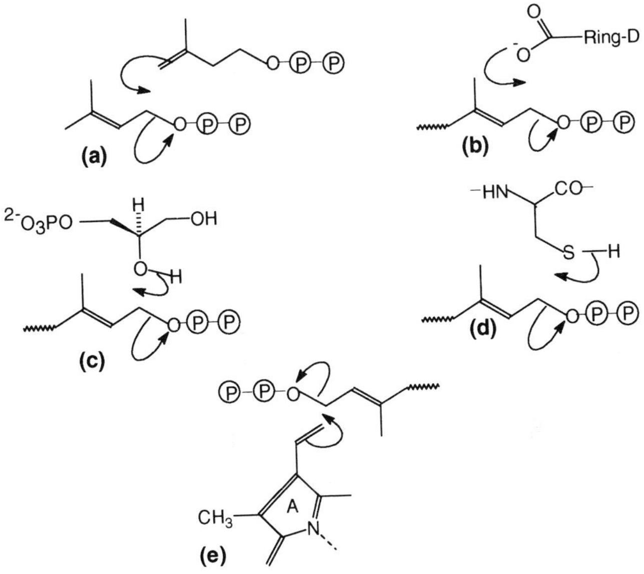

Scheme 11

References

Ajaz AA 1982 Mechanistic studies on the biosynthesis of bacteriochlorophylls in photosynthetic bacteria. PhD Thesis, University of Southampton, Southampton, UK

Ajaz AA, Corina DL, Akhtar M 1984 The mechanism of the C-13³ esterification step in the biosynthesis of bacteriochlorophyll *a*. Eur J Biochem 150:309–312

Akhtar M 1991 Mechanism and stereochemistry of enzymes involved in the conversion of uroporphyrinogen III into haem. In: Jordan PM (ed) Biosynthesis of tetrapyrroles. Elsevier, Amsterdam, p 67–69

Akhtar M, Jones C 1986 Preparation of stereospecifically-labeled porphobilinogens. Methods Enzymol 123:375–383

Akhtar M, Ajaz AA, Corina DL 1984 The mechanism of the attachment of esterifying alcohol in bacteriochlorophyll *a* biosynthesis. Biochem J 244:187–194

Barnard GF, Akhtar M 1979 Stereochemical and mechanistic studies on the decarboxylation of uroporphyrinogen III in haem biosynthesis. J Chem Soc Perkin Trans I, p 2354–2360

Battersby AR 1978 The discovery of Nature's biosynthetic pathways. Experientia 34:1–13

Battersby AR, Baldas J, Collins J, Grayson DH, James KJ, McDonald E 1972 Mechanism of biosynthesis of the vinyl groups of protoporphyrin IX. J Chem Soc Chem Commun, p 1265–1266

Battersby AR, Hunt E, Ihara M et al 1974 Structure of a heptacarboxylic porphyrin enzymically derived from porphobilinogen. J Chem Soc Chem Commun, p 994–995

Battersby AR, Staunton J, McDonald E, Redfern JR, Wightman RH 1976 Biosynthesis of porphyrins and related macrocycles. V. Structural integrity of the type III porphyrinogen macrocycle in an active biological system. Studies on the aromatisation of protoporphyrinogen IX. J Chem Soc Perkin Trans I, p 266–273

Battersby AR, Deutscher KR, Martinoni B 1983 Biosynthesis of vitamin B_{12}: stereochemistry of the decarboxylation step which generates the 12-*si*-methyl group. J Chem Soc Chem Commun, p 698–700

Borodin A 1882 Bot Ztg 40:608, quoted by Aronoff S 1966 An introductory survey. In: Vernon LP, Seely GR (eds) Chlorophylls. Academic Press, New York, p 4 and 19

Camadro JM, Chambon H, Jolles J, Labbe P 1986 Purification and properties of coproporphyrinogen oxidase from the yeast *Saccharomyces cerevisiae*. Eur J Biochem 156:579–587

Cavaleiro JAS, Kenner GW, Smith KM 1974 Pyrroles and related compounds. XXXII. Biosynthesis of protoporphyrin-IX from coproporphyrinogen III. J Chem Soc Perkin Trans I, p 1188–1194

Dailey HA (ed) 1990a Biosynthesis of heme and chlorophylls. McGraw-Hill, New York

Dailey HA 1990b Conversion of coproporphyrinogen to protoheme. In: Dailey HA (ed) Biosynthesis of heme and chlorophylls. McGraw-Hill, New York, p 123–161

del C Batlle AM, Grinstein M 1964 Porphyrin biosynthesis. II. Physiaporphyrinogen III, a normal intermediate in the biosynthesis of protoporphyrin 9. Biochim Biophys Acta 82:13-20

del C Batlle AM, Benson A, Rimington C 1965 Purification and properties of coproporphyrinogenase. Biochem J 97:731–740

Emery VC, Akhtar M 1987 Mechanistic studies on the phytylation and methylation steps in bacteriochlorophyll *a* biosynthesis: an application of the ^{18}O-induced isotope effect in ^{13}C NMR. Biochemistry 26:1200–1208

Jackson AH, Games DE, Couch PW, Jackson JR, Belcher RV, Smith SG 1974 Conversion of coproporphyrinogen III to protoporphyrin IX. Enzyme 17:81–87

Jackson AH, Sancovich HA, Ferramola AM et al 1976 Macrocyclic intermediates in the biosynthesis of porphyrins. Philos Trans R Soc Lond B Biol Sci 273:191–206

Jackson AH, Sancovich HA, Ferramola DE, Sancovich AM 1980 Synthetic and biosynthetic studies on porphyrins. III. Structures of intermediates between uroporphyrinogen III and coproporphyrinogen III: synthesis of fourteen heptacarboxylic, hexacarboxylic and pentacarboxylic porphyrins related to uroporphyrin III. Bioorg Chem 9:71–120

Jones C, Jordan PM, Akhtar M 1984 Mechanism and stereochemistry of the porphobilinogen deaminase and protoporphyrinogen IX oxidase reactions: stereospecific manipulation of hydrogen atoms at the four methylene bridges during the biosynthesis of haem. J Chem Soc Perkin Trans I, p 2625–2633

Jordan PM 1994 Porphobilinogen deaminase: mechanism of action and role in the biosynthesis of uroporphyrinogen III. In: The biosynthesis of the tetrapyrrole pigments. Wiley, Chichester (Ciba Found Symp 180) p 70–96

Kennedy GY, Jackson AH, Kenner GW, Suckling CJ 1970 Isolation, structure and synthesis of a tricarboxylic porphyrin from Harderian glands of the rat. FEBS (Fed Eur Biochem Soc) Lett 6:9–12 (correction, vol 7:205–206)

Kohno H, Furukawa T, Yoshinaga T, Tukunaga R, Taketani S 1993 Coproporphyrinogen oxidase. J Biol Chem 268:21359–21363

Labbe-Bois R, Labbe P 1990 Tetrapyrrole and heme biosynthesis in yeast. In: Daily HA (ed) Biosynthesis of heme and chlorophylls. McGraw-Hill, New York, p 235–286

Lash TD 1991 Action of uroporphyrinogen decarboxylation on uroporphyrinogen III: a reassessment of the clockwise decarboxylation hypothesis. Biochem J 278:901–902

Luo J, Lim CK 1993 Order of uroporphyrinogen III decarboxylation on incubation of porphobilinogen and uroporphyrinogen III with erythrocyte uroporphyrinogen decarboxylase. Biochem J 289:529–532

Mauzerall D, Granick S 1958 Porphyrin biosynthesis in erythrocytes. III. Uroporphyrinogen and its decarboxylase. J Biol Chem 232:1141–1162

Neve RA, Labbe RF, Aldrich RA 1956 Reduced porphyrin III in the biosynthesis of heme. J Am Chem Soc 78:691–692

Poulson R, Polglase WJ 1974 Aerobic and anaerobic coproporphyrinogenase activities in extracts from Saccharomyces cerevisiae. J Biol Chem 249:6367–6371

Rüdiger W, Benz J, Guthoff C 1980 Detection and partial characterization of activity of chlorophyll synthetase in etioplast membranes. Eur J Biochem 109:193–200

Sano S, Granick S 1961 Mitochondrial coproporphyrinogen oxidase and protoporphyrin formation. J Biol Chem 236:1173–1180

Seehra JS, Jordan PM, Akhtar M 1983 Anaerobic and aerobic coproporphyrinogen III oxidases of Rhodopseudomonas sphaeroides: mechanism and stereochemistry of vinyl group formation. Biochem J 209:709–718

Siepker LJ, Ford M, de Kock R, Kramer S 1987 Purification of bovine protoporphyrinogen oxidase: immunological cross-reactivity and structural relationship to ferrochelatase. Biochem Biophys Acta 913:349–358

Smith AG, Francis JE 1979 Decarboxylation of porphyrinogens by rat liver uroporphyrinogen decarboxylase. Biochem J 183:455–458

Tait GH 1972 Coproporphyrinogenase activity in extracts from Rhodopseudomonas sphaeroides and chromatium D. Biochem J 128:1159–1169

Tomio JM, Garcia RC, San Martin de Viale LC, Viale AA, Grinstein M 1970 Porphyrin biosynthesis. VII. Porphyrinogen carboxy-lyase from avian erythrocytes, purification and properties. Biochim Biophys Acta 198:353–363

Willstätter R, Stoll A 1913 Untersuchungen über chlorophyll. Springer-Verlag, Berlin (English translation by Shertz FM, Mertz GR 1928 Investigation on chlorophylls. Science Press, Lancaster, PA)

Yoshinaga T, Sano S 1980a Coproporphyrinogen oxidase. I. Purification, properties and activation by phospholipids. J Biol Chem 255:4722–4726

Yoshinaga T, Sano S 1980b Coproporphyrinogen oxidase. II. Reaction mechanism and role of tyrosine residues on the activity. J Biol Chem 255:4727–4731

Zaman Z, Akhtar M 1976 Mechanism and stereochemistry of vinyl-group formation in haem biosynthesis. Eur J Biochem 61:215–221

Zaman Z, Abboud MM, Akhtar M 1972 Mechanism and stereochemistry of vinyl group formation in haem biosynthesis. J Chem Soc Chem Commun, p 1263–1264

DISCUSSION

A. Smith: Is the stereospecificity of the non-enzymic conversion of protoporphyrinogen to protoporphyrin IX the same as that found in the enzymic removal of the hydrogens from the methene bridges?

Akhtar: No. Sir Alan Battersby and Tony Jackson found that in the non-enzymic reaction most of the tritium from non-stereospecifically labelled porphobilinogen is retained in protoporphyrin IX because of the operation of an isotope effect, whereas in the enzymic reaction, with $[11RS$-$^3H_2]$porphobilinogen as the substrate, there is exactly 50% retention of ^3H in protoporphyrin IX.

Arigoni: That does not mean that there is no isotope effect in the enzymic reaction. It's probably still there, but does not affect the distribution of isotopes, which is controlled exclusively by the stereochemistry.

Beale: In Scheme 11 you drew the farnesylation of haem *a*. It wasn't clear to me where the oxygen in the farnesyl hydroxyethyl group came from.

Akhtar: In (e), the nucleophile is the vinyl double bond. I merely wanted to show that the mechanism is identical in all cases. The only thing which is different in these isoprenylation reactions is the nature of the acceptor. In (a) and (e) the acceptors are the sp^2 carbon atoms. I should have put them next to each other.

Beale: Is the oxygen in the product derived from the farnesyl in (e)?

Akhtar: The partial mechanism shown in (e) will generate a carbonium ion which may be quenched by OH from water. Therefore, the hydroxyl group in haem *a* should be derived from H_2O, but I don't know whether this has been demonstrated.

Arigoni: What about (c)?

Akhtar: The problem is not relevant here because in this case we are dealing with the biosynthesis of isoprenyl ethers of archaebacteria.

Battersby: For haem *a*, it seems likely that the oxygen at the chiral centre is derived from water, which satisfies the electron density arising at this carbon as the vinyl group of the precursor undergoes farnesylation. Is the stereochemistry at that centre known?

K. Smith: It is not known for haem *a*, but it is for the *Chlorobium* chlorophylls, where it varies according to the molecular size of the side chains at the 4 position (Fischer nomenclature). In this case, I don't know whether or not the stereochemistry is known.

Chang: The secondary alcohol of haem *a* isomerizes so quickly in acidic demetallation procedures that I don't think you would be able to tell.

Arigoni: Professor Akhtar, you suggested three mechanisms for the oxidative decarboxylation of the propionic acid side chain (Scheme 5). You seemed to rule out Mechanism 1 on the basis that the reducing agent required by a mixed function oxidase is not present, and so favour Mechanism 3. If you quenched intermediate **16a** by addition of water, would you not generate the hydroxy compound **16b** of Mechanism 1?

Akhtar: I wave a white flag. You have produced the hydroxy compound by a different mechanism, using oxygen in the first step to generate **16a** (Scheme 5) which then gives the β-hydroxy compound (**16b**) by hydration. It is an interesting marriage of Mechanisms 2 and 3 in which a hydroxylated product is obtained without the need for a reducing agent.

Arigoni: You don't have to use oxygen. You could use NAD.

Akhtar: That may be true in anaerobic systems, but with the mammalian enzyme oxygen must be used because purified protoporphyrinogen-IX oxidase from several sources requires O_2 for the conversion.

K. Smith: The oxygen in the hydroxypropionate precursor in chlorophyll *a* biosynthesis definitely comes from O_2, not from water.

Kräutler: One should be careful when talking about the external reducing agent because protoporphyrinogen is itself a reducing agent.

Akhtar: In the conventional hydroxylation reaction two hydrogen atoms (two protons plus two electrons) and O_2 are involved. The fact that in Sano's experiment there was no requirement for a reducing agent means that if a β-hydroxypropionate derivative is an intermediate, it must be formed by an indirect path, perhaps through a dehydrogenation–hydration process (Sano 1966). Mechanism 3 permits this by providing a Michael acceptor which may be hydrated to give a β-hydroxypropionate side chain. In broad philosophical terms, the sequence **15→16a→16b** corresponds to the formation of β-hydroxyacyl-CoA derivatives in the fatty acid oxidation pathway.

K. Smith: The problem is the same in chlorophyll biosynthesis. Quite a lot of chlorophylls are produced in processes which are apparently totally anaerobic. In chlorophyll *a* the hydroxyl comes from molecular oxygen, so one wonders where the oxygen comes from in anaerobic systems. Perhaps it comes from water, as Duilio Arigoni suggested.

Akhtar: We preferred Mechanism 3 precisely because O_2 in the desaturation step can be easily replaced by another oxidant in anaerobes.

K. Smith: Chlorophyll biosynthesis is much easier to study because the oxygen remains in the product, in ring E, whereas it doesn't in the formation of vinyl from propionate, which means you cannot follow it by [18]O labelling.

Thauer: Professor Akhtar, you suggested that in anaerobes this decarboxylation might require *S*-adenosylmethionine.

Akhtar: Tait (1972), using a crude enzyme preparation from *Rhodobacter sphaeroides*, showed that the oxidant was $NADP^+$ but that methionine was also required. Although some groups were able to repeat Tait's finding, the anaerobic reaction has not yet been demonstrated with the purified enzyme.

Thauer: There are several examples in which *S*-adenosylmethionine replaces molecular oxygen as the initiator for radical formation. Is there any indication that the anaerobic mechanism involves radicals?

Akhtar: No. Only two groups have studied the anaerobic reaction in the past and recent attempts to purify the anaerobic enzyme have been unsuccessful.

Both protoporphyrinogen oxidase and coproporphyrinogen oxidase have yet to be isolated from anaerobes.

Thauer: E. coli should be able to catalyse the coproporphyrinogen dehydrogenation reaction, because it makes cytochromes anaerobically. It must have the enzyme, so one could try to overexpress it.

Jordan: We have isolated the coproporphyrinogen oxidase gene from *Rhodobacter sphaeroides*, but it is difficult to express.

Rebeiz: Is it possible to visualize a decarboxylation reaction which would result in a vinyl group at position 3 of the macrocycle and an ethyl group at position 8?

Akhtar: That is possible in principle, but there is good evidence that the ethyl group in ring B in chlorophylls comes from the corresponding vinyl group.

Rebeiz: There is unambiguous evidence for that in protochlorophyllide and chlorophyllide, but we haven't been able to demonstrate this in protoporphyrin IX. In my paper (p 186) I suggest that the monovinyl and divinylprotoporphyrins may be formed in parallel from coproporphyrinogen.

Akhtar: The type of mechanism I envisage would permit not the formation of a vinyl group at position 3 but an ethyl at position 8. However, you are correct that a free radical mechanism for the decarboxylation of the propionate side chain would give the ethyl radical, which could produce the vinyl group by the loss of a hydrogen atom. How you go from an ethyl radical to an ethyl group will obviously depend on whether Nature can produce a hydrogen donor.

Eschenmoser: I would like to comment on the relationship between Mechanism 2 and Mechanism 3 in Scheme 5. Constitutionally, the initiator in Mechanism 2 must be the acceptor. In such a reaction the process depicted in Mechanism 3 cannot but occur, because electrons leave faster than CO_2. Are you implying that Mechanism 2 differs from Mechanism 3 only in the conformation of the side chain?

Akhtar: No. Mechanism 2 was proposed by Granick, and I adapted it to get Mechanism 3, for the reason you gave, that electrons move faster than the decarboxylation process producing CO_2. Mechanism 3 would also allow an alternative way of producing the hydroxy compound which I had not realized until Duilio Arigoni pointed it out.

Eschenmoser: Mechanism 2 and Mechanism 3 are two distinctly different mechanisms, but one would have to express the difference between them in conformational terms.

Arigoni: There is a good precedent for the operation of Mechanism 2 in the formation of terminal olefins from saturated fatty acids in higher plants. Görgen & Boland (1989) showed that the stereochemistry of the elimination is exactly as observed during formation of the vinyl group in this case. The hydroxy compound is not an intermediate, and something like cytochrome P_{450} may be the cofactor. In this case the assistance of a nitrogen atom is not required.

Akhtar: If the mechanism is as you suggest, a carboxyl radical would have to be involved; hence, as you said, such a process will need to be modelled on the P_{450} system, which requires O_2 as well as two electrons.

Arigoni: That doesn't imply it is a radical mechanism.

Akhtar: It has to be patterned on the P_{450} system, for which iron, O_2 and two electrons are needed. The enzyme purified from three sources does not contain iron, nor is there any requirement for a reducing agent.

Arigoni: That is not what Albert Eschenmoser is challenging. We agree on the fact that on a formal basis the two mechanisms will merge.

Eschenmoser: If the process of hydrogen abstraction proceeds within the plane defined by the pyrrole ring, Mechanism 3 would not be expected to operate.

Jordan: During the formation of the formyl group at position 7 of chlorophyll *b* Professor Arigoni's idea could be used to make a 7-hydroxy intermediate that could be dehydrogenated to the formyl group.

Beale: There is an analogous transformation of a methyl to a formyl group in chlorophyll *b*. O_2 is incorporated in that reaction (Schneegurt & Beale 1992).

References

Görgen G, Boland W 1989 Biosynthesis of 1-alkenes in higher plants: stereochemical implications. A model study with *Carthamus tinctorius* (Asteraceae). Eur J Biochem 185:237–242

Sano S 1966 2,4-Bis-(β-hydroxypropionic acid) deuteroporphyrinogen IX a possible intermediate between coproporphyrinogen III and protoporphyrinogen IX. J Biol Chem 241:5276–5283

Schneegurt MA, Beale SI 1992 Origin of the chlorophyll *b* formyl oxygen in *Chlorella vulgaris*. Biochemistry 31:1677–1683

Tait GH 1972 Coproporphyrinogenase activity in extracts from *Rhodopseudomonas sphaeroides* and *chromatium D*. Biochem J 128:1159–1169

Biosynthesis of open-chain tetrapyrroles in plants, algae, and cyanobacteria

Samuel I. Beale

Division of Biology and Medicine, Brown University, Providence, RI 02912, USA

Abstract. Phycobilins are open-chain tetrapyrroles of plants and algae which act as the chromophores of phycobiliproteins where they function as light energy-harvesting pigments. Phytochromobilin, another open-chain tetrapyrrole, is the chromophore of phytochrome, which functions as a light-sensing pigment in plant development. These open-chain tetrapyrroles are biosynthetically derived from protohaem. Enzyme reactions that convert protohaem to biliverdin IXα, and biliverdin IXα to phycocyanobilin, have been detected and characterized in extracts of the unicellular rhodophyte *Cyanidium caldarium*. Algal haem oxygenase and algal biliverdin-IXα reductase are both soluble enzymes that use electrons derived from reduced ferredoxin. Biochemical intermediates in the conversion of biliverdin IXα to (3*E*)-phycocyanobilin were identified as 15,16-dihydrobiliverdin IXα, (3*Z*)-phycoerythrobilin and (3*Z*)-phycocyanobilin. Separate enzymes catalyse the two two-electron reduction steps in the conversion of biliverdin IXα to (3*Z*)-phycoerythrobilin. *Z*-to-*E* isomerization of the phycobilin ethylidine group is catalysed by an enzyme that requires glutathione for activity. Protein-bound phycoerythrobilin can be chemically converted to phytochromobilin which can then be released from the protein by methanolysis. This procedure was used to produce phytochromobilin in quantities sufficient to allow its chemical characterization and use in phytochrome reconstitution experiments. The results indicate that (2R,3*E*)-phytochromobilin spontaneously condenses with recombinant oat apophytochrome to form photoreversible holoprotein that is spectrally identical to native phytochrome.

1994 The biosynthesis of the tetrapyrrole pigments. Wiley, Chichester (Ciba Foundation Symposium 180) p 156–171

Phycobilins are open-chain tetrapyrroles that function as chromophores in light-harvesting chromoproteins in certain groups of photosynthetic organisms. In their functional state, phycobilins are covalently linked to specific proteins, to form pigment–protein complexes called phycobiliproteins, by one or two thioether bonds between cysteine residues in the protein and vinyl-derived bilin substituents (Fig. 1). The two major classes of phycobiliproteins are phycocyanins and phycoerythrins, which are blue and red, respectively. These pigments are largely responsible for the characteristic colours of organisms which

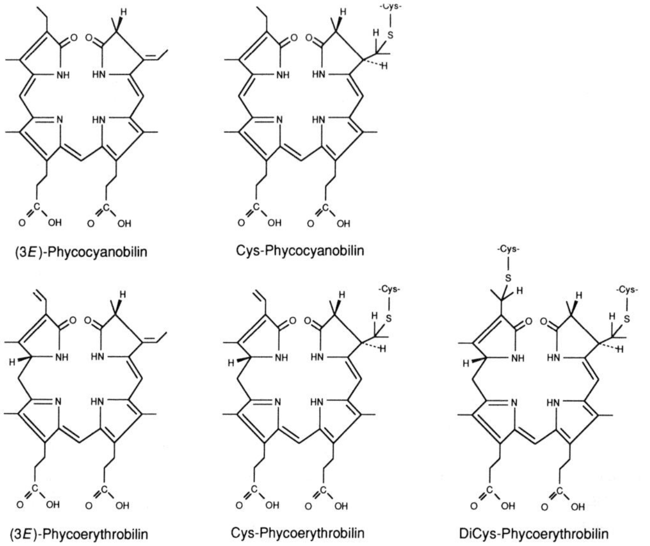

(3E)-Phycocyanobilin   Cys-Phycocyanobilin

(3E)-Phycoerythrobilin   Cys-Phycoerythrobilin   DiCys-Phycoerythrobilin

FIG. 1. Structures of free and protein-linked phycocyanobilin $(2,3^2,18^1,18^2$-tetrahydrobiliverdin IXα) and phycoerythrobilin $(2,3^2,15,16$-tetrahydrobiliverdin IXα). Chromophores that are linked to the proteins by a single thioether bond adjacent to a reduced pyrrole ring are released by heating in methanol, to yield primarily the (3E)-ethylidine isomers, plus smaller amounts of the (3Z)-isomers.

contain them. The different colours arise because the different phycobilins have different numbers of conjugated double bonds. In cyanobacteria (formerly blue-green algae) and rhodophytes (red algae), phycobiliproteins, together with several 'linker' polypeptides, form functional aggregates, called phycobilisomes, which are attached to thylakoid membranes and mediate efficient absorption of light and transfer of excitation energy to photosynthetic reaction centres. Crypto-monad algae also contain phycobiliproteins and utilize them for light harvesting, but the phycobiliproteins do not appear to be organized into phycobilisomes; instead, they are contained within the inner loculi of thylakoid membranes. Phycobilins are important because, on a global scale, light energy harvested by them is used to fix a substantial portion of the total carbon that is fixed in photosynthesis. Moreover, phytochrome, an important photomorphogenetic pigment that is found in plants and some algae, is a biliprotein whose chromophore closely resembles and shares biosynthetic features with phycobilins.

The first phycobilin for which the structure was determined was methanolysis-liberated phycocyanobilin (Fig. 1) (Cole et al 1967, Crespi et al 1967), followed closely by that of phycoerythrobilin (Chapman et al 1967, Crespi & Katz 1969). Structures of several additional phycobilins have been described, bringing the present total number of distinct chromophores to eight, four of which are known to occur both singly and doubly linked to apobiliproteins (Glazer & Hixson 1977, Bishop et al 1987, Wedemayer et al 1991, 1992). The pace of discovery of new phycobilins suggests that more structures will be uncovered as pigments of more organisms are examined in detail.

The field of phycobilin biosynthesis was the subject of a recent comprehensive review (Beale 1993). The following is a summary of recent work carried out primarily in the author's laboratory.

In vivo studies on phycobilin biosynthesis

Participation of haem in phycobilin formation is suggested by similarities of phycobilins to tetrapyrrole macrocycle ring-opening reaction products appearing in animal haem catabolism. Indirect evidence supporting the precursor status of haem was obtained by Beale & Chen (1983). N-Methylmesoporphyrin IX is a specific inhibitor of enzymic insertion of iron into protoporphyrin IX and therefore blocks haem formation (De Matteis et al 1980). Administration of N-methylmesoporphyrin IX to growing cells of the unicellular rhodophyte *Cyanidium caldarium* inhibited phycocyanin production, but not chlorophyll formation. This result suggested that haem is a direct precursor of phycobilins. The experiment was carried out in the dark using a mutant strain of *C. caldarium* that normally forms both pigments in the dark. In the light, synthesis of both phycocyanobilin and chlorophyll by wild-type cells was inhibited by N-methylprotoporphyrin IX (Brown et al 1982), but the effect on chlorophyll could have been an indirect one caused by the phototoxicity of the administered inhibitor, which is a photodynamically active porphyrin.

Direct evidence for the precursor status of haem was reported by Brown et al (1981) and Schuster et al (1983), who showed that exogenous [^{14}C]haem could contribute label to phycocyanobilin in greening *C. caldarium* cells (i.e., cells that are actively synthesizing and accumulating photosynthetic pigments). The specificity of haem incorporation was indicated by the fact that unlabelled chlorophyll was formed simultaneously with labelled phycocyanobilin in greening cells incubated with [^{14}C]haem. Further evidence supporting a role for haem was provided by the observation that non-radioactive haem decreased incorporation of ^{14}C-labelled 5-aminolaevulinic acid into phycocyanobilin (Brown et al 1981).

Beale & Cornejo (1983) found that the chromophores of phycocyanin became labelled when purified [^{14}C]biliverdin IXα was added to *C. caldarium* cells growing in the dark in the presence of N-methylmesoporphyrin IX, which was added to block endogenous haem formation. The strain of cells used in these experiments was capable of forming phycocyanin in the dark, so the experiment was done in the dark to eliminate possible phototoxic effects of administered N-methylmesoporphyrin or biliverdin. Cellular protohaem remained unlabelled, indicating that the incorporation of label into phycocyanobilin was direct, rather than via degradation of the administered labelled compound and subsequent re-utilization of the ^{14}C. Brown et al (1984) and Holroyd et al (1985) confirmed these results.

In vitro studies on phycobilin biosynthesis

By 1983, the involvement of haem and biliverdin in phycobilin biosynthesis had been established by the *in vivo* experiments described above. Cell-free extracts from phycobilin-forming organisms that are capable of catalysing biliverdin formation from haem, and phycocyanobilin formation from biliverdin, were first described in 1984 (Beale & Cornejo 1984a,b).

Algal haem oxygenase

Beale & Cornejo (1984b) detected haem oxygenase activity in extracts of *C. caldarium*. Like the animal system, the unfractionated algal haem oxygenase system required reduced pyridine nucleotide (NADPH was about twice as effective as NADH) as well as molecular oxygen. Ascorbate and other moderately strong reductants stimulated the reaction in unfractionated cell extracts, and were required after removal of low molecular mass materials from the enzyme system by gel filtration or dialysis. It is possible that these strong reductants reduce Fe of haem to the ferrous state, rather than serving as enzyme substrates (Cornejo & Beale 1988). Like the microsomal enzyme (Drummond & Kappas 1981), algal haem oxygenase is powerfully inhibited by Sn–protoporphyrin IX. In contrast, haem degradation via non-specific reactions, catalysed by haemoproteins such as myoglobin, is not inhibited by Sn–protoporphyrin (Cornejo & Beale 1988).

Originally, it was necessary to use mesohaem as a substrate in place of protohaem, the physiological substrate. The advantage of mesohaem is that the reaction product, mesobiliverdin, is not a natural product and can be distinguished from residual biliverdin that may be present in cell extracts. Moreover, because mesobiliverdin is not a substrate for phycobilin formation, it is more stable than biliverdin in cell extracts, and thus is more likely to accumulate to detectable levels during incubation. After the initial detection of haem oxygenase activity with mesohaem as a substrate, conversion of protohaem to biliverdin was also detected with optimized incubation conditions and partially purified enzymes (Cornejo & Beale 1988).

Products of the reaction with mesohaem and protohaem were identified as the IXα isomers of mesobiliverdin and biliverdin, respectively, indicating that the enzymic reaction specifically opens the macrocyclic ring at the same bridge carbon as does animal haem oxygenase (Beale & Cornejo 1984b, Cornejo & Beale 1988). Whereas the enzymic reaction exclusively produces IXα isomers, chemical attack on haems by ascorbate in aqueous pyridine produces a mixture of the four possible bilin isomers that can be formed by ring opening (Beale & Cornejo 1984b, Cornejo & Beale 1988), and coupled oxidation of myoglobin-bound haem also produces an isomer mixture (Beale & Cornejo 1984b).

The algal haem oxygenase system differs from the animal cell-derived microsomal system in that it is soluble, with virtually all activity appearing in the supernatant fraction of high speed-centrifuged cell homogenates (Beale & Cornejo 1984b). This finding is consistent with the fact that the reaction is thought to take place in plastids, and is also presumed to occur in prokaryotic cyanobacteria, because neither plastids nor prokaryotes have microsomes.

Microsomal heme oxygenase derived from animal cells requires two protein components for activity: haem oxygenase (EC 1.14.99.3) and NADPH–ferrihaemoprotein reductase (EC 1.6.2.4). In contrast, the soluble haem oxygenase system derived from C. caldarium can be separated (by fractionation) into three required protein components: ferredoxin-dependent NADPH–cytochrome-c reductase; a small Fe/S protein that appears to be ferredoxin; and a haem-binding oxygenase enzyme (Cornejo & Beale 1988). Reconstitution of haem oxygenase activity in vitro required all three protein components. Ferredoxin-dependent NADPH–cytochrome-c reductase could be replaced by spinach ferredoxin–NADP$^+$ reductase (EC 1.18.1.2), and ferredoxin could be replaced by commercial ferredoxin derived from spinach or the red alga Porphyra umbilicalis (Cornejo & Beale 1988, Rhie & Beale 1992).

A mechanistic question is whether the sole role of NADPH in the reaction is to serve as an electron source for ferredoxin reduction, or whether NADPH directly donates one or more electrons to haem oxygenase. To answer this question, Rhie & Beale (1992) assayed haem oxygenase activity in incubation mixtures containing C. caldarium haem oxygenase (separated from the other two protein components by fractionation) without added NADPH but supplemented with a light-driven ferredoxin reduction system derived from partially purified photosystem I from spinach leaves. In this mixed reconstitution assay, haem oxygenase activity was light dependent (Rhie & Beale 1992). In the dark, no activity was detected unless NADPH and ferredoxin–NADP$^+$ reductase were added to incubation mixtures. It can be concluded from these results that the sole, essential, role of NADPH and the ferredoxin-dependent NADPH–cytochrome-c reductase (ferredoxin–NADP$^+$ reductase) in the algal haem oxygenase system is to reduce ferredoxin, and that ferredoxin is the direct electron source for algal haem oxygenase.

It is likely that both photosystem I-derived and NADPH-derived electrons for ferredoxin reduction are important for phycobilin synthesis in vivo. In the light, ferredoxin is presumably reduced by the action of photosystem I. However, because the C. caldarium strain used for these studies can synthesize phycobilins in the dark as well as in the light (Beale & Chen 1983), the cells must be capable of reducing ferredoxin in the dark, using ferredoxin–NADP$^+$ reductase to transfer electrons from NADPH.

The haem-binding component of the algal haem oxygenase system has an apparent native relative molecular mass of approximately 38 000 and is resistant to inactivation by p-hydroxymercuribenzoate (Cornejo & Beale 1988). The enzyme is inactivated by diethylpyrocarbonate, and this inactivation is blocked by haem. In the reconstituted haem oxygenase system, the haem-binding component was the rate limiting one—addition of this component to unfractionated cell extracts increased the yield of biliverdin, whereas addition of ferredoxin or ferredoxin–NADP$^+$ reductase did not.

Transformation of biliverdin to phycocyanobilin

Because nearly all phycobilins that have been described so far contain at least two more hydrogen atoms than biliverdin, some form of biochemical reduction must be necessary for the transformation of biliverdin into these phycobilins. Although early attempts to detect biliverdin reduction activity in algal extracts were not successful (O'Carra & O'hEocha 1976), Beale & Cornejo (1984a) were able to measure enzymic conversion of biliverdin to free phycocyanobilin in extracts of *C. caldarium*. In addition to biliverdin IXα, the reaction required reduced pyridine nucleotide, NADPH being more effective than NADH. Activity was retained in the supernatant fraction after high-speed centrifugation and eluted with the protein fraction on gel filtration.

Incubation products included both the (3*Z*)- and the (3*E*)-ethylidine isomers of phycocyanobilin (Beale & Cornejo 1984a). Interestingly, both ethylidine isomers of phycocyanobilin are also formed upon methanolytic cleavage of the phycocyanin chromophore from the protein moiety (Fu et al 1979), but the *Z* isomer, being less stable (Weller & Gossauer 1980), isomerizes to the *E* isomer at the high temperatures at which methanolysis is carried out, and the equilibrium isomer ratio strongly favours the *E* form. The *Z*- and *E*-ethylidine isomers of phycocyanobilin can be interconverted by heating in acetic acid–methanol mixtures (Beale & Cornejo 1984a).

Preliminary evidence indicates that enzymic *cis–trans* isomerization of the ethylidine group is catalysed by *C. caldarium* cell extracts in the presence of reduced glutathione (GSH) (Beale & Cornejo 1991a). After removal of low molecular mass material from cell homogenates by Sephadex G-25 gel filtration, the major product formed enzymically was (3*Z*)-phycocyanobilin. Addition of GSH to incubations containing gel-filtered cell extracts restored the extracts' ability to form (3*E*)-phycocyanobilin. Moreover, purified (3*Z*)-phycocyanobilin was converted to (3*E*)-phycocyanobilin by cell extracts, and this conversion was greatly stimulated by GSH. The reverse conversion was not detected. In the absence of enzyme, GSH did not catalyse the conversion. These results indicate that extracts from *C. caldarium* cells contain a GSH-dependent enzyme activity that isomerizes the ethylidine group of (3*Z*)-phycocyanobilin to the *E* configuration.

This reaction falls into the category of *cis–trans* isomerizations that do not involve double bond migration (Seltzer 1972). Enzymes in this class of isomerases require sulphydryl groups for activity. One subclass of these enzymes uses integral enzyme sulphydryls, and a second subclass uses sulphydryls supplied by a cofactor (Seltzer 1972). Several members of the latter subclass, including maleate isomerase and several maleyl isomerases, specifically require GSH for activity (Knox 1960, Seltzer 1972). It is not known whether GSH or some other low molecular mass cofactor is the natural *in vivo* cofactor that is removed from *C. caldarium* cell extracts during purification of the protein fraction that catalyses ethylidine isomerization.

The existence of the (3Z)-phycobilins may be rationalized by proposing that the substrate for the pyrrole 2,3-reductase reaction is a bilin that has a 3-vinyl group and that the reaction produces, as the immediate product, a 3-vinyl-2,3-dihydrobilin. Gossauer et al (1989) have shown that synthetic 3-vinyl-2,3-dihydrobilins spontaneously isomerize to (3Z)-ethylidine-2,3-dihydrobilins. Thus, the enzymic pyrrole reduction of biliverdin or another 3-vinylbilin would initially yield the unstable intermediate 3-vinyl-2,3-dihydrobilin, which would isomerize non-enzymically to the (3Z)-ethylidine-2,3-dihydrobilin. However, if only the more stable (3E)-phycobilins are acceptable substrates for ligation of the chromophore to the apoprotein, there would be a need for a *Z–E* isomerase to transform the products of pyrrole ring reduction into the substrates for ligation.

In the course of fractionation and purification of proteins from *C. caldarium* extracts that catalyse transformation of biliverdin to phycocyanobilin, we observed that pigmented products were produced in our incubations in addition to phyco-cyanobilin. One of these products was identified as (3Z)-phycoerythrobilin by comparative absorption spectroscopy and HPLC elution, and the identification was confirmed by ^{1}H NMR (Beale & Cornejo 1991b). The occurrence of (3Z)-phycoerythrobilin as an incubation product was unexpected because *C. caldarium* does not contain phycoerythrin or other phycoerythrobilin-bearing phycobiliproteins. Further experiments indicated that *C. caldarium* extracts contain an enzyme that converts (3Z)-phycoerythrobilin to (3Z)-phycocyanobilin (Beale & Cornejo 1991b). The reverse reaction was not detected. The reaction does not require any substrate other than (3Z)-phycoerythrobilin. Because (3Z)-phycoerythrobilin and (3Z)-phycocyanobilin are isomeric, the enzyme can be considered to be an isomerase. In addition to being able to transform the (3Z) isomer, the protein fraction also catalysed conversion of (3E)-phycoerythrobilin to (3E)-phycocyanobilin. The enzyme activity was named phycoerythrobilin-to-phycocyanobilin (15,16-methylene-to-18^1,18^2-ethyl) isomerase. This unexpected conversion was anticipated by reports in the late 1960s that isolated phycoerthrobilin could be isomerized with 12 N hydrochloric acid to phycocyanobilin (Chapman et al 1967).

Proteins from cell extracts were partially fractionated by differential $(NH_4)_2SO_4$ precipitation, Blue-Sepharose affinity chromatography and

Sephadex G-75 gel filtration chromatography. The fraction containing proteins in the 30 000 to 40 000 relative molecular mass range could convert biliverdin to phycoerythrobilin but not to phycocyanobilin. The fraction containing proteins of relative molecular mass greater than 60 000 was inactive in biliverdin reduction, but catalysed isomerization of phycoerythrobilin to phycocyanobilin (Beale & Cornejo 1991b). These results suggest that phycoerythrobilin is an intermediate in the biosynthesis of phycocyanobilin from biliverdin.

Partially fractionated *C. caldarium* protein extract, when incubated with biliverdin IXα and reductant, produced, in addition to phycocyanobilin and phycoerythrobilin, a third pigment that was identified by comparative absorption spectrometry and ^{1}H NMR as 15,16-dihydrobiliverdin IXα (Beale & Cornejo 1991c). Further fractionation of the proteins by ferredoxin–Sepharose affinity chromatography yielded a fraction that formed 15,16-dihydrobiliverdin IXα as the sole product of biliverdin IXα reduction. Purified 15,16-dihydrobiliverdin IXα, when incubated with another protein fraction, was converted to (3Z)-phycoerythrobilin and (3Z)-phycocyanobilin. This conversion, as well as the conversion of biliverdin IXα to 15,16-dihydrobiliverdin IXα, required reductant in addition to the bilin substrate. These results suggest that 15,16-dihydrobiliverdin IXα is a partially reduced intermediate in the biosynthesis of phycoerythrobilin from biliverdin IXα. This finding was followed by the discovery that 15,16-dihydrobiliverdin functions as a phycobiliprotein chromophore (Wedemayer et al 1992).

Fractionation of proteins catalysing NADPH-dependent reduction of biliverdin to phycobilins in *C. caldarium* extracts revealed that two of the components are the same ones that are needed for the haem oxygenase system: ferredoxin and ferredoxin–NADP$^+$ reductase (Beale & Cornejo 1991a). As shown earlier for haem oxygenase (Cornejo & Beale 1988), these components could be replaced with commercial counterparts from spinach or a red alga. Also, as with haem oxygenase, the requirement for NADPH and ferredoxin–NADP$^+$ reductase could be supplanted by a light-driven ferredoxin-reducing system derived from spinach thylakoids (Rhie & Beale 1992). These results indicate that reduced ferredoxin is the only reductant needed for reduction of biliverdin to phycobilins.

Protein fractionation results also indicate that reduction of biliverdin to phycoerythrobilin proceeds by two two-electron steps, each of which is catalysed by a different enzyme. One enzyme, when supplied with biliverdin IXα plus a source of reduced ferredoxin, produces only 15,16-dihydrobiliverdin IXα (Beale & Cornejo 1991c). Further reduction of the bilin to phycoerythrobilin requires other proteins. The enzyme that catalyses the first reduction step can be separated from the other reductase by affinity chromatography on ferredoxin–Sepharose.

As indicated above, protein fractionation suggests that the reduction steps are catalysed by enzymes having apparent native relative molecular masses of

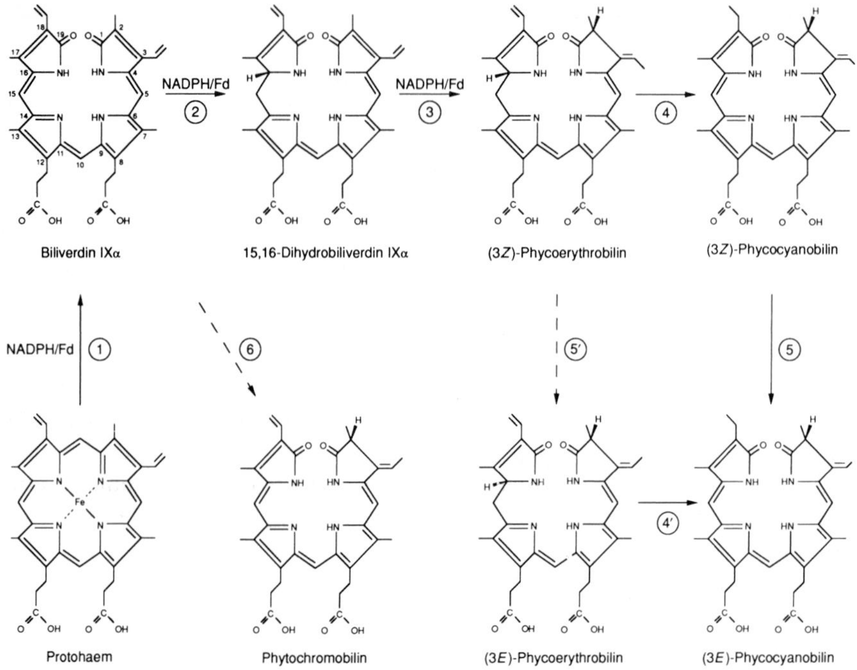

FIG. 2. The phycobilin biosynthetic pathway. Enzyme activities catalysing the reactions shown that have been characterized in *C. caldarium* extracts are: (1) haem oxygenase; (2) biliverdin IXα reductase; (3) 15,16-dihydrobiliverdin IXα reductase; (4) (3*Z*)-phycoerythrobilin-to-(3*Z*)-phycocyanobilin isomerase; and (5) phycobilin 3-ethylidine *cis–trans* isomerase. Isomerization of (3*E*)-phycoerythrobilin to (3*E*)-phycocyanobilin (step 4′) has also been detected *in vitro*. It is not known whether the ethylidine *cis–trans* isomerase can accept (3*Z*)-phycoerythrobilin as a substrate (step 5′). The conversion of biliverdin IXα to phytochromobilin (step 6) has not yet been demonstrated in cell-free or plastid-free extracts. The conventional bilin carbon numbering system is shown for biliverdin IXα.

about 30 000 to 40 000, and isomerization of phycoerythrobilin to phycocyanobilin is catalysed by an enzyme with an apparent native relative molecular mass greater than 60 000.

In summary, the biosynthesis of phycocyanobilin from haem in *C. caldarium* extracts is as follows: the transformation of haem to biliverdin IXα in a reaction catalysed by haem oxygenase is followed by two two-electron reductions, which yield, respectively, 15,16-dihydrobiliverdin IXα and phycoerythrobilin, which isomerizes into phycocyanobilin (Fig. 2). Each step is catalysed by a different enzyme. Both haem oxygenase and the bilin reduction steps are ferredoxin dependent. The extracts also contain an enzyme that isomerizes the 3-ethylidine group of phycoerythrobilin and phycocyanobilin from the *Z* configuration to the *E* configuration.

Relevance to phytochromobilin biosynthesis

The structure of the phytochrome chromophore closely resembles that of phycobilins (Smith & Kendrick 1976). Phytochromobilin is identical to phycocyanobilin except that the ethyl group is replaced by a vinyl group (Fig. 2). Phytochrome is present in higher and lower plants and at least some algae, including several green algae and the red algae *Porphyra tenera* (Dring 1974) and *Corallina elongata* (López-Figueroa et al 1989a,b). Elich & Lagarias (1987) obtained evidence that the phytochrome chromophore, like phycocyanobilin, can be synthesized *in vivo* from exogenously supplied biliverdin. These workers reported that phytochrome levels are substantially reduced in oat seedlings germinated in the presence of gabaculine (3-amino-2,3-dihydrobenzoic acid), a potent, specific inhibitor of 5-aminolaevulinic acid formation via the tRNA-dependent five-carbon pathway. When either 5-aminolaevulinic acid or biliverdin was administered to seedlings grown in the presence of gabaculine, there was a rapid increase in spectrophotometrically detected phytochrome. Label from [^{14}C]biliverdin was specifically incorporated into the phytochrome chromophore in oat leaves, thus establishing a clear biosynthetic link between this chromophore and phycobilins (Elich et al 1989). More recently, phytochrome-deficient mutant strains of *Arabidopsis thaliana* that are defective in chromophore biosynthesis have been shown to synthesize functional, phototransformable phytochrome from exogenous biliverdin IXα (Parks & Quail 1991). Thus, the experimental evidence indicates that phytochromobilin can be considered to be one of the end products of the general phycobilin biosynthetic pathway. It can be hypothesized that, in organisms that synthesize phytochromobilin, the pyrrole 2,3-reductase enzyme is able to accept biliverdin IXα as substrate, whereas in species that synthesize phycobilins but not phytochromobilin, this enzyme instead accepts 15,16-dihydrobiliverdin IXα as substrate.

It was recently discovered that phytochromobilin can be obtained from the unicellular red alga, *Porphyridium cruentum*, even though the alga does not contain detectable phytochrome (Cornejo et al 1992). The apparent source of the phytochromobilin is phycoerythrin chromophores, which can be oxidized at the 15,16-methine bridge while the reduced pyrrole ring remains protected from oxidation by the attachment to the protein. Subsequent cleavage of the protein–chromophore thioether link by methanolysis and solvent partitioning, DEAE–Sepharose ion-exchange chromatography and HPLC yielded purified (2R,3E)-phytochromobilin, whose structure was verified by comparative absorption spectroscopy, circular dichroism and ^{1}H NMR spectroscopy (Cornejo et al 1992). Phycoerythrin-derived phytochromobilin undergoes spontaneous covalent condensation with recombinant oat apophytochrome to yield photoreversible phytochrome that has spectroscopic properties identical to those of native oat phytochrome. The availability of relatively large quantities of phycoerythrin-derived phytochromobilin will aid future studies of phytochrome biosynthesis.

Conclusion

The results described here are beginning to shed light on the last major branch of tetrapyrrole biosynthesis to yield to experimental investigation. Now that the outline of phycobilin biosynthesis has been sketched and the principal biosynthetic intermediates and enzymes have been identified, future experiments will fill in important gaps in our understanding of enzyme properties, catalytic mechanisms and modes of metabolic regulation.

References

Beale SI 1993 Biosynthesis of phycobilins. Chem Rev 93:785–802

Beale SI, Chen NC 1983 N-Methyl mesoporphyrin IX inhibits phycocyanin, but not chlorophyll synthesis in Cyanidium caldarium. Plant Physiol 71:263–268

Beale SI, Cornejo J 1983 Biosynthesis of phycocyanobilin from exogenous labeled biliverdin in Cyanidium caldarium. Arch Biochem Biophys 227:279–286

Beale SI, Cornejo J 1984a Enzymic transformation of biliverdin to phycocyanobilin by extracts of the unicellular red alga, Cyanidium caldarium. Plant Physiol 76:7–15

Beale SI, Cornejo J 1984b Enzymatic haem oxygenase activity in soluble extracts of the unicellular red alga, Cyanidium caldarium. Arch Biochem Biophys 235:371–384

Beale SI, Cornejo J 1991a Biosynthesis of phycobilins: ferredoxin-mediated reduction of biliverdin catalyzed by extracts of Cyanidium caldarium. J Biol Chem 266:22328–22332

Beale SI, Cornejo J 1991b Biosynthesis of phycobilins: 3(Z)-phycoerythrobilin and 3(Z)-phycocyanobilin are intermediates in the formation of 3(E)-phycocyanobilin from biliverdin IXα. J Biol Chem 266:22333–22340

Beale SI, Cornejo J 1991c Biosynthesis of phycobilins: 15,16-dihydrobiliverdin IXα is a partially reduced intermediate in the formation of phycobilins from biliverdin IXα. J Biol Chem 266:22341–22345

Bishop JE, Rapoport H, Klotz AV et al 1987 Chromopeptides from phycoerythrocyanin: structure and linkage of the three bilin groups. J Am Chem Soc 109:875–881

Brown SB, Holroyd JA, Troxler RF, Offner GD 1981 Bile pigment synthesis in plants: incorporation of haem into phycocyanobilin and phycobiliproteins in Cyanidium caldarium. Biochem J 194:137–147

Brown SB, Holroyd JA, Vernon DI, Troxler RF, Smith KM 1982 The effect of N-methyl-protoporphyrin IX on the synthesis of photosynthetic pigments in Cyanidium caldarium: further evidence for the role of haem in the biosynthesis of plant bilins. Biochem J 208:487–491

Brown SB, Holroyd JA, Vernon DI 1984 Biosynthesis of phycobiliproteins: incorporation of biliverdin into phycocyanin of the red alga Cyanidium caldarium. Biochem J 219:905–909

Chapman DJ, Cole WJ, Siegelman HW 1967 The structure of phycoerythrobilin. J Am Chem Soc 89:5976–5977

Cole WJ, Chapman DJ, Siegelman HW 1967 The structure of phycocyanobilin. J Am Chem Soc 89:3643–3645

Cornejo J, Beale SI 1988 Algal heme oxygenase from *Cyanidium caldarium*: partial purification and fractionation into three required protein components. J Biol Chem 263:11915–11921

Cornejo J, Beale SI, Terry MJ, Lagarias JC 1992 Phytochrome assembly: the structure and biological activity of 2(R),3(*E*)-phytochromobilin derived from phycobiliproteins. J Biol Chem 67:14790–14798

Crespi HL, Katz JJ 1969 Exchangeable hydrogen in phycoerythrobilin. Phytochemistry 8:759–761

Crespi HL, Boucher LJ, Norman GD, Katz JJ, Dougherty RC 1967 Structure of phycocyanobilin. J Am Chem Soc 89:3642–3643

De Matteis F, Gibbs AH, Smith AG 1980 Inhibition of protohaem ferro-lyase by *N*-substituted porphyrins: structural requirements for the inhibitory effect. Biochem J 189:645–648

Dring MJ 1974 Reproduction. In: Stewart WDP (ed) Algal physiology and biochemistry. University of California Press, Berkeley, CA, p 814–837

Drummond GS, Kappas A 1981 Prevention of neonatal hyperbilirubinemia by tin protoporphyrin IX, a potent competitive inhibitor of heme oxidation. Proc Natl Acad Sci USA 78:6466–6470

Elich TD, Lagarias JC 1987 Phytochrome chromophore biosynthesis: both 5-aminolevulinic acid and biliverdin overcome inhibition by gabaculine in etiolated *Avena sativa* L. seedlings. Plant Physiol 84:304–310

Elich TD, McDonagh AF, Palma LA, Lagarias JC 1989 Phytochrome chromophore biosynthesis: treatment of tetrapyrrole-deficient *Avena* explants with natural and non-natural bilatrienes leads to formation of spectrally active holoproteins. J Biol Chem 264:183–189

Fu E, Friedman L, Siegelman HW 1979 Mass-spectral identification and purification of phycoerythrobilin and phycocyanobilin. Biochem J 179:1–6

Glazer AN, Hixson CS 1977 Subunit structure and chromophore composition of Rhodophytan phycoerythrins: *Porphyridium cruentum* B-phycoerythrin and b-phycoerythrin. J Biol Chem 252:32–42

Gossauer A, Nydegger F, Benedikt E, Köst H-P 1989 Synthesis of bile pigments. XVI. Synthesis of a vinyl substituted 2,3-dihydrobilindione: possible role of this new class of bile pigments in phycobilin biosynthesis. Helv Chim Acta 72:518–529

Holroyd JA, Vernon DI, Brown SB 1985 Biliverdin, an intermediate in the biosynthesis of plant pigments. Biochem Soc Trans 13:209–210

Knox WE 1960 Glutathione. In: Boyer PD, Lardy H, Myrbäck K (eds) The enzymes. Academic Press, New York, vol 2A:253–294

López-Figueroa F, Lindemann P, Braslavsky SE, Schaffner K, Schneider-Poetsch HAW, Rüdiger W 1989a Detection of a phytochrome-like protein in macroalgae. Bot Acta 102:178–180

López-Figueroa F, Perez R, Niell FX 1989b Effects of red and far-red light pulses on the chlorophyll and biliprotein accumulation in the red alga *Corallina elongata*. J Photochem Photobiol B Biol 4:185–193

O'Carra P, O'hEocha C 1976 Algal biliproteins and phycobilins. In: Goodwin TW (ed) Chemistry and biochemistry of plant pigments, 2nd edn. Academic Press, New York, vol 1:328–376

Parks BM, Quail PH 1991 Phytochrome-deficient hy1 and hy2 long hypocotyl mutants of *Arabidopsis* are defective in phytochrome chromophore biosynthesis. Plant Cell 3:1177–1186

Rhie G, Beale SI 1992 Biosynthesis of phycobilins: ferredoxin-supported, NADPH-independent heme oxygenase and phycobilin-forming activities from *Cyanidium caldarium*. J Biol Chem 67:16088–16093

Schuster A, Köst H-P, Rüdiger W, Holroyd JA, Brown SB 1983 Incorporation of haem into phycocyanobilin in levulinic acid treated *Cyanidium caldarium*. Plant Cell Rep 2:85–87

Seltzer S 1972 Cis-trans isomerization. In: Boyer PD (ed) The enzymes, 3rd edn. Academic Press, New York vol 6:381–406

Smith H, Kendrick RE 1976 The structure and properties of phytochrome. In: Goodwin TW (ed) Chemistry and biochemistry of plant pigments, 2nd edn. Academic Press, New York, vol 1:377–424

Wedemayer GJ, Wemmer DE, Glazer AN !991 Phycobilins of Cryptophytan algae: structures of novel bilins with acryloyl substituents from phycoerythrin 566. J Biol Chem 266:4731–4741

Wedemayer GJ, Kidd DG, Wemmer DE, Glazer AN 1992 Phycobilins of Cryptophytan algae: occurrence of dihydrobiliverdin and mesobiliverdin in cryptomonad biliproteins. J Biol Chem 267:7315–7331

Weller J-P, Gossauer A 1980 Synthesen von Gallenfarbstoffen X: Synthese und Photoisomerisierung des racem. Phytochromobilin-dimethylesters. Chem Ber 113:1603–1611

DISCUSSION

Scott: Is there an enzyme that connects the protein with the vinylic side chain, or is this spontaneous in the same way as the cysteine residue we talked about earlier (Cys-242, p 90) in reconstituted porphobilinogen deaminase reacts autocatalytically with porphobilinogen?

Beale: In phytochrome, the apoprotein appears to condense with phyto-chromobilin spontaneously. With the phycobiliproteins, however, there is a different set of results, mostly from Glazer's lab (Fairchild et al 1992). In cyanobacteria specific gene products are required for each condensation. The only one that has been studied in any detail is the one involved in the specific attachment of phycocyanobilin to the α-c-phycocyanin site. Several proteins seem to be required for full conjugation of phycobiliproteins.

Timkovich: All types of cytochrome *c* have the amino acid cysteine added across a vinyl group at that position on the tetrapyrrole, and the process requires an elaborate enzyme system. I'm surprised that there is a non-enzymic way.

Beale: Phytochrome is a very large protein which might carry its own conjugating enzyme.

Kräutler: Did I understand correctly that some of the phycobilins are attached to the protein by two linkages?

Beale: Some of the phycobilins—not those containing the ethyl precursor, and only some of those containing vinyl precursors, such as phycoerythrobilin—are attached at this site as well as the ethylidine site.

Leeper: Chemically, one would expect glutathione to work in the same way, by adding to the double bond first before being eliminated, which is how the *E*-to-*Z* isomerization could be brought about. I can't see why an enzyme is necessary.

Beale: The process requires an enzyme. Glutathione alone is not sufficient.

K. Smith: Vinylporphyrins (**1**, Scheme 1) can be photoreduced in the vinyl-containing ring to give a reduced ring with a vinyl group (**2**). Non-enzymically, the *E* isomer is always formed, whereas in your work you get the *Z* form which is then converted to the *E* form. Chemically, the thermodynamically favoured isomer is the *E* form.

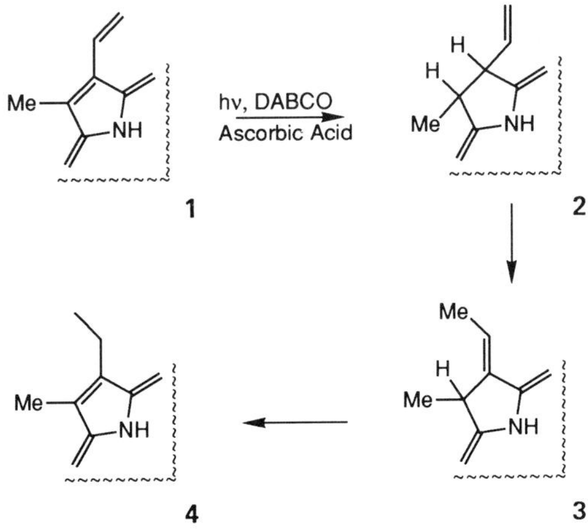

Scheme 1 (*K. Smith*)

Beale: That's been demonstrated, and it's also been shown that you can photochemically isomerize *Z* to *E* in a non-aqueous solvent using the methyl esters of phycocyanobilin.

K. Smith: That also supports your finding that an enzyme is required to get the *Z* form first.

Beale: That is not necessarily true, because in the photoreduction a lot of energy is being put in and two reactions may be being carried out at once.

K. Smith: You can isolate the vinylchlorin (**2**) and watch its conversion to the ethylidene chlorin (**3**) with clean isobestic points. It goes only to the *E* form, then migrates again to give the ethyl unsaturated porphyrin (**4**).

Beale: However, the model vinylchlorins studied by Gossauer et al (1989) isomerized to the *Z*-ethylidine isomer rather than the *E* form.

Arigoni: What is the difference in energy between the *Z* and *E* isomers?

Beale: Spectrophotometry suggests the energy difference is slight. There must be some large-scale structural, conformational, differences between the isomers, however, because they are easily separable by HPLC.

Griffiths: Does the purified oxygenase take Mg–protoporphyrin IX?

Beale: No, nor does it use protoporphyrin. It does not use cobalt–haem either, and we are interested to know whether it can be made to use cobalt–haem if we provide a strong reductant, because there have been reports that mammalian haem oxygenase can use cobalt–protoporphyrin as a substrate under some conditions (Maines & Kappas 1977, Vernon & Brown 1984).

Akhtar: There are many enzymes which catalyse oxygenase-type reactions, such as collagen hydroxylase, which carry out the hydroxylation process in the Fe(II) resting state. However, ascorbate is required for the enzyme to stay in the Fe(II) form. This seems to be a peculiarity of a large group of enzymes.

Beale: A requirement for ascorbate *in vitro* may simply be a reflection of the fact that the enzyme normally exists in an environment, the chloroplast stroma, which is sufficiently reducing to keep the iron reduced.

Akhtar: Ascorbate is not truly enzymically involved. Once the Fe(III) is converted to Fe(II), one assumes that ascorbate is irrelevant until the enzyme has to be reduced again.

A. Smith: Dr Beale, could you speculate on phytochrome synthesis, in particular on how the two compounds, phytochrome and phycobilin, might have evolved and whether they are connected in any way.

Beale: There are two two-electron reduction steps leading from biliverdin to phycocyanobilin. If you directly reduce the biliverdin pyrrole, you end up with phytochromobilin in a one-step reaction. It seems eminently reasonable that that's the way phytochromobilin synthesis would occur. No enzyme capable of that has yet been isolated, but biliverdin is known to be the precursor of the phytochrome chromophore. Parks & Quail (1991) studied *Arabidopsis* mutants that are unable to make the phytochrome chromophore. These mutants have elongated hypocotyls because they cannot sense light and keep growing vertically. If you spray biliverdin onto these plantlets they will undergo a phytochrome response, which shows that they can produce phytochrome from biliverdin.

Ilag: Euglena does not seem to have phytochrome, yet it responds to light. What do you think its functional equivalent of phytochrome is?

Beale: Euglena has not been reported to contain phytochrome, but it is known to have two, perhaps three, photosensing systems. One is based on proto-chlorophyllide reductase, with sensitivity in the red and blue region, and the other is a cytoplasmically localized blue light photoreceptor.

Ilag: Do you think any of its phycobilins serve the same function as phytochrome, acting as a photomorphogenetic pigment?

Beale: Apparently not. Neither of those systems has an action spectrum that matches any phycobilin spectra.

Arigoni: Is there any consensus on the mechanism responsible for the extrusion of the *meso* position as carbon monoxide?

Beale: You can imagine haem oxygenase working by three rounds of a similar reaction wherein ferrohaem binds oxygen and accepts an electron to produce a hydroperoxide that disproportionates to a ferric superoxide anion radical, which attacks the bridge region (Beale 1993). The first round would generate a hydroxyl group at the α *meso* position. The second would generate what's called a verdohaem, in which there is an oxygen bridge in place of the bridge carbon. CO is liberated at this stage.

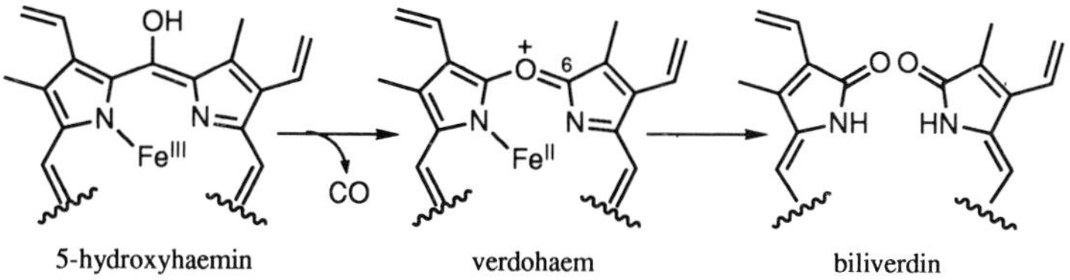

Scheme 2 (*Leeper*)

Leeper: The mechanism is very complicated and I don't think it is known. The 5-hydroxyhaemin is definitely an intermediate (Scheme 2; Yoshinaga et al 1990). The next established intermediate, verdohaem, has oxygen at position 5. If the oxygen is positively charged it is an aromatic system; alternatively, water addition at C-6 produces a neutral but non-aromatic compound.

Arigoni: What happens between verdohaem and biliverdin?

Leeper: You would think that verdohaem could be simply hydrolysed to biliverdin, but the enzyme actually cleaves it oxidatively in some way.

Beale: Each of the lactam oxygens in biliverdin, and the oxygen of CO, are derived from different O_2 molecules. The most detailed proposals for this mechanism are given by Sano et al (1986), Leeper (1987) and Beale (1993).

Chang: The CO extrusion step is not necessarily enzymically catalysed. If you take a hydroxyporphyrin with Fe(II) and expose it to air, you instaneously get verdohaem. It is known that *meso*-hydroxyporphyrin has a tendency to delocalize spin density on the pyrrole α-position, which might react with oxygen spontaneously.

References

Beale SI 1993 Biosynthesis of phycobilins. Chem Rev 93:785–802

Fairchild CD, Zhao J, Zhou J, Colson SE, Bryant DA, Glazer AN 1992 Phycocyanin α-subunit phycocyanobilin lyase. Proc Natl Acad Sci USA 89:7017–7021

Gossauer A, Nydegger F, Benedikt E, Köst H-P 1989 Synthesis of bile pigments. XVI. Synthesis of a vinyl substituted 2,3-dihydrobilindione: possible role of this new class of bile pigments in phycobilin biosynthesis. Helv Chim Acta 72:518–529

Leeper FJ 1987 The biosynthesis of porphyrins, chlorophylls and vitamin B_{12}. Nat Prod Rep 4:441–468

Maines MD, Kappas A 1977 Enzymatic oxidation of cobalt protoporphyrin IX: observations on the mechanism of heme oxygenase action. Biochemistry 16:419–423

Parks BM, Quail PH 1991 Phytochrome-deficient hy1 and hy2 long hypocotyl mutants of *Arabidopsis* are defective in phytochrome chromophore biosynthesis. Plant Cell 3:1177–1186

Sano S, Sano T, Morishima I, Shiro Y, Maeda Y 1986 On the mechanism of the chemical and enzymic oxygenations of α-oxyprotohaemin IX to Fe-biliverdin IXα. Proc Natl Acad Sci USA 83:531–535

Vernon DI, Brown SB 1984 Formation of bile pigments by coupled oxidation of cobalt-susbtituted haemoglobin and myoglobin. Biochem J 223:205–209

Yoshinaga T, Sudo Y, Sano S 1990 Enzymic conversion of α-oxyprotohaem IX into biliverdin IXα by haem oxygenase. Biochem J 270:659–664

General discussion I

Chlorophyll degradation

Arigoni: As we all know, when autumn comes, leaves turn yellow and red. What happens to the chlorophyll? Dr Kräutler has a partial answer to that question.

Kräutler: This work is the result of a collaboration with Professor Ph. Matile and his group at the Institute of Plant Biology at the University of Zürich. They isolated a colourless material from senescing plants which appeared to be derived from a porphyrin, presumably chlorophyll. This colourless material, which was first isolated from barley, *Hordeum vulgare*, turned red on exposure to air when they attempted to purify it by chromatography on silica gel plates. It was therefore called a 'rusty pigment' originally.

In the UV spectrum of the colourless material, the major absorption maximum is at 315 nm. The material is optically active. By FAB spectroscopy, we determined the molecular mass and found the molecular formula to be $C_{35}H_{41}N_4O_{10}K$. Using about 100 g senescent barley primary leaves (aged by being kept in the dark for seven days) and extensive HPLC chromatography, we isolated about 10 mg of the pure catabolite RP-14 (*Hv*-NCC-1). With the help of a series of one- and two-dimensional NMR spectra, using various techniques for homonuclear and heteronuclear correlations, we were able to establish the constitution of the catabolite as that of a 1-formyl-19-oxobilane, a linear tetrapyrrole (Kräutler et al 1991, 1992), as shown in Fig. 1. Comparison with the structure of chlorophyllide *a*, derived from chlorophyll *a* by dephytylation (see Fig. 1), shows that the chromophore of this catabolite is characterized by loss of the magnesium ion, opening of the macrocyclic ring at the C-5 *meso* position and saturation at the other *meso* positions.

Recent work by Professor Matile's group has revealed the presence of colourless and presumably structurally related chlorophyll catabolites in the cotyledons of rape (*Brassica napus*) that become senescent under natural growth conditions and under photoperiodic conditions. A similar analytical strategy to that described above enabled us to determine the molecular formula, $C_{37}H_{37}N_4O_{11}K_3$, and constitution of the colourless catabolite *Bn*-NCC-1 from the cotyledons of this plant. The compound again proved to be a 1-formyl-19-oxobilane, with the same basic skeleton as but with a different pattern of substituents from *Hv*-NCC-1 (Fig. 2; Mühlecker et al 1993).

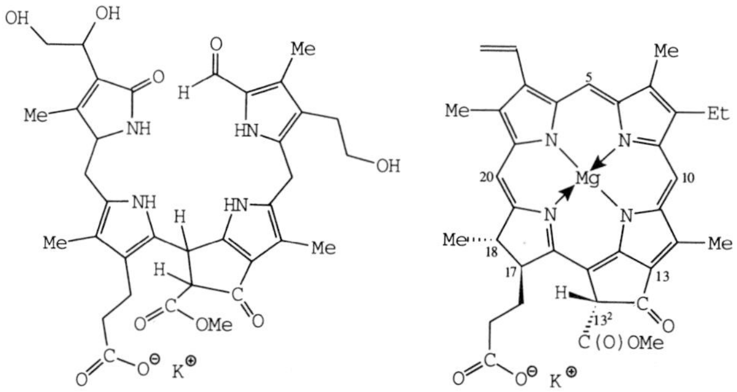

FIG. 1. (*Kräutler*) Constitutional formula of *Hv*-NCC-1 (*left*), the chlorophyll catabolite from barley (*Hordeum vulgaris*), and the structural formula of chlorophyllide *a* (*right*).

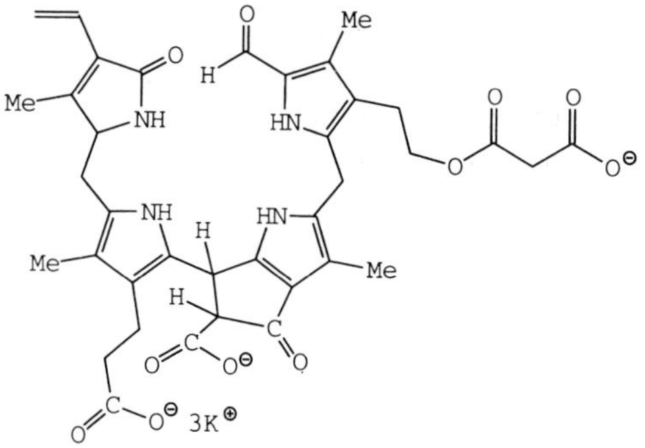

FIG. 2. (*Kräutler*) Constitutional formula of *Bn*-NCC-1, the chlorophyll catabolite from rape (*Brassica napus*).

Beale: There have been several reports that, among the products that are found in green or greening algal cells put back into the dark, in addition to pheophorbide (chlorophyll minus the magnesium and phytol), which would be a precursor to your compound, pyropheophytin and pyropheophorbide are found (Schoch et al 1981, Ziegler et al 1988). The pyro compound lacks the methoxycarbonyl moiety, yet your compound has that moiety. Where does that place pyro compounds in relation to your compound on the chlorophyll degradation pathway?

Kräutler: That is a difficult question. More reactions are needed to arrive from chlorophyll *a* at one of the pyropheophorbide degradation products than at the corresponding chlorophyll degradation products which still have that functional group.

Eschenmoser: Is the recovery of the magnesium the *raison d'être* for this pathway?

Kräutler: When naturally ageing plants lose their leaves the leaves still contain more than 50% of their original Mg, but about 80% of the leaves' nitrogen is relocated. I am told that the *raison d'être* behind chlorophyll breakdown is the rapid destruction of a potential photosensitizing agent that can produce, in the presence of light and oxygen, a non-specific cell poison, singlet oxygen, which would disturb the natural enzyme-catalysed processes of senescence. The purpose of natural senescence of leaves is for valuable nutrients, nitrogen in particular, to be taken from proteins and other constituents of the leaves into that part of the plant which will survive into the next year.

Spencer: Does the degradation of chlorophyll stop once the ring is opened?

Kräutler: No; these catabolites are more complex than simply chlorophyll with the ring opened. The lack of colour is associated with the saturation of the *meso* positions. If you simply opened the porphyrin macrocycle, by oxygen insertion, for example, you would obtain coloured compounds; the chromophore system must be further interrupted to get colourless compounds.

Spencer: Do you think that compound is the end point?

Kräutler: We don't know. After about five days, the concentration of RP-14 (*Hv*-NCC-1) in these artificially senescing plant leaves starts to decrease slowly.

Griffiths: Do you place any significance on the fact that the in the degradation product the C-20 *meso* position is reduced whereas ring D is oxidized? During chlorophyll formation the reduction of ring D is thought to be mediated by initial reduction at C-20 followed by migration of hydrogen to ring D, i.e., an apparent reversal of the breakdown process.

Kräutler: We haven't placed any significance on that particular aspect. The interruption of the chromophore at C-20 is the reason why it is colourless.

Rebeiz: What is the yield of this degradation product in the light? You are inducing senescence by seven days of darkness, which is not a natural phenomenon; most senescence takes place under photoperiodic conditions. Does the intermediate accumulate if you subject the tissue to light? Under natural conditions light may have important effects on chlorophyll degradation that you would miss in your system.

Kräutler: Of course, these leaves are aged in an artificial way, which is why we were interested in getting similar catabolites of chlorophyll from rape cotyledons, which lose their colour under natural growth conditions. These senescent cotyledons do contain large quantities of the chlorophyll catabolite, *Bn*-NCC-1.

Evolution of the series III porphyrinogens

Rebeiz: Because the world experts on porphyrin chemistry are here, I am tempted to ask the following question. Finian Leeper has shown us that Nature has gone to great lengths to flip ring D to end up with coproporphyrinogen III, instead of coproporphyrinogen I (Leeper 1994, this volume). Why has the type III series evolved rather than the type I series?

Eschenmoser: This is a classic question and an important one. I remember that this question was also asked at the last Ciba Foundation Symposium I attended, *Further perspectives in organic chemistry* in 1977, where it was addressed by R.B. Woodward (Battersby 1978). Of the four constitutional isomers of uroporphyrinogen, type III is the most stable (statistically preferred) isomer thermodynamically, representing 50% of the equilibrium mixture of the four. If porphinoids became biologically functional, that is, were engaged by enzymes to fulfil a biological function, before their biosynthesis had evolved, it is reasonable (though no more than reasonable) to assume that the adaptation of the functional enzymes referred to the isomer which was most abundant. If this were so, the evolving biosynthetic pathway would have had to stick to type III.

Rebeiz: Asymmetry is important for vectorial electron transport in photosynthesis.

Arigoni: We are mixing up two different things.

Rebeiz: No we aren't. Coproporphyrinogen III is much more asymmetric than coproporphyrinogen I and generates a more asymmetric chlorophyll molecule which then has to fulfil an asymmetric function in electron transport.

Arigoni: I do not follow your argument. If type I had evolved into a chlorophyll this would be as asymmetric as the type III-derived chlorophyll.

Rebeiz: Would it? I don't think so, as a more symmetrical coproporphyrinogen I may yield a more symmetrical type I chlorophyll.

K. Smith: The type III isomer is the one in which all of the substituents are uniquely different. Once Nature had chosen type III, because that was the predominant form, and then developed its enzymes and biosynthetic pathways to produce that type, it would be difficult for type I to be used because it is symmetrical and all the enzymes have been generated to be asymmetric.

Arigoni: There is something rather uncanny about the idea that the proteins were adapted to the pre-existing, low molecular mass building blocks.

Castelfranco: Are there any examples of natural type I tetrapyrroles?

Battersby: Uroporphyrin I is present in the shell of the oyster, *Pinctada vulgaris* (Comfort 1950).

Scott: There is a species of fox squirrel, *Sciurus niger*, which suffers from porphyria, that lives with type I tetrapyrroles, mostly (Levin 1975). It manages somehow. If you reconstitute haem with type I porphyrin it still works. There is also a snail that has almost entirely type I tetrapyrroles. I don't know what its blood is like or what it does with its excess uroporphyrin I.

Castelfranco: Are there any examples from clinically known porphyrias?

Elder: There are type I tetrapyrroles but they are formed as by-products from hydroxymethylbilane. In congenital erythropoietic porphyria, the large amounts of uroporphyrinogen I which are produced presumably come non-enzymically from accumulated hydroxymethylbilane.

Jordan: Surely these type I forms are merely a small amount that is formed spuriously. I can't believe that all the enzymes in these organisms use type I. They must have type III porphyrins as the physiologically important isomers.

Arigoni: Ian Scott's work has shown that some enzymes which normally specialize in type III can be fooled into taking type I.

Rebeiz: The reaction is usually slower.

Scott: Type I tetrapyrroles do in fact occur in Nature. The coupling of porphobilinogen deaminase and cosynthase is somewhat leaky, so a small amount of type I uroporphyrinogen is always formed, even in humans. In *Propionibacterium shermanii* there are at least five compounds derived from uroporphyrinogen I in the natural lifestyle. These are heavily processed, with extra methyl groups and so on. It is interesting that uroporpyrinogen I is a much better substrate for the methyltransferases than the 'natural' uroporphyrinogen III, which is inhibitory at high concentration.

Jordan: Type I porphyrins do not get beyond the coproporphyrinogen oxidase stage in haem biosynthesis.

Scott: These type I compounds are not decarboxylated, i.e., towards haem, but are related to sirohaem and B_{12}, containing up to four extra C-methyl groups.

Rebeiz: This discussion suggests there are no thermodynamic advantages for the enzymes in handling type III rather than type I tetrapyrroles.

Arigoni: There is certainly an entropic advantage.

References

Battersby AR 1978 Ideas and experiments in biosynthesis. In: Further perspectives in organic chemistry. Elsevier Science Publishers, Amsterdam (Ciba Found Symp 53) p 25–51

Comfort A 1950 Molluscan shells as a practical source of uroporphyrin I. Science 112:279–280

Kräutler B, Jaun B, Bortlik K, Schellenberg M, Matile Ph 1991 On the enigma of chlorophyll degradation: the constitution of a secoporphinoid catabolite. Angew Chem Int Ed Engl 30:1315–1318

Kräutler B, Jaun B, Amrein W, Bortlick K, Schellenberg M, Matile P 1992 Breakdown of chlorophyll: constitution of a secoporphinoid chlorophyll catabolite isolated from senescent barley leaves. Plant Physiol Biochem 30:333–346

Leeper FJ 1994 The evidence for a spirocyclic intermediate in the formation of uroporphyrinogen III by cosynthase. In: The biosynthesis of the tetrapyrrole pigments. Wiley, Chichester (Ciba Found Symp 180) p 111–130

Levin EY 1975 Porphyria in man and animals. Ann NY Acad Sci 244:481–495

Mühlecker W, Kräutler B, Ginsburg S, Matile Ph 1993 Breakdown of chlorophyll: a tetrapyrrolic chlorophyll catabolite from senescent rape leaves. Helv Chim Acta 76:2976–2980

Schoch S, Scheer H, Schiff JA, Rüdiger W, Siegelman HW 1981 Pyropheophytin *a* accompanies pheophytin *a* in darkened light grown cells of *Euglena*. Z Naturforsch Sect C Biosci 36:827–833

Ziegler R, Blaheta A, Guha N, Schönegge B 1988 Enzymatic formation of pheophorbide and pyropheophorbide during chlorophyll degradation in a mutant *Chlorella fusca* Shihira et Kraus. J Plant Physiol 132:327–332

Chlorophyll *a* biosynthetic heterogeneity

Constantin A. Rebeiz, Ramin Parham, Dionysia A. Fasoula and Ioannis M. Ioannides

Laboratory of Plant Pigment Biochemistry and Photobiology, University of Illinois, 1201 West Gregory Street, Urbana, IL-61801-3838, USA

Abstract. Chlorophyll *a* biosynthesis is presently interpreted in terms of two different biochemical pathways. According to one pathway, chlorophyll *a* is made via a single linear chain of reactions starting with divinylprotoporphyrin IX and ending with monovinylchlorophyll *a*. The experimental evidence for this pathway is marred by incompletely characterized intermediates that were detected in *Chlorella* mutants. The second pathway considers chlorophyll *a* to be made via multiple and parallel biosynthetic routes that result in the formation and accumulation of monovinyl- and divinylchlorophyll *a* chemical species. Two of these routes, namely the di/monocarboxylic monovinyl and divinyl routes, are responsible for the biosynthesis of most of the chlorophyll *a* in green plants. The experimental evidence for these two routes consists of: (a) the detection and spectroscopic characterization of intermediates and end products; (b) the demonstration of precursor–product relationships between various intermediates *in vivo* and *in vitro*; and (c) the detection of 4-vinylreductases that appear to be mainly responsible for the observed biosynthetic heterogeneity. The biological significance of chlorophyll *a* biosynthetic heterogeneity is becoming better understood. On the basis of the prevalence of the di/monocarboxylic monovinyl- and divinylchlorophyll *a* biosynthetic routes, green plants have been classified into three different greening groups. It now appears that the major chlorophylls in the euphotic zone of tropical waters are divinylchlorophyll *a* and *b*. It also appears that the di/monocarboxylic monovinyl and divinyl biosynthetic routes lead to the formation of different pigment proteins in different greening groups of plants, and that the more highly evolved monovinylchlorophyll *a* biosynthetic route is associated with higher field productivity in wheat.

1994 The biosynthesis of the tetrapyrrole pigments. Wiley, Chichester (Ciba Foundation Symposium 180) p 177–193

Tetrapyrroles are natural products which are essential for organic life in the biosphere. Haems are involved in oxygen and electron transport, chlorophylls and bacteriochlorophylls catalyse the conversion of solar energy to chemical energy, and vitamin B_{12} participates in important biochemical reactions. As a consequence, tetrapyrrole chemistry and biochemistry have attracted research talent for over a hundred years. Our focus here is on the biosynthesis of chlorophyll *a*, but we shall also explore the experimental basis of what constitutes a biosynthetic pathway. Current experimental evidence is no longer compatible with that which served as a basis for the conventional linear chlorophyll *a* biosynthetic pathway. Indeed, chlorophyll *a* biosynthesis is presently interpreted by different research groups in terms of two different biochemical pathways. One camp believes that chlorophyll *a* is made via a single linear chain of reactions starting with divinylprotoporphyrin IX and ending with monovinylchlorophyll *a* (2-vinylchlorophyll *a*), whereas the other camp considers chlorophyll *a* to be the product of multiple and/or parallel biosynthetic routes that result in the formation and accumulation of monovinyl- (2-vinyl-) or divinylchlorophyll *a* (2,4-divinylchlorophyll *a*) chemical species, or both. The biological importance of this biosynthetic heterogeneity is becoming better understood.

The linear chlorophyll *a* biosynthetic pathway

The classical chlorophyll *a* biosynthetic pathway is visualized as a linear sequence of biochemical reactions which convert divinylprotoporphyrin to monovinyl-chlorophyll *a* via Mg–divinylprotoporphyrin, Mg–divinylprotoporphyrin

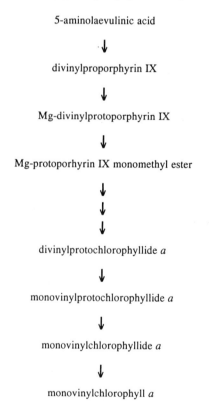

FIG. 1.   The classical linear chlorophyll *a* biosynthetic pathway.

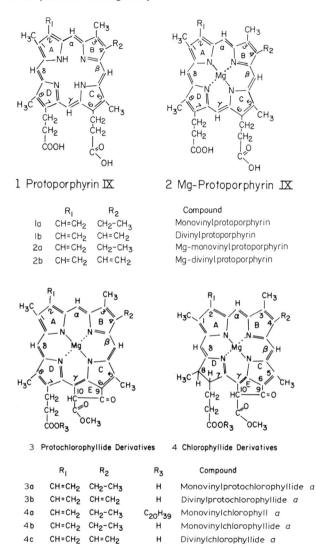

	R_1	R_2		Compound
1a	$CH=CH_2$	CH_2-CH_3		Monovinylprotoporphyrin
1b	$CH=CH_2$	$CH=CH_2$		Divinylprotoporphyrin
2a	$CH=CH_2$	CH_2-CH_3		Mg-monovinylprotoporphyrin
2b	$CH=CH_2$	$CH=CH_2$		Mg-divinylprotoporphyrin

	R_1	R_2	R_3	Compound
3a	$CH=CH_2$	CH_2-CH_3	H	Monovinylprotochlorophyllide *a*
3b	$CH=CH_2$	$CH=CH_2$	H	Divinylprotochlorophyllide *a*
4a	$CH=CH_2$	CH_2-CH_3	$C_{20}H_{39}$	Monovinylchlorophyll *a*
4b	$CH=CH_2$	CH_2-CH_3	H	Monovinylchlorophyllide *a*
4c	$CH=CH_2$	$CH=CH_2$	H	Divinylchlorophyllide *a*

FIG. 2. Chemical structures of some of the tetrapyrroles involved in chlorophyll *a* biosynthesis. The structures are numbered according to the Fischer nomenclature system.

monomethyl ester, divinylprotochlorophyllide *a*, monovinylprotochlorophyllide *a* and monovinylchlorophyllide *a* (Fig. 1, Fig. 2). The salient features of this pathway are: (a) that divinylprotochlorophyllide *a* does not accumulate in higher plants, but is a transient metabolite which is rapidly converted to monovinylchlorophyll *a* via monovinylprotochlorophyllide *a*; and (b) that there is no detectable monovinyltetrapyrrole between divinylprotoporphyrin and Mg–divinylprotoporphyrin monomethyl ester or divinyltetrapyrroles between monovinylprotochlorophyllide *a* and monovinylchlorophyll *a* (Fig. 1).

Experimental evidence in support of the linear chlorophyll a biosynthetic pathway

The demonstration of metabolic pathways is a multistep process involving at least three stages: (a) detection and characterization of metabolic intermediates; (b) demonstration of precursor–product relationships between putative intermediates; and (c) purification and characterization of the enzymes involved in metabolic interconversions. These criteria will be applied in our evaluation of the experimental evidence supporting the linear chlorophyll *a* biosynthetic pathway.

Detection and characterization of metabolic intermediates. The linear chlorophyll *a* biosynthetic pathway was essentially proposed by Granick (1950) and modified by Wolff & Price (1957) and Jones (1963). Granick detected the accumulation of divinylprotoporphyrin (Granick 1948a), Mg–divinylproto-porphyrin (Granick 1948b) and monovinylprotochlorophyllide *a* (Granick 1950) in *Chlorella* with X-ray-induced mutations which impaired chlorophyll biosynthetic capabilities. Granick organized the detected tetrapyrroles in order of increasing chemical complexity into a paper chemistry pathway. In this scheme, divinylprotoporphyrin was presumed to be convertible to Mg–divinylprotoporphyrin, which in turn was assumed to be convertible to monovinylprotochlorophyllide *a*. Monovinylprotochlorophyllide *a* was presumed to be convertible to monovinylprotochlorophyllide *a* phytyl ester. At that time it was believed that the immediate precursor of monovinyl-chlorophyll *a* was monovinylprotochlorophyllide *a* phytyl ester (Koski 1950, Granick 1950); a few years later Wolff & Price (1957) demonstrated that the immediate precursor of monovinylchlorophyll *a* was in fact monovinyl-chlorophyllide *a* which was formed by photoreduction of monovinylproto-chlorophyllide *a*. Then Granick (1961) detected the formation and accumulation of esterified Mg–divinylprotoporphyrin (Mg–protoporphyrin monomethyl ester) in another *Chlorella* mutant, and in etiolated barley leaves treated with 5-aminolaevulinic acid (ALA) and 2,2′-dipyridyl; he proposed that Mg–divinyl-protoporphyrin monomethyl ester was another intermediate of chlorophyll *a* biosynthesis formed by esterification of Mg–divinylprotoporphyrin at position 6 (Fischer nomenclature, equivalent to position 13 in IUPAC–IUB JCBN nomenclature) of the macrocycle. Jones (1963) identified divinylprotochlo-rophyllide *a* in *Rhodobacter sphaeroides* treated with 8-hydroxyquinoline and proposed that divinylprotochlorophyllide *a* was the immediate precursor of monovinylprotochlorophyllide *a* in bacteriochlorophyll and chlorophyll *a* biosynthesis. Because divinylprotochlorophyllide *a* had not been detected in higher plants it was assumed to be a transient intermediate which was rapidly converted to monovinylprotochlorophyllide *a* by reduction of the vinyl group at position 4 (8) of the macrocycle to an ethyl group. With the inclusion of divinylprotochlorophyllide *a* in the proposed linear chlorophyll *a* biosynthetic pathway, the pathway took on a form (Fig. 1) which for the next two decades was universally accepted by the scientific community.

Demonstration of precursor–product relationships between putative intermediates of the linear chlorophyll a biosynthetic pathway. One of the first partial reactions to be demonstrated *in vitro* was the conversion of Mg–divinyl-protoporphyrin to Mg–divinylprotoporphyrin monomethyl ester by methylation of the propionic acid residue at position 6 (13) of the macrocycle (Radmer & Bogorad 1967). The reaction mechanism was later investigated extensively by Ellsworth et al (1974) and Hinchigeri et al (1984).

Further progress in the demonstration of precursor–product relationships between putative intermediates of the pathway was hindered by the lack of cell-free systems capable of catalysing the remaining postulated reactions of the linear biosynthetic pathway. This obstacle was overcome when Rebeiz & Castelfranco (1971a,b) described a cell-free system which converted exogenous [^{14}C]ALA to [^{14}C]protochlorophyllide *a*, [^{14}C]protochlorophyllide *a* ester and [^{14}C]chlorophyll *a* and *b*. The performance of this cell-free system was improved over the years, with significant enhancements including: (a) the development of analytical spectrofluorimetric techniques that allowed net synthetic rates to be monitored *in organello* (Rebeiz et al 1975); (b) the discovery that high ATP concentrations in the incubation medium considerably improved synthetic performance (Castelfranco et al 1979, Rebeiz et al 1982); and (c) the demonstration that optimally operating cell-free systems could support thylakoid membrane assembly in isolated etioplasts (Rebeiz et al 1984).

For a time, studies of precursor–product relationships *in organello* and in suborganellar preparations capable of supporting total or partial monovinyl-chlorophyll *a* biosynthetic reactions appeared to confirm the operation of the linear chlorophyll *a* biosynthetic pathway between divinylprotoporphyrin and monovinylchlorophyll *a*. One partial reaction successfully demonstrated *in organello* was the photoconversion of exogenous protochlorophyllide *a* to chlorophyll(ide) *a* (Griffiths 1974, Mattheis & Rebeiz 1978). The putative metabolic role of divinylprotochlorophyllide *a* as a precursor of monovinyl-protochlorophyllide *a* appeared to be proven when Griffiths & Jones (1975) demonstrated the conversion of exogenous divinylprotochlorophyllide *a* to chlorophyll(ide) *a* by isolated barley etioplasts, presumably after conversion of divinylprotochlorophyllide *a* to monovinylprotochlorophyllide *a*. Protochlo-rophyllide-*a* oxidoreductase, the enzyme that catalyses the photoconversion of monovinylprotochlorophyllide *a* to monovinylchlorophyllide *a*, was investigated extensively by Griffiths, who showed it was NADPH dependent (Griffiths 1991). The biosynthetic role of divinylprotoporphyrin was demonstrated *in organello* when its conversion to protochlorophyllide *a* was shown (Mattheis & Rebeiz 1977a). Smith & Rebeiz (1977) then observed that insertion of Mg into exogenous divinylprotoporphyrin *in organello* yields Mg–protoporphyrins; the ATP requirement for this reaction was demonstrated by Castelfranco et al (1979). Conversion of Mg–divinylprotoporphyrin monomethyl ester to protochlorophyllide *a* and formation of the cyclopentanone ring was demonstrated *in organello* by Mattheis & Rebeiz (1977b). Finally, Rüdiger and co-workers established that in etiolated tissues, the esterification of chlorophyllide *a* proceeds via geranylgeraniol which is reduced step-wise to phytol (3,7,11,15-tetramethyl-2-hexadecen-1-ol, hexahydrogeranylgeraniol), via dihydro and tetrahydrogeranylgeraniol intermediates, in the presence of ATP and NADPH, and named the activity chlorophyll synthetase (Rüdiger et al 1980).

   In the above studies, tetrapyrroles were characterized mainly by chemical derivatization combined with electronic absorbance spectroscopy, an approach that did not reveal the full chemical complexity of the pools under investigation. It was therefore assumed wrongly that all the metabolic pools between divinylprotoporphyrin and divinylprotochlorophyllide *a* were populated by divinyltetrapyrroles while all the phorbin pools between monovinylproto-chlorophyllide *a* and chlorophyll *a* were populated by monovinyltetrapyrroles. Experimental evidence contrary to this prompted a re-examination of the validity of the linear chlorophyll *a* biosynthetic pathway.

The multibranched chlorophyll *a* biosynthetic pathway

At the root of the discovery of chlorophyll *a* biosynthetic heterogeneity in plants was the development of sensitive spectrofluorimetric analytical techniques. These techniques revealed that the monocarboxylic and fully esterified metabolic pools between protoporphyrin and chlorophyll *a* were chemically more complex than earlier studies had suggested, and were populated by tetrapyrroles that could not be accounted for by operation of the linear chlorophyll *a* biosynthetic pathway.

   A working hypothesis of the multibranched chlorophyll *a* biosynthetic pathway (Fig. 3) was proposed a decade ago (Rebeiz et al 1983). Chlorophyll *a* biosynthesis was considered to consist of six possible biosynthetic routes which led to the formation of multiple chlorophyll *a* chemical species (Fig. 3). Routes 1 and 6 (Fig. 3) are two minor routes that involve fully esterified divinyl and monovinyltetrapyrroles, respectively. Routes 2 and 5, on which we shall concentrate here, are two major routes that involve di- and monocarboxylic divinyl and monovinyltetrapyrroles. These two routes account for the biosynthesis of most of the chlorophyll *a* in Nature and are interconnected by 4-vinylreductases (Fischer numbering). Their functionality is based on

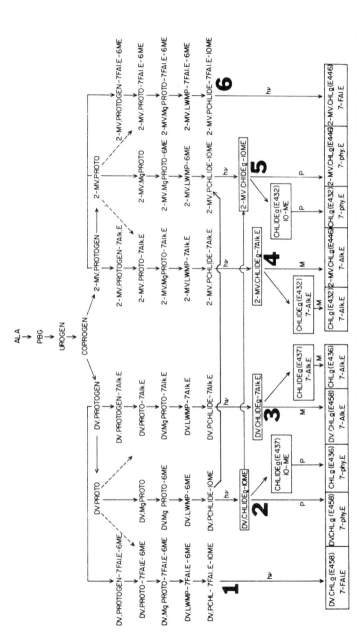

FIG. 3. The multibranched, six-route, chlorophyll *a* biosynthetic pathway proposed by Rebeiz et al (1983). ALA, 5-aminolaevulinic acid; PBG, porphobilinogen; UROGEN, uroporphyrinogen III; COPROGEN, coproporphyrinogen III; PROTOGEN, protoporphyrinogen IX; PROTO, protoporphyrin IX; PCHLIDE, protochlorophyllide; CHLIDE, chlorophyllide; PCHL, protochlorophyll; CHL, chlorophyll; DV, divinyl; MV, monovinyl; FAl, fatty alcohol; LWMP, longer-wavelength metalloporphyrins (putative intermediates of cyclopentanone ring formation); E, ester; ME, methyl ester; Alk, alkyl group of unknown chain length; phy, phytol; P, esterification with geranylgeraniol followed by step-wise reduction to phytol, or direct esterification with phytol; M, methylation. Adapted from Rebeiz et al (1983).

well-established experimental evidence. Routes 3 and 4 are highly hypothetical routes, supposedly populated by geometrical isomers of routes 2 and 5.

Experimental evidence in support of the di/monocarboxylic biosynthetic routes (routes 2 and 5)

Detection and characterization of di/monocarboxylic intermediates. Most of the tetrapyrroles that populate routes 2 and 5 have been purified and their chemical structure ascertained by electronic, field desorption and NMR spectroscopy. Divinylprotochlorophyllide *a* and monovinylprotochlorophyllide *a* have been purified from several plant tissues and their electronic spectroscopic properties have been described (Belanger & Rebeiz 1980a). Divinylprotochlorophyllide *a* was further characterized by field desorption and NMR spectroscopy (Wu & Rebeiz 1984). The identity of purified divinylchlorophyllide *a* was ascertained by extensive chemical derivatization combined with electronic spectroscopy (Belanger et al 1982); its mass spectroscopic and NMR profiles were reported by Wu & Rebeiz (1984). The molar extinction properties of purified divinylchlorophyll *a* and its NMR profile have been described by Shedbalkar & Rebeiz (1992). Mg–divinylprotoporphyrin and Mg–monovinylprotoporphyrin and their monomethyl esters were purified from etiolated cucumber cotyledons and were characterized by extensive chemical derivatization combined with electronic spectroscopic analysis (Belanger & Rebeiz 1982). Monovinylprotoporphyrin was detected spectrofluorimetrically in extracts of etiolated cucumber cotyledons treated with 2,2′-dipyridyl (Rebeiz & Lascelles 1982), and also in the protoporphyrin pool formed by etioplasts incubated with ALA. The divinylprotoporphyrin and monovinylprotoporphyrin components were purified by chromatography on thin layers of polyethylene after methylation and identified by fluorescence spectroscopy before and after conversion to Mg analogues (Kuhajda 1984). Insufficient samples were available for further characterization of monovinylprotoporphyrin by NMR and mass spectroscopy. Finally, chlorophyll *a* (E432), one of the end products of route 5, was characterized as a monovinyl-10-hydroxychlorophyll *a* lactone, by field desorption mass spectrometry and NMR spectroscopy (Wu & Rebeiz 1988).

Demonstration of precursor–product relationships among the di/monocarboxylic intermediates of routes 2 and 5. Precursor–product relationships between the intermediates of routes 2 and 5 were established either *in vivo* or *in organello*. The following partial reactions of route 2 were demonstrated *in organello*: divinylprotoporphyrin to divinylprotochlorophyllide *a*, divinylprotoporphyrin to Mg–divinylprotoporphyrin, Mg–divinylprotoporphyrin to divinylprotochlorophyllide *a*, and Mg–divinylprotoporphyrin monomethyl ester to divinylprotochlorophyllide *a* (Tripathy & Rebeiz 1986). The conversion of divinylprotochlorophyllide *a* to divinylchlorophyllide *a* (Duggan & Rebeiz 1982) and of divinylchlorophyllide *a* to divinylchlorophyll *a* (Rebeiz et al 1983) have also been described. Likewise, the following partial reactions of route 5 were demonstrated *in organello*: monovinylprotoporphyrin to monovinylprotochlorophyllide *a*, monovinylprotoporphyrin to Mg–monovinylprotoporphyrin, Mg–monovinyl-protoporphyrin to monovinylprotochlorophyllide *a*, and Mg–monovinylproto-porphyrin monomethyl ester to monovinylprotochlorophyllide *a* (Tripathy & Rebeiz 1986). The conversion of monovinylprotochlorophyllide *a* to monovinyl-chlorophyllide *a* has also been reported (Belanger & Rebeiz 1980b).

Interconnection of routes 2 and 5. Routes 2 and 5 are connected by 4-vinyl reductases (8-vinyl under IUPAC–IUB JCBN nomenclature), which convert a vinyl group at position 4 (8) of the macrocycle to an ethyl group, at two different levels at least. In dark-monovinyl/light-divinyl plants such as barley, during the shift from the divinyl to the monovinyl di/monocarboxylic chlorophyll *a* biosynthetic mode, about 50% of the monovinylprotochlorophyllide *a* may be formed through reduction of the 4-vinyl group of divinylprotochlorophyllide *a* to ethyl (Tripathy & Rebeiz 1988). This conversion rate drops drastically as the plant settles into the monovinylchlorophyll *a* biosynthetic mode. 4-Vinylprotochlorophyllide-*a* reductase, the enzyme which catalyses the vinyl reduction, has been demonstrated *in organello* (Tripathy & Rebeiz 1988). Its purification and the study of its properties are in progress. In dark-divinyl/light-divinyl plants such as cucumber, routes 2 and 5 are strongly interconnected at the level of divinylchlorophyllide *a* (Duggan & Rebeiz 1982). 4-Vinylchlorophyllide-*a* reductase, the enzyme that catalyses this reaction, is NADPH-dependent and is bound to the plastid membranes (Parham & Rebeiz 1992). This enzyme appears to be different from the 4-vinylprotochlorophyllide-*a* reductase and is inactive towards divinylprotochlorophyllide *a*. It is a very potent enzyme, the conversion it catalyses being about 65% complete after one minute. It has pH and temperature optima at 6.3 and 30 °C respectively, and is present in both etiolated and greening tissues; it is less active in dark-monovinyl/light-divinyl plants such as corn (R. Parham & C. A. Rebeiz, unpublished). Purification of 4-vinylchlorophyllide-*a* reductase is in progress.

Routes 2 and 5 may also be interconnected at other biosynthetic levels. Divinylchlorophyll *a*, which is formed rapidly by esterification of newly formed divinylchlorophyllide *a* (Rebeiz & Lascelles 1982), disappears rapidly after a few minutes in darkness as monovinylchlorophyll *a* accumulates (Belanger & Rebeiz 1980b). The disappearance of divinylchlorophyll *a* may be due to its rapid conversion to monovinylchlorophyll *a* by 4-vinylchlorophyllide-*a* reductase or by a novel 4-vinylchlorophyll-*a* reductase.

Tripathy & Rebeiz (1986) observed that Mg–divinylprotoporphyrin monomethyl ester was rapidly converted to monovinylprotochlorophyllide *a* in isolated barley etioplasts, under conditions in which 4-vinylprotochlorophyllide-*a* reductase was inactive. They proposed that a Mg–4-vinylporphyrin reductase operates somewhere at the level of an intermediate of the cyclopentanone ring formation. This may well be at the level of Mg-divinylprotoporphyrin monomethyl ester ketopropionate, as suggested by the observation that Mg-monovinylprotoporphyrin monomethyl ester ketopropionate was transformed into monovinylprotochlorophyllide *a* at four times the rate of conversion of the divinyl analogue (Walker et al 1988). Finally, Ellsworth & Hsing (1974) reported that Mg-divinylprotoporphyrin monomethyl ester could be converted to Mg-monovinylprotoporphyrin monomethyl ester by an NADH-dependent enzyme. To our knowledge, no one, ourselves included, has been able to confirm these results.

Present status of routes 2 and 5. Although experimental evidence in support of the operation of the di/monocarboxylic routes (2 and 5) is comprehensive, some synthetic details are still unresolved. It would be reassuring to purify monovinylprotoporphyrin from a variety of plant tissues and to confirm its monovinyl identity by NMR spectroscopy. Furthermore, the biosynthetic origin of this putative intermediate has not yet been resolved. It is not clear, for example, whether monovinylprotoporphyrinogen is formed from divinyl-protoporphyrinogen by a 4-vinyl reductase, or whether it is formed from coproporphyrinogen III via an unknown reaction, as depicted in Fig. 3. It is also not known whether one 4-vinyl reductase of broad specificity or several 4-vinyl reductases of narrow specificity are involved in the conversion of divinyltetrapyrroles to monovinyltetrapyrroles, nor is it clear whether one set of enzymes of broad specificity or two sets of enzymes of narrower specificity catalyse the reactions between divinylprotoporphyrin and divinylprotochlo-rophyllide *a* on the one hand, and the reactions between monovinylproto-porphyrin and monovinylprotochlorophyllide *a* on the other. It should be emphasized that although the outcome of investigations into enzyme specificity will deepen our understanding of the regulation of the di/monocarboxylic biosynthetic routes, it will have no bearing on the functionality of routes 2 and 5 as two distinct and separate biosynthetic routes.

Significance of the multibranched chlorophyll *a* biosynthetic pathway

A complete understanding of chlorophyll *a* biosynthetic heterogeneity cannot be achieved without a thorough understanding of its biological significance and of the role it plays in Nature. The discovery of the di/monocarboxylic biosynthetic routes, routes 2 and 5, has led to the realization that green plants can be classified into one of three greening groups according to which di/monocarboxylic chlorophyll *a* biosynthetic route is predominant during the dark and light phases of the photoperiod (Ioannides et al 1992). Other important findings that have followed the discovery of chlorophyll *a* biosynthetic heterogeneity are: (a) that the major chlorophyll species in the euphotic zone of tropical oceans (the zone that supports plankton growth) are divinylchlorophyll *a* and divinylchlorophyll *b* (Chisholm et al 1992); (b) that the susceptibility of various plant species to ALA-dependent photodynamic herbicides depends, among other things, on the greening group into which the plant falls (Rebeiz et al 1990); (c) that the chlorophyll *a* formed via biosynthetic routes 2 and 5 is preferentially incorporated into different pigment protein complexes (D. Fasoula & C. A. Rebeiz, unpublished); and (d) that in spring-wheat cultivars which belong to the dark-monovinyl/light-divinyl greening group, the accumulation of higher proportions of monovinylprotochlorophyllide *a* appears to be associated with higher yields under field conditions (D. Fasoula & C. A. Rebeiz, unpublished work).

Summary and conclusions

Available experimental evidence indicates that chlorophyll *a* is formed via multiple biosynthetic routes. Although additional biosynthetic details need to be established, most chlorophyll *a* appears to be formed via two routes populated by divinyl and monovinyl di/monocarboxylic tetrapyrroles. These conclusions are based on: (a) purification of key metabolic intermediates and end-products; (b) thorough chemical characterization of purified intermediates and end-products by various spectroscopic techniques including NMR and field desorption mass spectroscopy; and (c) demonstration of precursor–product relationship among various intermediates *in vivo* and *in organello*. With our present state of understanding, chlorophyll *a* biosynthetic heterogeneity can be used as a basis for classifying plants into three different greening groups, according to the prevalence of a particular divinyl or monovinyl di/monocarboxylic biosynthetic route at night or in daytime. Chlorophyll *a* biosynthetic heterogeneity also explains the prevalence of divinylchlorophyll *a* and divinylchlorophyll *b* in the euphotic zone of tropical oceans and the differential susceptibility of green plants to photodynamic herbicides, and also appears to be related to crop yields under field conditions.

Acknowledgement

The preparation of this manuscript was supported by the John P. Trebellas Photobiotechnology Research endowment to C. A. Rebeiz.

References

Belanger FC, Rebeiz CA 1980a Chloroplast biogenesis. Detection of divinyl protochlorophyllide in higher plants. J Biol Chem 255:1266–1272
Belanger FC, Rebeiz CA 1980b Chloroplast biogenesis. 30. Chlorophyll(ide) (E459F675) and chlorophyll(ide) (E449F675) the first detectable products of divinyl and monovinyl protochlorophyll photo-reduction. Plant Sci Lett 18:343–350

Belanger FC, Rebeiz CA 1982 Chloroplast biogenesis. Detection of monovinyl magnesium-protoporphyrin monoester and other magnesium-porphyrins in higher plants. J Biol Chem 257:1360–1371

Belanger FC, Duggan JX, Rebeiz CA 1982 Chloroplast biogenesis. Identification of chlorophyllide *a* (E458F674) as a divinyl chlorophyllide *a*. J Biol Chem 257: 4849–4858

Castelfranco PA, Weinstein JD, Schwarcz S, Pardo AD, Wezelman BE 1979 The Mg insertion step in chlorophyll biosynthesis. Arch Biochem Biophys 192:592–598

Chisholm SW, Frankel SL, Goericke R et al 1992 *Prochlorococcus marinus* nov gen nov sp: an oxyphototrophic marine prokaryote containing divinyl chlorophyll *a* and chlorophyll *b*. Arch Microbiol 157:297–300

Duggan JX, Rebeiz CA 1982 Chloroplast biogenesis. 42. Conversion of divinyl chlorophyllide *a* to monovinyl chlorophyllide *a* in vivo and in vitro. Plant Sci Lett 27:137–145

Ellsworth RK, Hsing AS 1974 Activity and some properties of Mg-4-ethyl-(4-desvinyl)-protoporphyrin IX monomethyl ester:NAD^+ oxidoreductase in crude homogenates from etiolated wheat seedlings. Photosynthetica 8:228–234

Ellsworth RK, Dullaghan JP, St Pierre ME 1974 The reaction mechanism of S-adenosyl-L-methionine:magnesium protoporphyrin IX methyltransferase of wheat. Photosynthetica 8:375–383

Granick S 1948a Protoporphyrin 9 as a precursor of chlorophyll. J Biol Chem 172:717–727

Granick S 1948b Magnesium protoporphyrin as a precursor of chlorophyll in Chlorella. J Biol Chem 175:333–342

Granick S 1950 Magnesium vinyl pheoporphyrin a_5, another intermediate in the biological synthesis of chlorophyll. J Biol Chem 183:713–730

Granick S 1961 Magnesium protoporphyrin monoester and protoporphyrin monomethyl ester in chlorophyll biosynthesis. J Biol Chem 236:1168–1172

Griffiths WT 1974 Protochlorophyll and protochlorophyllide as precursors for chlorophyll synthesis in vitro. FEBS (Fed Eur Biochem Soc) Lett 49:196–200

Griffiths WT 1991 Protochlorophyllide photoreduction. In: Scheer H (ed) Chlorophylls. CRC Press, Boca Raton, FL, p 433–449

Griffiths WT, Jones OTG 1975 Magnesium 2,4-divinylphaeoporphyrin a_5 as a substrate for chlorophyll biosynthesis in vitro. FEBS (Fed Eur Biochem Soc) Lett 50: 355–358

Hinchigeri SB, Nelson DW, Richards WR 1984 The purification and reaction mechanism of S-adenosyl-L-methionine: magnesium protoporphyrin methyltransferase from *Rhodopseudomonas sphaeroides*. Photosynthetica 18:168–178

Ioannides IM, Robertson KR, Rebeiz CA 1992 Chloroplast biogenesis. 65. Study of the greening group affiliation of green plants. Plant Physiol (Life Sci Adv) 11:79–88

Jones OTG 1963 Magnesium 2,4-divinylphaeoporphyrin a_5 monomethyl ester, a protochlorophyll-like pigment produced by *Rhodopseudomonas sphaeroides*. Biochem J 89:182–189

Koski VM 1950 Chlorophyll formation in seedlings of *Zea mays* L. Arch Biochem 29:339–343

Kuhajda MS 1984 The identification and site of biosynthesis of monovinyl protoporphyrin divinyl protoporphyrin, and protoporphyrin monoester intermediates in the biosynthetic pathway of chlorophyll *a*. MSc thesis, University of Illinois, Urbana, IL, p 122

Mattheis JR, Rebeiz CA 1977a Chloroplast biogenesis. Net synthesis of protochlorophyllide from protoporphyrin IX by developing chloroplasts. J Biol Chem 252:8347–8349

Mattheis JR, Rebeiz CA 1977b Chloroplast biogenesis. Net synthesis of protochlorophyllide from magnesium-protoporphyrin monoester by developing chloroplasts. J Biol Chem 252:4022–4024

Mattheis JR, Rebeiz CA 1978 Chloroplast biogenesis. The conversion of exogenous protochlorophyllide into phototransformable protochlorophyllide in vitro. Photochem Photobiol 28:55–60

Parham R, Rebeiz CA 1992 Chloroplast biogenesis. [4-Vinyl]chlorophyllide *a* reductase is a divinyl chlorophyllide *a* specific, NADPH-dependent enzyme. Biochemistry 31:8460–8464

Radmer RJ, Bogorad L 1967 (–)S-Adenosyl-L-methionine-magnesium protoporphyrin methyl transferase, an enzyme in the biosynthetic pathway of chlorophyll in *Zea mays*. Plant Physiol 42:463–465

Rebeiz CA, Castelfranco P 1971a Protochlorophyll biosynthesis in a cell-free system from higher plants. Plant Physiol 47:24–32

Rebeiz CA, Castelfranco P 1971b Chlorophyll biosynthesis in a cell-free system from higher plants. Plant Physiol 47:33–37

Rebeiz CA, Lascelles J 1982 Biosynthesis of pigments in plants and bacteria. In: Govindjee (ed) Photosynthesis: energy conversion by plants and bacteria. Academic Press, New York, vol 1:699–780

Rebeiz CA, Mattheis JR, Smith BB, Rebeiz CC, Dayton D 1975 Chloroplast biogenesis. Biosynthesis and accumulation of protochlorophyll by isolated etioplasts and developing chloroplasts. Arch Biochem Biophys 171:549–567

Rebeiz CA, Daniell H, Mattheis JR 1982 Chloroplast bioengineering. The greening of chloroplasts in vitro. Biotechnol Bioeng Symp 12:413–439

Rebeiz CA, Wu SM, Kuhadja M, Daniell H, Perkins EJ 1983 Chlorophyll *a* biosynthetic routes and chlorophyll *a* chemical heterogeneity in plants. Mol Cell Biochem 58:97–125

Rebeiz CA, Montazer-Zouhoor A, Daniell H 1984 Chloroplast culture. 10. Thylakoid assembly in vitro. Isr J Bot 33:225–235

Rebeiz CA, Reddy KN, Nandihalli UB, Velu J 1990 Tetrapyrrole-dependent photodynamic herbicides. Photochem Photobiol 52:1099–1117

Rüdiger WR, Benz J, Guthoff C 1980 Detection and partial characterization of activity of chlorophyll synthetase in etioplast membranes. Eur J Biochem 109:193–200

Shedbalkar VP, Rebeiz CA 1992 Chloroplast biogenesis. Determination of the molar extinction coefficients of divinyl chlorophyll *a* and *b* and their pheophytins. Anal Biochem 207:261–266

Smith BB, Rebeiz CA 1977 Chloroplast biogenesis: detection of Mg-protoporphyrin chelatase in vitro. Arch Biochem Biophys 180:178–185

Tripathy BC, Rebeiz CA 1986 Chloroplast biogenesis. Demonstration of the monovinyl and divinyl monocarboxylic routes of chlorophyll biosynthesis in higher plants. J Biol Chem 261:13556–13564

Tripathy BC, Rebeiz CA 1988 Chloroplast biogenesis. 60. Conversion of divinyl protochlorophyllide to monovinyl protochlorophyllide in green(ing) barley, a dark monovinyl/light divinyl plant species. Plant Physiol 87:89–94

Walker CJ, Mansfield KE, Rezzano IN, Hanamoto CM, Smith KM, Castelfranco PA 1988 The magnesium-protoporphyrin IX (oxidative) cyclase system. Studies of the mechanism and specificity of the reaction sequence. Biochem J 255:685–692

Wolff JB, Price L 1957 Terminal steps of chlorophyll *a* biosynthesis in higher plants. Arch Biochem Biophys 72:293–301

Wu SM, Rebeiz CA 1984 Chloroplast biogenesis. 45. Molecular structure of protochlorophyllide (E443 F625) and of chlorophyllide *a* (E458 F674). Tetrahedron 40:659–664

Wu SM, Rebeiz CA 1988 Chloroplast biogenesis: molecular structure of short wavelength chlorophyll *a* (E432 F662). Phytochemistry 27:353–356

DISCUSSION

Kräutler: I have a nomenclature question. Couldn't one call the 'monovinyl-protoporphyrins' and so on dihydroprotoporphyrins?

Rebeiz: Yes. We are catering mainly for biochemists, and feel that divinyl or monovinyl is much simpler.

Akhtar: This was an important and interesting wide-ranging analysis, which reminds me of similar situations in the steroid field. As you can with steroids, you can divide the porphyrin molecule into the side chains and the nucleus. The sequence of steroid biosynthesis is rigid up to the cyclization stage which produces that nucleus, just as it is in porphyrin biosynthesis, but there is exactly the same kind of heterogeneity with respect to nuclear demethylations and modifications. Consider the C-24–25 double bond modifications in various organisms. Here, one sees a great deal of diversity, which may depend on the level of the side chain-modifying enzyme(s) relative to the enzymes involved in nuclear modifications. Of course, you have raised a more important point, that in some cases the ethyl group on ring B has come into being by a mechanism entirely different from that involved in the formation of the vinyl group.

Rebeiz: We have definite evidence that the divinylprotochlorophyllide reductase is a different enzyme from the divinylchlorophyllide reductase. The divinylchlorophyllide reductase is an NADPH-dependent enzyme; it is much faster than the divinylprotochlorophyllide reductase and is totally inactive towards 4-vinylprotochlorophyllide.

Akhtar: Are you saying that the enzyme which produces the vinyl group in ring A is different from that which introduces the vinyl group into ring B?

Rebeiz: No. I'm saying that the enzyme that catalyses the reduction in ring B of divinylchlorophyllide is different from the enzyme that catalyses the reaction in divinylprotochlorophyllide. Nature has developed two different enzyme systems for the reduction of the vinyl group to an ethyl group, one for chlorophyllide and one for protochlorophyllide.

Arigoni: Is each specific for its substrate?

Rebeiz: Yes; specificity is absolute.

Arigoni: As Professor Akhtar pointed out, a similar situation prevails in steroid biosynthesis. We still do not understand the need for multiple pathways.

Rebeiz: The monovinyl route, which is the more recently evolved route, has an advantage in that it appears to be associated with higher productivity in green plants (T. Tanaka K. Takanashi, Y. Hotta, Y. Takeushi & M. Konnai, unpublished paper, 19th Meet PGR Soc Am, July 1992, and unpublished paper, Molec Regul Chloroplast Funct Meet, August 1992). In plants sprayed with low concentrations of ALA field productivity is increased and the monovinyl route is enhanced (D. A. Fasoula & C. A. Rebeiz, unpublished work 1993).

Akhtar: When you refer to the monovinyl pathway, do you mean a direct decarboxylation of the propionate to the ethyl group, or are you talking about two different reductases?

Rebeiz: I am saying that the monovinyl route can originate in two different ways, either at the level of protoporphyrin IX or at the level of the 4-vinyl reductases, of which there may be several with strict specificities.

Castelfranco: It is now possible to separate the various chlorophyll–protein complexes. Have you checked these complexes for monovinylchlorophyll and divinylchlorophyll? Some of them might contain divinylchlorophyll and others monovinylchlorophyll.

Rebeiz: The divinylchlorophyll *a* which is produced disappears after a few minutes in higher plants. We don't know where it goes, or whether it is masked. We were unable to find any trace of divinylchlorophyll *a* or *b* in higher plants. In the phytoplankton of tropical oceans, in contrast, divinylchlorophyll *a* and *b* are predominant. Figure 3 is a paper chemistry composite pathway that involves both higher and lower plants. Divinylchlorophyllide *a* is formed in higher plants.

Jordan: You just mentioned that spraying plants with low concentrations of ALA changes the pathway operating. Do you think it's possible that some of your results are somewhat artefactual, arising because you overload the system with ALA which is normally produced in a regulated manner? We once did some experiments on the incorporation of ALA into red blood cells, and found that if we used no carrier at all, but just added the ^{14}C label, the results were very different from those obtained when we added carrier to the $[^{14}C]$ALA. This was because we were overloading the enzyme systems, causing the build up of intermediates which normally don't accumulate.

Rebeiz: Spraying low concentrations of ALA activates the monovinyl route. Many of the reported intermediates in that pathway (Fig. 3) have been observed *in vivo* without any addition of ALA.

Griffiths: The one enzyme on the chlorophyll synthesis pathway that has been identified unambiguously, i.e., the photoenzyme which catalyses protochlorophyllide reduction, works perfectly well with both the divinyl and monovinyl compounds.

Rebeiz: That's correct. The enzymes in the forward direction don't seem to discriminate between the divinyl and monovinyl components. In their work on magnesium-chelating activity, Walker & Weinstein (1991) reported that chelation was faster with monovinylprotoporphyrin than with divinylprotoporphyrin. It is the side bridge reactions which are specific. In other words, we are dealing with a biosynthetic heterogeneity brought about by enzymes which don't care what they are given in the forward reaction. What is given to them is dictated, however, by the specific 4-vinyl reductases.

Beale: Where does maize fit in this scheme?

Rebeiz: Maize is in between barley and cucumber.

Beale: So you would expect it to have two 4-vinyl reductases.

Rebeiz: Yes. The divinylprotochlorophyllide reductase in maize is less powerful than that in barley, but it is present, in addition to the divinyl-chlorophyllide reductase.

Beale: There is a mutant strain of maize, which presumably lacks a single gene, which has only divinylchlorophyllide *a* and *b*. A single lesion can prevent the synthesis of monovinylchlorophylls. If there were two enzymes, one working at the protochlorophyllide level and the other at the chlorophyllide level, one would presumably be able to replace the other, to give a monovinyl product, yet there is none in this strain.

Rebeiz: That is only true if loss of a single gene dictates the outcome, and I don't know if that is the case. We have looked at that mutant carefully. It does produce divinylchlorophyllide *a* and divinylchlorophyllide *b*, but we haven't seen any monovinylchlorophyllide. All I know is that *in vitro*, the divinyl-protochlorophyllide and chlorophyllide enzymes are strictly specific for their substrates.

A. Smith: A single gene mutation could knock out a common factor required by two enzymes with differing specificities. In other words, another gene could encode a subunit that determines the specificity of the enzyme. I don't think this could be ruled out, particularly in view of the likelihood that most of these enzymes are multisubunit enzymes.

Jordan: How was this mutant maize spotted? Does it make less chlorophyll than normal?

Rebeiz: The mutant was found in Govindjee's lab by a graduate student, M. B. Bazzaz, who came upon this mutant and couldn't explain what was going on. After getting her PhD she joined my lab at the time when we were beginning to unravel the monovinyl/divinyl story. She wondered if the compound in the maize mutant was a divinyl compound, and went on to demonstrate this. The mutant strain is small and non-viable. Phytoplankton which have only divinylchlorophyll *a* and *b* survive perfectly well, so the maize mutant must lack something more than just the monovinyl branch.

Battersby: Several things have had to happen to the molecule to give 2-monovinylprotoporphyrinogen.

Rebeiz: This is paper chemistry, in which I assume that coproporphyrinogen is giving rise to two reactions, one of which produces divinylprotoporphyrinogen and the other monovinylprotoporphyrinogen. In other words, in one reaction the vinyl groups at positions 2 and 4 (Fischer nomenclature) are being produced by the oxidative decarboxylation of the propionate residue, whereas in the other, the ethyl group at position 4 is somehow produced from coproporphyrinogen. It's quite likely that this scheme is not valid, and that divinylprotoporphyrinogen is really being converted by a 4-vinyl reductase to monovinylprotoporphyrinogen. We looked for that reaction four years ago, but couldn't find it. We now have a more potent cell-free system with which we shall re-examine this question.

Arigoni: You could test this by using a substrate labelled with tritium in the methylene group adjacent to the pyrrole ring. Half of this label should be lost if formation of the double bond is a prerequisite.

Rebeiz: That is a possible approach. We find it easier to look for the enzyme.

Arigoni: Do I understand correctly that you are looking for an enzyme that will decarboxylate the propionate side chains to leave an ethyl group?

Rebeiz: No; I'm looking for an enzyme that converts the 4-vinyl group to a 4-ethyl.

Leeper: Is there any chance that the monovinylprotoporphyrin that you isolate comes from some other compound which has lost its magnesium, or an ester which could have been hydrolysed by an esterase?

Rebeiz: That is a good question. This problem arises if you try to accelerate the process by incubating the tissue with ALA and a modulator such as 2,2′-dipyridyl. Under these conditions, considerable amounts of Mg–monovinyl-protoporphyrin and Mg–divinylprotoporphyrin are formed; some protoporphyrin may then be formed from the demetallation of Mg–protoporphyrin. However, we have demonstrated the conversion of ALA to protoporphyrin in a cell-free system in which Mg^{2+} cannot be inserted. Of the protoporphyrin formed in that system, 25% was in the monovinyl form.

Scott: In Fig. 3, do all the vertical arrows that are laterally aligned correspond to the same enzyme?

Rebeiz: I would need an hour to go through all this. I concentrated on routes 2 and 5 because 95% of chlorophylls are made via these two routes. Routes 1 and 6 are fully esterified routes. I feel that different enzymes are involved here. The photoreduction of protochlorophyllide ester, for example, is weak; only about 1% of the protochlorophyllide ester pool can be photoconverted to chlorophyll, whereas about 95–98% of the protochlorophyllide pool is photoconverted under the same conditions. Routes 1 and 6 probably use different enzymes, but account for only about 2% of the total chlorophyll. Routes 3 and 4 are highly hypothetical, and based only on the occurrence of HPLC doublets (Rebeiz et al 1983). We have some experimental evidence for routes 1 and 6, but most of the work has been confined to routes 2 and 5, because these routes make most of the chlorophyll.

References

Rebeiz CA, Wu SM, Kuhadja M, Daniell H, Perkins EF 1983 Chlorophyll *a* biosynthetic routes and chlorophyll *a* chemical heterogeneity in plants. Mol Cell Biochem 58:97–125
Walker CJ, Weinstein JD 1991 Further characterization of the magnesium chelatase in isolated developing cucumber chloroplasts. Plant Physiol 95:1189–1196

Biosynthetic studies on chlorophylls: from protoporphyrin IX to protochlorophyllide

Paul A. Castelfranco, Caroline J. Walker† and Jon D. Weinstein†

Division of Biological Sciences, Section of Botany, University of California, Davis, CA 95616 and †Department of Biological Sciences, Clemson University, Clemson, SC 29634-1903, USA

Abstract. The series of reactions leading from protoporphyrin IX to protochlorophyllide have been studied over the last 15 years in the authors' laboratories at Davis and Clemson. Here, two crucial steps are emphasized, the discovery of the ATP requirement for Mg^{2+} chelation, and the oxidative cyclization of Mg–protoporphyrin IX monomethyl ester to protochlorophyllide. The *in vitro* systems for the chelation of Mg^{2+} and for the oxidative cyclization of Mg–protoporphyrin IX monomethyl ester both require membrane-associated and soluble heat-labile components. We speculate about the enzymological mechanisms of these important reactions, their sub-plastidic localization and the relationship of these individual steps to the broader questions of chloroplast and cell development.

1994 The biosynthesis of the tetrapyrrole pigments. Wiley, Chichester (Ciba Foundation Symposium 180) p 194–209

The multistep transformation of protoporphyrin IX to protochloropyllide* (Fig. 1) is fundamental to the chlorophyll biosynthetic pathway. In this process: (a) the central Mg ion is introduced—this is an essential feature of all chlorophylls, regardless of function or taxonomic distribution; (b) ring E is formed—this is also a feature of all chlorophylls; (c) during this process, the first asymmetric centre is formed at the C-10 of ring E (Fischer formula; position 13^2, IUPAC–IUB JCBN), which may play a part in directing subsequent asymmetric steps of the chlorophyll pathway. Here, we have limited ourselves to a review of the research from our own laboratories. Readers wishing a more complete coverage should consult the recent review article by Beale & Weinstein (1990).

---

*The term protochlorophyllide is used without consideration of whether it has an ethyl or vinyl substituent at position 4 (Fischer formula; position 8, IUPAC–IUB JCBN). The reactions proceed regardless of the oxidation state of this substituent. The question of this substituent is addressed elsewhere (Rebeiz et al 1994, this volume).

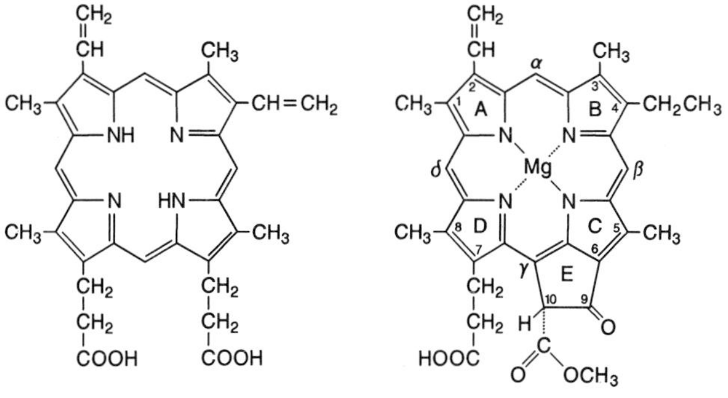

Protoporphyrin IX   **Protochlorophyllide**

FIG. 1.   Structures of protoporphyrin IX and protochlorophyllide (see footnote on p 194 about protochlorophyllide). The latter is shown with the ethyl substituent in the 4-position and is numbered according to the Fischer system.

Biological materials and preparation of subcellular fractions

In our experiments, the biological material consisted of either: cotyledons of cucumber seedlings germinated in the dark for 6–7 days and exposed to white fluorescent light for 12–20 hours; or the primary leaves of pea seedlings germinated under photoperiodic conditions and harvested just before the leaflets were fully expanded. Both tissues are soft and easily yield intact developing chloroplasts (Fuesler et al 1984a, Walker & Weinstein 1991a). Sub-plastidic fractions were obtained by various treatments involving osmotic shocks, sonication and freeze-thawing (Wong & Castelfranco 1984, Walker et al 1991, Walker & Weinstein 1991a, Walker et al 1992, Whyte & Castelfranco 1993a). In all experiments, reaction products were detected spectrofluorimetrically, spectrophotometrically or chromatographically (Castelfranco et al 1979, Fuesler et al 1982, Vijayan et al 1992, Walker et al 1992, Walker & Weinstein 1991b).

Results and discussion

Serious progress on Mg^{2+} chelation began at Davis when we discovered that in a crude preparation of cucumber chloroplasts containing 1–2 mM ATP, protoporphyrin IX was converted to Mg–protoporphyrin IX in the presence of glutamate or tricarboxylic acid cycle intermediates (Castelfranco et al 1979). Because Mg^{2+} chelation also required O_2 and was inhibited by malonate, the involvement of mitochondria was suggested (Pardo et al 1980). However, as the ATP concentration was raised, glutamate stimulation became less prominent, and both the O_2 requirement and the inhibition by malonate disappeared. When we purified this crude chloroplast pellet to remove contaminating mitochondria, it could still sustain Mg^{2+} chelation, but only in the presence

TABLE 1   Specificity of the ATP requirement for chelation of Mg^{2+} by protoporphyrin IX

Additions to the standard system[a]	Activity (pmol Mg–protoporphyrin IX)
Experiment 1	
1.5 mM ATP	3.5 ± 0.2
1.5 mM ATP + PEP + PK	820 ± 68
10 mM ATP	625 ± 5
10 mM ATP + PEP + PK	922 ± 20
Experiment 2	
1.5 mM ATP + PEP + PK	566 ± 66
PEP + PK	6.4 ± 0.8
1.5 mM ATP + PK	1.5 ± 0.1
1.5 mM ATP + PEP	65.6 ± 7
Experiment 3	
10 mM ATP	937 ± 63
10 mM GTP	2.3 ± 0.7
10 mM UTP	1.5 ± 0.0
10 mM CTP	1.7 ± 0.1
10 mM ITP	1.9 ± 0.3
Experiment 4	
10 mM ATP	982 ± 7.5
20 mM ADP	1.9 ± 0.5

Reproduced from Pardo et al (1980) with the permission of the American Society of Plant Physiologists.
[a]For experimental details, see Pardo et al (1980).
PEP, phosphoenolpyruvate.
PK, pyruvate kinase (EC 2.7.1.40)

of 10–15 mM ATP (Pardo et al 1980) (Table 1). The ATP could not be replaced by ADP or by other trinucleotides (Pardo et al 1980) and AMP was strongly inhibitory (Pardo et al 1980); non-hydrolysable ATP analogues did not support Mg^{2+}-chelating activity and were, in fact, inhibitory (Walker & Weinstein 1991b). Deutero- and mesoporphyrins could replace protoporphyrin in the chelation assay. Haematoporphyrin was only slightly active. Other porphyrins tested failed to chelate Mg^{2+} (Pardo et al 1980).

The next major step forward in characterizing this enzyme came much later at Clemson when, after several unsuccessful attempts (Walker & Weinstein 1991b), we succeeded in obtaining plastid-free magnesium-chelating activity (Walker & Weinstein 1991a). To our surprise, the activity required two heat-labile

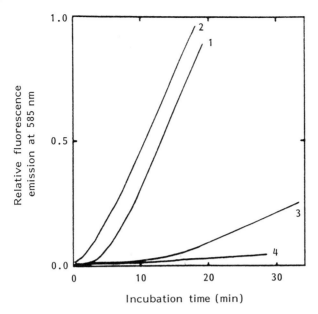

FIG. 2. Effect of preincubating a pea plastid fraction, a mixture of light membranes plus supernatant, with the substrates of Mg-chelatase. The emission at 585 nm is a measure of the Mg–deuteroporphyrin formed (see Walker et al 1992 for experimental details). Curve 1, no preincubation; curve 2, a six-minute preincubation with ATP; curve 3, a six-minute preincubation with deuteroporphyrin; curve 4, a six-minute preincubation without substrates. Reproduced from Walker et al (1992) with the permission of the Editor.

fractions, one soluble and one membrane-associated. (The analogous reaction, insertion of iron to form haem, is catalysed by ferrochelatase [EC 4.99.1.1], a single membrane-bound protein.) Initially, our chloroplast fractionation experiments were done with peas, but we later applied our technique to cucumbers with the same results (Hartwell & Weinstein 1992).

Using a pea plastid fraction which is low in chlorophyll (consisting of 'light' membranes and stroma) we developed a continuous fluorimetric assay (Walker et al 1992). Figure 2, curve 1, shows a typical activity profile which starts with an initial lag which is followed by a linear phase of Mg–deuteroporphyrin accumulation. If the sample was preincubated with ATP, and the reaction initiated by the addition of porphyrin (curve 2), the initial lag was almost abolished. Preincubation without substrates destroyed the activity (curve 4) and preincubation with porphyrin (curve 3) afforded only a slight protection. These results indicated that ATP might be involved in an 'activation' process and might also provide protection from inactivation.

After the insertion of Mg^{2+}, the C-6 propionic acid side chain must be esterified before the oxidative cyclization can take place (Fig. 3). This is because,

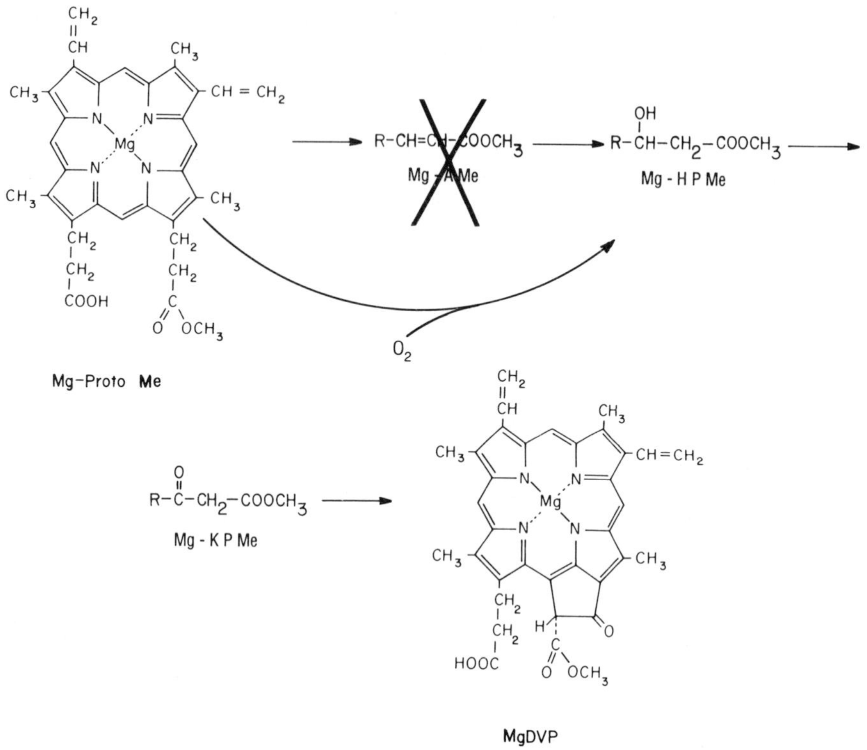

FIG. 3. Mechanism of conversion of Mg–protoporphyrin IX monomethyl ester to protochlorophyllide. The notations Mg-A Me, Mg-HP Me and Mg-KP Me represent, respectively, the acrylate, β-hydroxypropionate and β-oxopropionate derivatives of Mg–protoporphyrin IX monomethyl ester; MgDVP is divinylprotochlorophyllide (Mg–divinylphaeoporphyrin a_5). The acrylate derivative, first proposed by Granick (1951), is not involved in the conversion. Modified from Wong et al (1985), with the permission of the American Society of Plant Physiologists.

in the absence of the methyl ester, the β-oxopropionic acid is decarboxylated too readily for ring closure to occur. Therefore, the next enzyme after the Mg-chelatase is magnesium-protoporphyrin O-methyltransferase (EC 2.1.1.11), which uses S-adenosyl-L-methionine (SAM) as a methyl donor. Although intact chloroplasts do not appear to have any SAM of their own, exogenous SAM will enable the methylation of Mg–protoporphyrin (Fuesler et al 1982, 1984a) or other chloroplast constituents (Wallsgrove et al 1983).

Chloroplasts possess a methyl esterase (or esterases) that convert Mg–proto-porphyrin IX monomethyl ester back to Mg–protoporphyrin IX (Chereskin et al 1982, Walker et al 1988, Shiau et al 1991, Walker et al 1991). This enzyme appears to be mostly soluble. Its function might be to degrade any

Mg–protoporphyrin IX monomethyl ester that falls off the growing photosynthetic membrane. The soluble fraction of etioplasts and developing chloroplasts also contains an O_2-dependent oxidase that destroys plastidic pigments, including exogenous Mg–protoporphyrin IX monomethyl ester (Gassman & Ramanujam 1986, Whyte & Castelfranco 1993b).

Mg–protoporphyrin IX monomethyl ester is converted to protochlorophyllide by the magnesium–protoporphyrin monomethyl ester (oxidative) cyclase, which requires O_2 and NADPH. This conversion is a six-electron oxidation and involves the incorporation of one atom of oxygen in position 9 (position 13^1, IUPAC–IUB JCBN nomenclature) (Fig. 3). We have shown, by an $^{18}O_2$-labelling experiment, that this oxygen atom comes from atmospheric O_2 (Walker et al 1989). A mechanism for this process with acrylate, hydroxy and oxo intermediates was proposed by Granick (1951) by analogy to β-oxidation of fatty acids (Fig. 3). However, synthetic acrylate could not be cyclized by intact cucumber plastids (Walker et al 1988) or a plastid-free reconstituted cyclase system (see below) (Y.-S. Wong & P. A. Castelfranco, unpublished results), indicating that this compound is not an intermediate. However, the β-hydroxy and the β-oxo derivatives are definitely intermediates in the cyclization process. The hydroxy compound was isolated from *in vitro* reaction mixtures and characterized by comparison with synthetic standard (Wong et al 1985) after it was observed that the accumulation of this compound followed the kinetic behaviour expected of an intermediate. The two enantiomers of the synthetic β-hydroxy compound were resolved, and we found that only one of these could be converted to protochlorophyllide (Walker et al 1988) by our plant extracts. However, because we have not yet been able to grow a crystal of sufficient size and purity for X-ray crystallography, the chirality of the β-hydroxy intermediate is still in doubt (Walker et al 1988). The synthetic β-oxo derivative was cyclized very readily by developing chloroplasts and by our plastid-free reconstituted system (see below) (Wong et al 1985, Walker et al 1988).

The cyclase system can be resolved into sub-plastidic components if the intact chloroplasts are submitted to sonication (Wong & Castelfranco 1984) or osmotic shock (Walker et al 1991, Whyte & Castelfranco 1993a). Centrifugation of the broken chloroplasts yielded two fractions, a supernatant and a pellet, which were inactive alone, but active upon recombination. The protochlorophyllide produced *in vitro* was phototransformable, being photoreduced to chlorophyllide, which is gradually converted to chlorophyll in the light or in the dark (Nasrulhaq-Boyce et al 1987, Vijayan et al 1992). We could not, however, simplify our incubation mixtures with the reconstituted system by using the β-hydroxy or the β-oxo intermediates instead of Mg–protoporphyrin IX monomethyl ester, because supernatant, pellet, NADPH and O_2 were still required (Wong et al 1985). In no case could $NADP^+$ replace NADPH plus O_2. The cyclase activity of the reconstituted system was stimulated by Mg^{2+}, which could be replaced by a greater concentration of Na^+ or K^+. At high

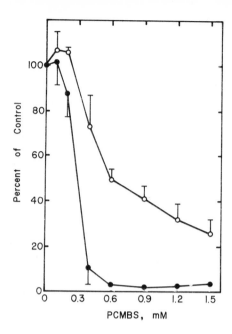

FIG. 4. The effect of PCMBS (*p*-chloromercuribenzene sulphonate) on Mg-chelatase (●) and Mg–protoporphyrin IX monomethyl ester (oxidative) cyclase (○). Activity is expressed as a percentage of that measured in the absence of PCMBS. The bars represent the range of duplicate samples. Reproduced from Fuesler et al 1984b, with the permission of the American Society of Plant Physiologists.

concentrations, all three cations were inhibitory, as are all other divalent and trivalent cations that have been tested (Whyte et al 1992).

Cyclase activity is inhibited by SH-complexing agents such as *p*-chloro-mercuribenzoate (PCMB) and *N*-ethylmaleimide (NEM). Some SH-containing compounds themselves inhibit the oxidative cyclization of Mg–protoporphyrin IX monomethyl ester (Wong & Castelfranco 1985). Many compounds that can act as artificial electron carriers, such as methyl viologen (paraquat), phenazine methosulphate, methylene blue, benzoquinone and hydroquinone, are inhibitory (Chereskin et al 1982, Whyte & Castelfranco 1993a); CN^- and N_3^- are also inhibitory, particularly in the reconstituted cyclase system (Whyte & Castelfranco 1993a), but CO does not appear to inhibit the activity (Chereskin et al 1982), and neither do other compounds that are commonly reported to inhibit cytochrome *P*-450-type mixed function oxidases (Whyte & Castelfranco 1993a).

A study on the localization of the Mg-chelatase and cyclase was done at Davis. At that time, both enzyme systems could be assayed in intact chloroplasts, but only the cyclase could be resolved into sub-plastidic fractions. Because both enzymes were inhibited by sulphydryl-complexing reagents, we compared the effect of PCMB and its non-penetrant analogue, PCMBS (*p*-chloromercuribenzene sulphonate), on the two systems in intact chloroplasts (Fig. 4) (Fuesler et al 1984b). PCMB inhibited both systems to a similar extent, but PCMBS inhibited the chelatase much more than it did the cyclase. The reconstituted cyclase was inhibited about equally by PCMB and PCMBS. From such data, we inferred that the chelatase is located superficially on the chloroplast, whereas the cyclase is deep inside the chloroplast and, therefore, not accessible to non-penetrant reagents like PCMBS. We have not yet confirmed the localization of either enzyme using the plastid-free assay systems and known markers.

Conclusions, speculations and new directions

The enzymology in this section of the chlorophyll biosynthetic pathway is far from being fully worked out. Some questions come immediately to mind.

(1) What is the role of ATP in the Mg-chelatase? The chelatase is activated by preincubation with ATP. We have suggested that the components must interact in an ATP-dependent manner before an active Mg-chelatase is formed. One of several possible mechanisms would be protein phosphorylation. This would explain why, in contrast to the situation with ferrochelatase, at least two proteins are required for activity. Because Mg-chelatase is at the branch-point of haem and chlorophyll synthesis, a sophisticated mechanism of regulation might be expected.

(2) *S*-Adenosyl-L-methionine appears to be limiting in isolated chloroplasts. Is it also limiting *in vivo*? If SAM is synthesized in the cytosol, how is it transported into the chloroplasts? Is there a specific carrier? More generally, do the various carriers that mediate the transport of the building blocks of the chlorophyll molecule in and out of the chloroplasts have a regulatory function in this biosynthetic pathway?

(3) What is the role of NADPH and O_2 in the cyclization? Do these co-substrates participate in a mixed function oxidase? Studies with inhibitors are not suggestive of a *typical* mixed function oxidase (Whyte & Castelfranco 1993a), but little is known about the mixed function oxidases found in chloroplasts; in fact, the best characterized plant mixed function oxidases are localized in the endoplasmic reticulum (Kleinig 1989). Is a common mechanism responsible for all, or many, of the O_2-requiring steps which occur during chloroplast development and which involve a number of different synthetic pathways (e.g. those to haems, chlorophylls, unsaturated lipids, carotenes, xanthophylls, etc.)?

(4) The overall cyclization process involves three steps (Fig. 3). These steps need to be separated and characterized individually, either by the use of specific metabolic inhibitors or by mutations. The intermediate substrates, for which synthetic protocols are now available, will be invaluable in this research.

(5) How do anaerobic photosynthetic bacteria (e.g., *Chlorobium* sp.) carry out the cyclization, which in chloroplasts has an absolute requirement for O_2?

(6) The *in vitro* cyclase system consists of pellet and soluble components. What is the enzymological contribution of each? Preliminary data from affinity chromatography suggest that the soluble fraction may bind porphyrins (Mg–protoporphyrin IX dimethyl ester–Sepharose) but may not bind NADPH (Blue Sepharose); the pellet fraction is irreversibly inhibited by 8-hydroxyquinoline and desferrioxamine mesylate and may therefore contain an essential heavy metal (Walker et al 1991). The origin of each component of our *in vitro* system, in terms of sub-plastidic structures, is also unknown. Conceivably, the soluble component(s) could be derived from the stroma or the intermembrane space of the envelope, or could be extrinsic proteins that are sheared from the membranes. Likewise, the pellet component(s) could come from the envelope or from the thylakoid membranes.

(7) Developing chloroplasts contain an active haem-synthesizing system. During light exposure of dark-grown barley seedlings chlorophyll accumulates rapidly, while chloroplast haems and haemoproteins are accumulated more slowly. However, haem is turned over in greening plant tissue (Castelfranco & Jones 1975) and also may be exported to other cell compartments (Thomas & Weinstein 1990). How are these two branches of the tetrapyrrole pathway related to each other? Which pools of intermediates are shared and which pools are duplicated? In one limiting case, the two pathways could share all the intermediates and all the enzymes from glutamate to protoporphyrin IX; at the other extreme, the syntheses of haem and Mg–protoporphyrin IX could be mediated by two totally separate complexes, with the two pathways diverging at the level of ALA formation. In this context, it should be noted that protoporphyrinogen oxidase (EC 1.3.3.4) has been found in two sub-plastidic locations (Matringe et al 1992).

(8) Finally, we must consider the way in which the all-important environmental factor, light, regulates the magnesium branch of the tetrapyrrole pathway. Despite a great amount of research, the mechanics of this regulation has not been worked out (Huang et al 1989).

Variability between species, stage of development and heterogeneity at the organ level will complicate further the experimental analysis of this important pathway, which is intimately connected to the developmental processes which lead to the formation of a mature chloroplast and a functioning photosynthetic cell.

Acknowledgements

The generous support of the National Science Foundation USA (P.A.C.) and the US Department of Energy (J.D.W.) through the years is gratefully acknowledged. Dr K. M. Smith greatly contributed to this research by providing synthetic substrates and reaction intermediates that are not commercially available. Dr S. M. Theg read this manuscript and made many helpful suggestions.

References

Beale SI, Weinstein JD 1990 Tetrapyrrole metabolism in photosynthetic organisms. In: Dailey HA (ed) Biosynthesis of heme and chlorophylls. McGraw-Hill, New York, p 287–391

Castelfranco PA, Jones OTG 1975 Protoheme turnover and chlorophyll synthesis in greening barley tissue. Plant Physiol (Bethesda) 55:485–490

Castelfranco PA, Weinstein JD, Schwarcz S, Pardo AD, Wezelman BE 1979 The Mg insertion step in chlorophyll biosynthesis. Arch Biochem Biophys 192:592–598

Chereskin BM, Wong Y, Castelfranco PA 1982 In vitro synthesis of the chlorophyll isocyclic ring. Transformation of magnesium-protoporphyrin IX and magnesium-protoporphyrin IX monomethyl ester into magnesium-2,4-divinyl pheoporphyrin A_5. Plant Physiol (Bethesda) 70:987–993

Fuesler TP, Hanamoto CM, Castelfranco PA 1982 Separation of Mg-protoporphyrin IX and Mg-protoporphyrin IX monomethyl ester synthesized de novo by developing cucumber etioplasts. Plant Physiol (Bethesda) 69:421–423

Fuesler TP, Castelfranco PA, Wong Y-S 1984a Formation of Mg-containing chlorophyll precursors from protoporphyrin IX, δ-aminolevulinic acid, and glutamate in isolated, photosynthetically competent, developing chloroplasts. Plant Physiol (Bethesda) 74:928–933

Fuesler TP, Wong Y-S, Castelfranco PA 1984b Localization of the Mg-chelatase and Mg-protoporphyrin IX monomethyl ester (oxidative) cyclase activites within isolated, developing cucumber chloroplasts. Plant Physiol (Bethesda) 75:662–664

Gassman M, Ramanujam P 1986 Relation between enzymatic destruction of magnesium porphyrins and chloroplast development. In: Akoyunouglou G, Senger H (eds) Regulation of chloroplast differentiation. Alan R Liss, New York, p 115–123

Granick S 1951 The structural and functional relationships between heme and chlorophyll. Harvey Lect 44:220–245

Hartwell JG, Weinstein JD 1992 Cucumber chloroplast magnesium-chelatase. Plant Physiol (Bethesda) (suppl) 99:4

Huang L, Bonner BA, Castelfranco PA 1989 Regulation of 5-aminolevulinic acid (ALA) synthesis in developing chloroplasts. II. Regulation of ALA-synthesizing capacity by phytochrome. Plant Physiol (Bethesda) 90:1003–1008

Kleinig H 1989 The role of plastids in isoprenoid biosynthesis. Annu Rev Plant Physiol Plant Mol Biol 40:39–59

Matringe M, Camadro J-M, Block MA, Scalla R, Labbe P, Douce R 1992 Localization within chloroplasts of protoporphyrinogen oxidase, the target enzyme for diphenylether-like herbicides. J Biol Chem 267:4646–4651

Nasrulhaq-Boyce A, Griffiths WT, Jones OTG 1987 The use of continuous assays to characterize the oxidative cyclase that synthesizes the chlorophyll isocyclic ring. Biochem J 243:23–29

Pardo AD, Chereskin BM, Castelfranco PA, Franceschi VR, Wezelman BE 1980 ATP requirement for Mg chelatase in developing chloroplasts. Plant Physiol (Bethesda) 65:956–960

Rebeiz CA, Parham R, Fasoula DA, Ioannides IM 1994 Chlorophyll a biosynthetic heterogeneity. In: The biosynthesis of the tetrapyrrole pigments. Wiley, Chichester (Ciba Found Symp 180) p 177–193

Shiau F, Whyte BJ, Castelfranco PA, Smith KM 1991 Partial syntheses of the isomerically pure magnesium (II) protoporphyrin IX monomethyl esters and their identification. J Chem Soc Perkins Trans I 1781–1785

Thomas J, Weinstein JD 1990 Measurement of heme efflux and heme content in isolated developing chloroplasts. Plant Physiol (Bethesda) 94:1414–1423

Vijayan P, Whyte B, Castelfranco P 1992 A spectrophotometric analysis of the Mg-protoporphyrin monomethyl ester (oxidative) cyclase. Plant Physiol & Biochem 30:271–278

Walker CJ, Weinstein JD 1991a *In vitro* assay of the chlorophyll biosynthetic enzyme Mg-chelatase: resolution of the activity into soluble and membrane-bound fractions. Proc Natl Acad Sci USA 88:5789–5793

Walker CJ, Weinstein JD 1991b Further characterization of the magnesium chelatase in isolated developing cucumber chloroplasts: substrate specificity, regulation, intactness and ATP requirements. Plant Physiol (Bethesda) 95:1189–1196

Walker CJ, Mansfield KE, Rezzano IN, Hanamoto CM, Smith KM, Castelfranco PA 1988 The magnesium-protoporphyrn IX (oxidative) cyclase system. Studies on the mechanism and specificity of the reaction sequence. Biochem J 255:685–692

Walker CJ, Mansfield KE, Smith KM, Castelfranco PA 1989 Incorporation of atmospheric oxygen into the carbonyl functionality of the protochlorophyllide isocyclic ring. Biochem J 257:599–602

Walker CJ, Castelfranco PA, Whyte BJ 1991 Synthesis of divinyl protochlorophyllide: enzymological properties of the Mg-protoporphyrin IX monomethyl ester oxidative cyclase system. Biochem J 276:691–697

Walker CJ, Hupp LR, Weinstein JD 1992 Activation and stabilization of Mg-chelatase activity by ATP as revealed by a novel *in vitro* continuous assay. Plant Physiol & Biochem 30:263–269

Wallsgrove R, Lea P, Miflin B 1983 Intracellular localization of aspartate kinase and the enzymes of threonine and methionine biosynthesis in green leaves. Plant Physiol (Bethesda) 71:780–784

Whyte BJ, Castelfranco PA 1993a Further observations on the Mg-protoporphyrin IX (oxidative) cyclase system. Biochem J 290:355–359

Whyte BJ, Castelfranco PA 1993b Breakdown of thylakoid pigments by soluble proteins of developing chloroplasts. Biochem J 290:361–367

Whyte BJ, Vijayan P, Castelfranco PA 1992 *In vitro* synthesis of protochlorophyllide: effects of Mg^{2+} and other cations on the reconstituted (oxidative) cyclase. Plant Physiol & Biochem 30:279–284

Wong Y-S, Castelfranco PA 1984 Resolution and reconstitution of Mg-protoporphyrin IX monomethyl ester (oxidative) cyclase, the enzyme system responsible for the formation of the chlorophyll isocyclic ring. Plant Physiol (Bethesda) 75:658–661

Wong Y-S, Castelfranco PA 1985 Properties of the Mg-protoporphyrin IX monomethyl ester (oxidative) cyclase system. Plant Physiol (Bethesda) 79:730–733

Wong Y-S, Castelfranco PA, Goff DA, Smith KM 1985 Intermediates in the formation of the chlorophyll isocyclic ring. Plant Physiol (Bethesda) 79:725–729

Note added in proof: The Mg-chelatase-catalysed reaction has now been convincingly separated into activation and catalysis. Both steps require ATP; however, the former has an apparent $K_{m(ATP)}$ of 0.3 mM and the latter one of < 0.2 mM. The analogue ATPγS substitutes for ATP in the activation step, but not in the catalysis. (Walker CJ, Weinstein JD 1994 The magnesium insertion step of chlorophyll biosynthesis is a two-stage reaction. Biochem J, in press)

DISCUSSION

Rebeiz: In your mechanism (Fig. 3) you showed the β-hydroxypropionate, which you have seen in HPLC profiles of cucumber extracts. Have you also seen the β-oxopropionate?

Castelfranco: No; the synthetic oxopropionate derivative acts as a substrate, but we have not isolated it from reaction mixtures.

Rebeiz: It would be reassuring to isolate it, to prove that your results are not due to non-specificity of the cyclase. As you know, these forward enzymes are not very specific.

Castelfranco: It would be reassuring, but I think that our isolation of the β-hydroxypropionate was a stroke of luck.

Rebeiz: By the way, S. M. Wu in my group detected the β-hydroxypropionate two years before your publication of its mass spectrum, but we didn't publish this finding because I didn't believe it.

Castelfranco: Isn't this always the way with preliminary results? If you rush to publish them, they turn out to be wrong; if you sit on them, someone else comes along and publishes them before you.

Arigoni: You saw a peak in the mass spectrum corresponding to the molecular mass expected for the hydroxy compound.

Rebeiz: Yes. We were actually looking for magnesium–monovinylprotoporphyrin.

Arigoni: All you saw was a peak at this molecular mass. You didn't know that the compound had that structure.

Rebeiz: We knew it had the molecular ion mass of a hydroxypropionate, but we had more pressing things to follow up at that time.

Beale: We have just confirmed Professor Castelfranco's result on the incorporation of O_2 into the keto group, by determining that O_2 also goes into the formyl group of chlorophyll *b* (Schneegurt & Beale 1992). In incubations with $^{18}O_2$, chlorophyll *a* was heavy by two atomic units and chlorophyll *b* by four units.

What do you suppose is the role of the O_2 in the conversion of the keto compound to the isocyclic ring?

Castelfranco: The conversion of the keto compound to the divinylprotochlorophyllide is still an oxidation—two hydrogens have to be removed, one from the *meso* carbon and the other from the side chain. Presumably, the same mechanism as used to remove the preceding four hydrogens would operate. The cyclization is a six-electron oxidation; when you get to the keto compound you have removed four electrons, so there are two more to go in the form of hydrogens. Whatever mechanism is used by the chloroplast to remove the first four reducing equivalents must also be used for the last two reducing equivalents.

Arigoni: Your reaction looks formally like a P_{450}-catalysed insertion of oxygen into a C—H bond.

Castelfranco: There are many analogies between what we have seen and P_{450}-catalysed reactions, but P_{450} reactions in chloroplasts are not well characterized.

Arigoni: I am not suggesting that P_{450} is actually involved, but something similar to it may well be. As for the ring closure step, there is some analogy with non-enzymic work by the late George Kenner (Kenner et al 1974).

K. Smith: Yes, there is. There is a photochemical thallium(III)-promoted one-electron oxidation of the porphyrin nucleus to make a cation radical, followed by a one-electron oxidation of the keto ester to make a keto ester radical, and the porphyrin cation radical can react with the keto ester radical to form the five-membered ring after loss of a proton.

Arigoni: You could also use the cationic centre to attack the enol form of the keto ester.

K. Smith: Basically, you can either make a dication of porphyrin and cyclize with the anion, or a cation radical and cyclize with the keto ester radical. This involves a peculiar photochemical thallium reaction, but in principle the biological cyclization could be just the same, going through a magnesium(II)–porphyrin cation radical which could cyclize with the keto ester radical and then lose a proton.

Castelfranco: Didn't you also get a cyclization with iodine in methanol in the presence of Na_2Co_3?

K. Smith: That's right, but there you finish up with a methoxy in position 10 (Fischer nomenclature).

Akhtar: Granick's suggestion that the ring E keto group is formed by dehydrogenation followed by addition of water and finally oxidation is likely to be correct in anaerobes, where I doubt that $^{18}O_2$ would label the ring E keto oxygen.

Castelfranco: Sam Granick postulated that scheme entirely by analogy with such things as β-oxidation, without any experimental evidence. I don't know how anaerobic organisms make that isocyclic ring. We've been thinking about extending this work to *Chlorobium*, but somehow haven't had the courage.

Rebeiz: Granick didn't even publish a scheme, only a sentence saying that there is probably a β-oxidation (Granick 1967).

Arigoni: How would the saturated side chain be converted to the β-keto acid?

Akhtar: There is no problem with an anaerobic organism having a non-oxygen oxidant. The Granick mechanism merely says that if the two hydrogen atoms are removed by a suitable oxidant, the hydroxylated side chain may be produced by hydration. Anaerobic organisms have buckets of non-oxygen oxidants.

Arigoni: Does the reaction operate on the free carboxylic acid?

Akhtar: No. The methyl ester is known to be made earlier in the biosynthesis, so the desaturation will occur on the methylated propionate side chain.

Leeper: The step from the hydroxypropionate to the oxopropionate appears rather simple, and might require just one alcohol dehydrogenase, perhaps. Have you looked for that activity in more fractionated systems?

Castelfranco: No.

Arigoni: In the biosynthesis of oestrogens the C-19 methyl group of the precursor is converted into an aldehydo group in two P_{450}-catalysed steps which first generate the hydrate of the aldehyde. There is a chance that something similar happens here.

Castelfranco: Formally, that second step is much like the reaction catalysed by lactate dehydrogenase, but we don't know its mechanism.

Rebeiz: Professor Akhtar has pointed out the similarities between the decarboxylation of the propionate residue and the formation of the cyclopentane ring. One difference between the two processes is that the acrylate in the decarboxylation may be an intermediate; it has been detected in meconium and its conversion to a vinyl group has been proposed (French et al 1970). During cyclopentane ring formation the acrylate substrate is completely inactive.

K. Smith: The acrylate equivalent in the coproporphyrin series, dehydrocoproporphyrin, or S-411 porphyrin, is found in meconium (Couch et al 1976) but I don't recall its conversion to vinyl being reported.

Akhtar: Isotopic experiments done by my group and Sir Alan Battersby's group in four different systems—duck and chicken erythrocytes, *Rhodobacter sphaeroides* and *Euglena*—rule out acrylate as an intermediate in the formation of the vinyl group. Both α-hydrogen atoms are retained in haem as well as in chlorophylls.

Beale: If you expose an etiolated system or a greening system to an iron chelator, you block isocyclic ring formation. What is the role of iron? Does iron have any effect *in vitro*?

Castelfranco: Iron has much less effect *in vitro*. If you treat etiolated tissue with 2,2'-dipyridyl or *O*-phenanthroline overnight, you get accumulation of several intermediates, including Mg–protoporphyrin, some protochlorophyllide and some protoporphyrin. The array of compounds depends on the tissue type. Short-term exposure of developing chloroplasts to iron or iron chelators, in our hands, has little effect; high concentrations inhibit everything, and low concentrations do nothing. We have not been able to make the transition from long-term incubations *in vivo* to short-term incubations with isolated chloroplasts.

Jordan: Professor Arigoni, what you said (p 206) about using hydroxylation to make the ketone worried me, because there's no hydroxylation proposed to assist in the formation of ring E.

Arigoni: I was merely suggesting that the keto compound could be formed as the result of a double hydroxylation.

Jordan: But in oestrogen biosynthesis the hydroxylation is necessary to activate the system for C–C bond cleavage.

Arigoni: It's true that in oestrogen biosynthesis there's a C–C bond cleavage, but not every oxidation has to be followed by a C–C bond cleavage.

Jordan: There are two steps in chlorophyll ring E synthesis—formation of the ketone from the hydroxy intermediate and cyclization to make the E ring.

Arigoni: Yes, and each step requires two electrons.

Jordan: I wasn't clear whether oxygen was required for each of those two steps.

Castelfranco: We cannot say whether oxygen is required for each of those steps, because we have not followed the conversion from Mg–protoporphyrin monomethyl ester to the hydroxy, or from the hydroxy to the ketone, or from the ketone to the final product. We always go from a substrate to the final product. We have not been able to abolish the oxygen requirement, nor have we shown that NADP can replace NADPH plus oxygen.

Arigoni: Does carbon monoxide inhibit your system?

Castelfranco: No. Cyanide and azide are fairly good inhibitors though.

Arigoni: If you start with the hydroxy compound, what effect do these inhibitors have?

Castelfranco: We have not tried that with carbon monoxide. That conversion is inhibited by methylene blue.

Jordan: This route has implications for how the formyl group at position 7 (IUPAC–IUB JCBN numbering, equivalent to position 3 in Fischer nomenclature) is generated and whether it also proceeds through a hydroxy intermediate.

Eschenmoser: In non-enzymic chemistry it is quite difficult to transport magnesium from a hydrated magnesium salt into a porphinoid ring. One efficient way is to put the magnesium into a lipid environment first. I wondered whether something like this might be involved in the chelating process. I get the impression that the number of carboxylate groups in the porphinoid substrate correlates with the chelatase's ability to insert Mg^{2+}.

Castelfranco: Protoporphyrin, mesoporphyrin and deuteroporphyrin all have two carboxylic acid groups, whereas haematoporphyrin also has two hydroxyethyl groups. The chelatase is considerably less active towards haematoporphyrin. Magnesium is not inserted into uroporphyrin or coproporphyrin in our system.

Eschenmoser: Is it conceivable from a biochemical point of view that the ATP could help transport magnesium from a hydrophilic environment into a lipid environment?

Castelfranco: If that were the case, chelation should be inhibited by uncouplers or ionophores, but it is not.

Eschenmoser: If the ATP acted by esterifying the magnesium hydrate with a lipid group, would uncouplers be expected to be inhibitory?

Castelfranco: I don't know.

Griffiths: I would suggest that Mg^{2+}–ATP might be the substrate.

Castelfranco: It probably is, because both are required, at about the same concentration.

References

Couch PW, Games DE, Jackson AH 1976 Synthetic and biosynthetic studies of porphyrins. I. Synthesis of the 'S-411' porphyrin obtained from meconium. J Chem Soc Perkin Trans I, p 2492–2501

French J, Nicholson DC, Rimington C 1970 Identification of acrylate porphyrin S-411 from meconium. Biochem J 120:393–397

Granick S 1967 The heme and chlorophyll biosynthetic chain. In: Goodin TW (ed) Biochemistry of chloroplasts. Academic Press, New York, vol 2:373–410

Kenner GW, McCombie SW, Smith KM 1974 Pyrroles and related compounds. 30. Cyclisation of porphyrin β-keto-esters to phaeoporphyrins. J Chem Soc Perkin Trans I, p 527–530

Schneegurt MA, Beale SI 1992 Origin of the chlorophyll b formyl oxygen atom in *Chlorella vulgaris*. Biochemistry 31:11677–11683

Biosynthesis of coenzyme F$_{430}$, a nickel porphinoid involved in methanogenesis

Rudolf K. Thauer and Lutz G. Bonacker

Laboratorium für Mikrobiologie des Fachbereichs Biologie der Philipps-Universität Marburg, Karl-von-Frisch-Strasse, Postfach 1929, and Max-Planck-Institut für Terrestrische Mikrobiologie Marburg, D-35043 Marburg, Germany

Abstract. Coenzyme F$_{430}$ is the prosthetic group of methyl-coenzyme-M reductase, which catalyses the final step of methane formation in methanogenic bacteria. The coenzyme is a nickel-containing macrocyclic tetrapyrrole of unique structure. We describe the biosynthesis of this nickel porphinoid from L-glutamate via 5-aminolaevulinic acid, uroporphyrinogen III and dihydrosirohydrochlorin, the binding of the coenzyme to methyl-coenzyme-M reductase and the regulation of coenzyme F$_{430}$ biosynthesis. We end with some evolutionary considerations on the biosynthesis of macrocyclic tetrapyrroles and remarks on the degradation of these compounds under anaerobic conditions.

1994 The biosynthesis of the tetrapyrrole pigments. Wiley, Chichester (Ciba Foundation Symposium 180) p 210–227

An overview of methanogenesis and the role of coenzyme F$_{430}$

Most methanogenic bacteria grow on CO_2 and H_2 as their sole energy sources ($CO_2 + 4H_2 \rightarrow CH_4 + 2H_2O$; $\Delta G^{\circ}{}' = -131$ kJ/mol). CO_2 is reduced to CH_4 via carrier-bound, one-carbon intermediates. CO_2 first reacts with methanofuran to yield formylmethanofuran; this endergonic reaction is catalysed by formylmethanofuran dehydrogenase (EC 1.2.99.5), a membrane-bound iron–sulphur protein which contains a molybdopterin prosthetic group and molybdenum or tungsten (extreme thermophiles tend to have tungsten). The formyl group is then transferred to tetrahydromethanopterin, and the resultant N^5-formyltetrahydromethanopterin is cyclized to N^5,N^{10}-methenyltetrahydromethanopterin, which is then reduced in two two-electron steps to N^5-methyltetrahydromethanopterin. The enzymes catalysing these steps are soluble and contain no prosthetic groups. The methyl group of N^5-methyltetrahydromethanopterin is transferred to coenzyme M (2-mercaptoethanesulphonate) by N^5-methyltetrahydromethanopterin:coenzyme-

210

M methyltransferase (tetrahydromethanopterin S-methyltransferase, EC 2.1.1.86), a membrane-bound, seven-subunit, iron–sulphur, corrinoid (5-hydroxy-benzimidazolyl cobamide) enzyme. This step is the first in the pathway to be exergonic enough to involve conservation of energy ($\Delta G°' = -30\,kJ/mol$), and the conversion is thought to be coupled to the generation of a sodium-motive force (Becher et al 1992). This sodium gradient drives the endergonic CO_2 reduction to formylmethanofuran and can be converted to a proton-motive force.

The final step in methane formation is the reductive demethylation of methyl-coenzyme M (2-[methylthio]ethane sulphonate, CH_3-S-CoM) with N^7(mercaptoheptanoyl)-L-threonine-O^3-phosphate (H-S-HTP) to yield methane and the heterodisulphide of coenzyme M, as shown in 1 (Ellerman et al 1988). Ellefson et al (1982) discovered that F_{430} is the prosthetic group of the enzyme which catalyses this reaction, methyl-coenzyme-M reductase.

$$CH_3\text{-S-CoM} + \text{H-S-HTP} \rightarrow CH_4 + \text{CoM-S-S-HTP} \quad \Delta G°' = -45\,kJ/mol \quad \mathbf{1}$$

This step is exergonic, but there is no evidence for energy conservation. However, reduction of the heterodisulphide with H_2 ($\Delta G°' = -40\,kJ/mol$) is associated with the generation of a proton motive force which can drive formation of ATP from ADP and P_i (Blaut et al 1992).

Coenzyme F_{430} (Fig. 1) is a nickel porphinoid found only in methanogenic bacteria, in which it is always present. The yellow, non-fluorescent, low molecular mass substance was first isolated in 1978 by Gunsalus & Wolfe, who named it factor F_{430} because of its absorption maximum at 430 nm. In 1980 the compound was shown to contain nickel (Diekert et al 1980a, Whitman & Wolfe 1980). Several biosynthetic experiments indicated that the newly isolated nickel-containing compound was in fact a macrocyclic tetrapyrrole (Diekert et al 1980b, Jaenchen et al 1981). Its structure was unravelled in the laboratory of Eschenmoser at the ETH, Zürich (Pfaltz et al 1982, 1985, Livingston et al 1984, Fässler et al 1985, Färber et al 1991). Battersby's laboratory had a part in the elucidation of the absolute configuration in the region of rings A and B (Fässler et al 1985).

Evidence is available from electron paramagnetic resonance studies that active methyl-coenzyme-M reductase contains coenzyme F_{430} in a reduced Ni(I) form (Rospert et al 1992). Studies with models suggest that cleavage of the methyl–sulphur bond of CH_3–S–CoM by attacking reduced Ni(I)–F_{430} leads to transfer of the methyl group to the nickel such that methane then arises by protonolysis of the methylated cofactor (Lin & Jaun 1991, Berkessel 1991). Consistent with this mechanism is the finding that the reduction of the methyl group to methane involves inversion of configuration (Ahn et al 1991).

Presently, about 10^9 tonnes of methane are generated every year in anoxic environments on earth by the action of methanogenic bacteria using coenzyme

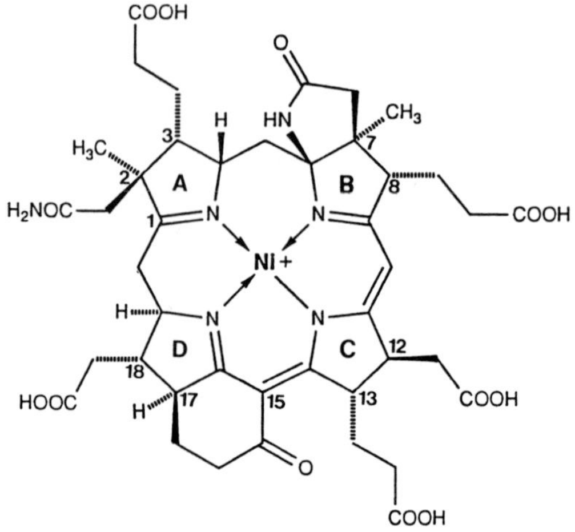

FIG. 1.   The structure of coenzyme F_{430}.

F_{430} as the catalyst. About 40% of the methane is released into the atmosphere, where its concentration has been increasing continuously for many years. This is an important issue, because methane is a greenhouse gas which contributes significantly to the anthropogenic greenhouse effect. Thus, coenzyme F_{430} is involved in the catalysis of a process of global importance.

Here, we concentrate on the biosynthesis of coenzyme F_{430}. For detailed treatises on the chemistry and biochemistry of coenzyme F_{430}, see reviews by DiMarco et al (1990), Friedmann et al (1991) and Jaun (1993). For reviews on the biochemistry of methanogenesis in general, see Ferry (1992), Blaut et al (1992) and Weiss & Thauer (1993). For a historical review, see Wolfe (1991).

Biosynthesis of coenzyme F_{430} from 5-aminolaevulinic acid via uroporphyrinogen III and dihydrosirohydrochlorin

Biosynthetic experiments, as already mentioned, indicated soon after coenzyme F_{430}'s discovery that it is a macrocyclic tetrapyrrole. Growing methanogenic bacteria were shown to incorporate ^{14}C-labelled 5-aminolaevulinic acid into coenzyme F_{430}. The ^{14}C-labelled coenzyme had a specific radioactivity eight times higher than that of the 5-aminolaevulinic acid used (Diekert et al 1980b). Labelling studies with L-[*methyl*-^{14}C]methionine and L-[*methyl*-^{3}H]methionine then established the presence in coenzyme F_{430} of two methyl groups derived from methionine (Jaenchen et al 1981). At this point NMR spectroscopic analysis was performed with a variety of differently labelled coenzyme F_{430} preparations isolated from methanogenic bacteria grown in the presence of 5-amino[2-^{13}C]-laevulinic acid, 5-amino[3-^{13}C]laevulinic acid, 5-amino[4-^{13}C]laevulinc acid,

5-amino[5-^{13}C]laevulinc acid or L-[*methyl*-^{13}C]methionine. The results of these and other studies led to the structure shown in Fig. 1 (Pfaltz et al 1982, 1985, Livingston et al 1984, Fässler et al 1985, Färber et al 1991).

This structure has side chains analagous to those of uroporphyrinogen III and two carbon–carbon methyl groups identical to those in dihydrosirohydrochlorin, strongly suggesting that one step in coenzyme F_{430} biosynthesis is the

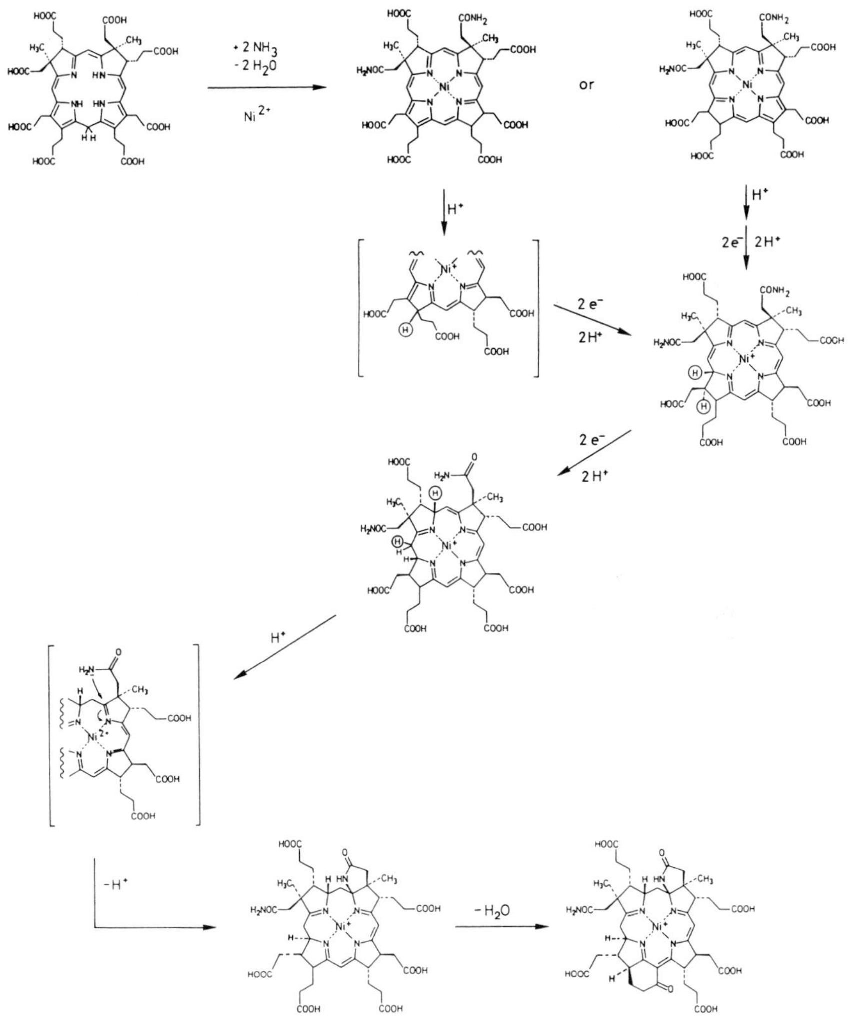

FIG. 2. Proposed steps in the biosynthesis of coenzyme F_{430} from dihydrosirohydrochlorin. The structure immediately preceding coenzyme F_{430} is 15,17^3-seco-F_{430}-17^3-acid (Pfaltz et al 1987).

conversion of uroporphyrinogen III to dihydrosirohydrochlorin, a change also encountered in the course of B_{12} and sirohaem formation. The occurence of this conversion was substantiated by the subsequent observation that coenzyme F_{430} is formed both from uroporphyrinogen III in growing methanogenic bacteria (Gilles & Thauer 1983) and from sirohydrochlorin in cell extracts of these microorganisms (Mucha et al 1985).

A more detailed inspection of the structure of coenzyme F_{430} suggests that dihydrosirohydrochlorin is converted to coenzyme F_{430} via the five steps shown in Fig. 2 (not necessarily in the following order): (i) insertion of nickel; (ii) amidation of the ring A and ring B acetate groups; (iii) reduction of two double

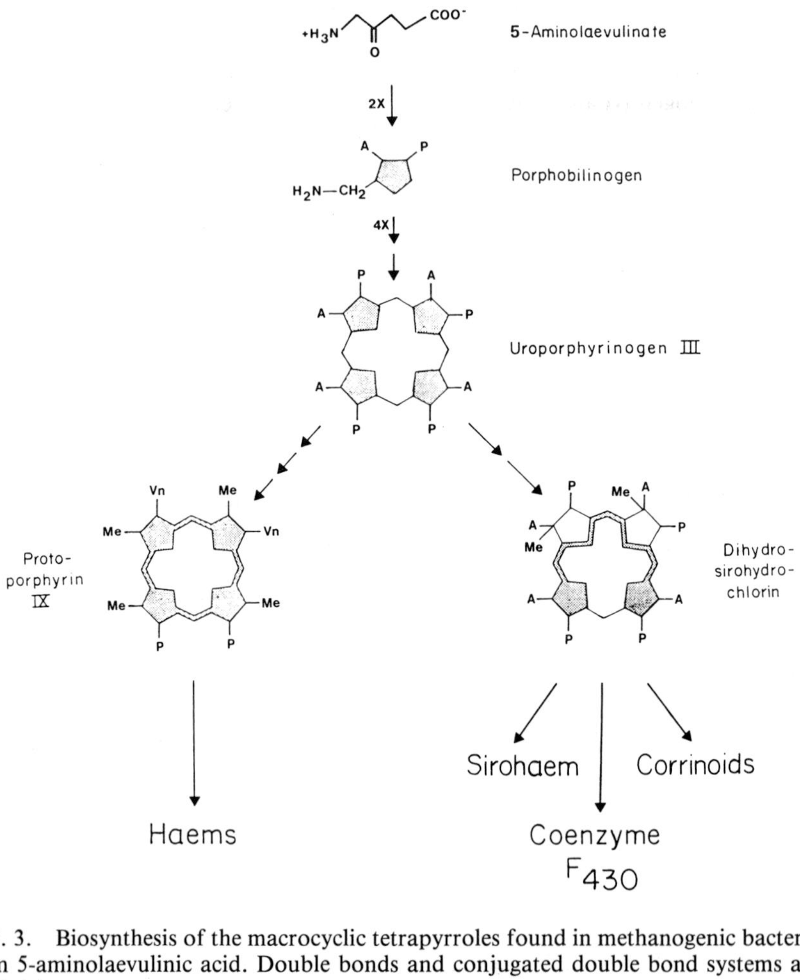

FIG. 3.   Biosynthesis of the macrocyclic tetrapyrroles found in methanogenic bacteria from 5-aminolaevulinic acid. Double bonds and conjugated double bond systems are shaded in grey.

bonds; (iv) cyclization of the acetamide of ring B; and (v) cyclization of the propionic acid groups of ring D (Pfaltz et al 1987). In this reaction sequence, only the formation of $15,17^3$-seco-F_{430}-17^3 acid (Fig. 2) and its conversion to coenzyme F_{430} have been demonstrated experimentally (Pfaltz et al 1987). Seco-F_{430} contains an uncyclized propionic acid group at ring D. Before cyclization can occur, this group must almost certainly be activated, possibly via acetyl-CoA thioester.

Methanogenic bacteria also all contain a corrinoid, generally 5-hydroxy-benzimidazolyl cobamide (Stupperich et al 1990). Some additionally contain haem (cytochromes b and c) (Blaut et al 1992) or sirohaem (sulphite reductase) or both (Moura et al 1982). From studies with other organisms, uroporphyrinogen III is known to be the common precursor in the biosynthesis of corrinoids, haem and sirohaem, and dihydrosirohydrochlorin is known to be the common intermediate in both sirohaem and corrinoid biosynthesis. Thus uroporphyrinogen III and dihydrosirohydrochlorin are required in methanogenic bacteria for more than just the biosynthesis of coenzyme F_{430} (Fig. 3).

Biosynthesis of 5-aminolaevulinic acid from L-glutamate

5-Aminolaevulinic acid is made in methanogenic bacteria from L-glutamate (Gilles et al 1983). The involvement of a tRNA^{Glu} has been demonstrated (Friedmann & Thauer 1986, Friedmann et al 1987), as shown in (2). The pathway of 5-aminolaevulinic acid formation for coenzyme F_{430} biosynthesis is thus the same as that in plants and in most eubacteria.

$$\text{Glu} + \text{tRNA}^{Glu} + \text{ATP} \rightarrow \text{Glu-tRNA}^{Glu} + \text{AMP} + \text{PP}_i$$

$$\text{Glu-tRNA}^{Glu} + \text{NAD(P)H} \rightarrow \text{glutamate 1-semialdehyde} + \text{NAD(P)} + \text{tRNA}^{Glu}$$

$$\text{Glutamate 1-semialdehyde} \rightarrow \text{5-aminolaevulinic acid} \qquad \textbf{2}$$

Biosynthesis of glutamate from CO_2

Most methanogenic bacteria can grow using CO_2 as a sole carbon source from which all cell components thus have to be made. CO_2 fixation in these microorganisms is not via the Calvin cycle, but through a pathway in which acetyl-CoA is the first fixation product with a carbon–carbon bond. Formylmethanofuran, N^5-formyltetrahydromethanopterin, N^5,N^{10}-methenyl-tetrahydromethanopterin, N^5,N^{10}-methylenetetrahydromethanopterin and N^5-methyltetrahydromethanopterin are most probably intermediates in the biosynthesis of the methyl group of acetyl-CoA, and carbon monoxide is an intermediate in the formation of the carbonyl group of acetyl-CoA. The final step is the formation of acetyl-CoA and tetrahydromethanopterin from N^5-methyltetrahydromethanopterin, CO and CoA. Pyruvate is then generated from acetyl-CoA by reductive carboxylation. Pyruvate is then converted to oxaloacetate via phosphoenolpyruvate (Länge & Fuchs 1987).

There are two pathways from oxaloacetate to glutamate. In the order *Methanobacteriales*, glutamate is synthesized from oxaloacetate and CO_2 via malate, fumarate, succinate, succinyl-CoA and 2-oxoglutarate. In the order *Methanomicrobiales*, glutamate is synthesized from oxaloacetate and acetyl-CoA via citrate, aconitate, isocitrate and 2-oxoglutarate (for a review see Jones et al 1987).

Binding of coenzyme F_{430} to methyl-coenzyme-M reductase

Coenzyme F_{430} is present in methanogenic bacteria both in the free state and bound to methyl-coenzyme-M reductase (Ankel-Fuchs et al 1984). The bound form is released only on denaturation of the enzyme and denaturation appears to be irreversible, at least *in vitro*. Originally it was thought that the two forms differed from each other, but it has now been shown that the free and the bound coenzyme F_{430} (after release from the protein) are chemically identical (Hüster et al 1985).

There is evidence that coenzyme F_{430} participates in methane formation only when bound to methyl-coenzyme-M reductase. The free form apparently functions only as a 'precursor' of the bound form (Ankel-Fuchs et al 1984).

Methyl-coenzyme-M reductase is a hexamer of three different subunits with an $\alpha_2\beta_2\gamma_2$ composition and an apparent molecular mass of 300 kDa. Each molecule of the holoenzyme contains two molecules of tightly bound coenzyme F_{430}. There is some evidence that coenzyme F_{430} is associated with the α subunit (DiMarco et al 1990).

Methyl-coenzyme-M reductase has absorption maxima at 278 nm and at 420 nm with a shoulder at 445 nm. The $A_{278}:A_{420}$ ratio is six. The extinction coefficient at 420 nm, 22 000 cm^{-1} M^{-1}, is practically the same as that of the free coenzyme at 430 nm (DiMarco et al 1990). The shift in the absorption maximum and the generation of the 445 nm absorbance shoulder are not understood. It has been postulated that the absorbance changes are due to the formation of a Schiff's base between the carbonyl group of the carbocyclic ring of coenzyme F_{430} and an amino group in the α subunit. However, attempts to reduce the putative base and thus to link the coenzyme covalently to the peptide chain have been unsuccessful (Friedmann et al 1991).

The primary structures of the three subunits of methyl-coenzyme-M reductase have been determined by analysis of the encoding genes; methyl-coenzyme-M reductase genes from five different methanogens have been completely sequenced (Friedmann et al 1991, Reeve 1992). These sequences show a high degree of conservation, with consensus sequences composed of identical or functionally homologous amino acids. These consensus sequences may be considered to be part of functionally essential domains determining, for example, cofactor binding, nucleation of protein folding, or subunit interactions. A binding site for coenzyme F_{430} was not apparent.

Regulation of coenzyme F_{430} biosynthesis

The concentration of free coenzyme F_{430} in methanogenic bacteria is dependent on the concentration of nickel in the growth medium. At low nickel concentrations only coenzyme F_{430} bound to methyl-coenzyme-M reductase is found, and at growth-limiting nickel concentrations the concentration of methyl-coenzyme-M reductase and thus of bound coenzyme F_{430} is decreased (Ankel-Fuchs et al 1984). These findings indicate that the biosynthesis of coenzyme F_{430} is regulated mainly by the availability of nickel rather than by feedback control.

Cells grown under conditions of nickel limitation appear not to contain the apoenzyme of methyl-coenzyme-M reductase. The biosynthesis of the apoenzyme is therefore probably positively regulated either by nickel or by coenzyme F_{430}. Alternatively, the apoenzyme could be degraded when synthesized in the absence of coenzyme F_{430}.

It has been found recently that some methanogenic bacteria which grow on H_2 and CO_2 contain two isoenzymes of methyl-coenzyme-M reductase which are differentially expressed, depending on the growth conditions (Bonacker et al 1992). Isoenzyme I is more abundant in cells grown under conditions of H_2 limitation and isoenzyme II in cells grown under non-H_2-limiting conditions. Evidently, biosynthesis of these isoenzymes is subject to regulatory control by other factors in addition to the availability of nickel and coenzyme F_{430}.

The three genes encoding the three subunits of methyl-coenzyme-M reductase are organized in a transcription unit in which two other genes have been found in all cases analysed (Friedmann et al 1991, Reeve 1992) (Fig. 4). These two open reading frames lie between the genes coding for the β and γ subunits and are conserved between species to different degrees, the smaller one being more heterogeneous both in size and in the amino acid sequence it encodes. From the structure of their translational signal and their codon usage, and from

FIG. 4. The common arrangement of the methyl-coenzyme-M reductase gene clusters in methanogenic bacteria. Genes *mcrA*, *mcrB*, and *mcrG* encode subunits α, β and γ of the enzyme, respectively. The functions of the *mcrC* and *mcrD* gene products are not known. The lower and upper limits of the molecular masses (in kDa) of each type of polypeptide encoded are shown below the genes. The molecular masses were derived from the five known nucleotide sequences from *Methanobacterium thermoautotrophicum*, *Methanothermus fervidus*, *Methanococcus voltae*, *Methanococcus vannielii* and *Methanosarcina barkeri*.

218

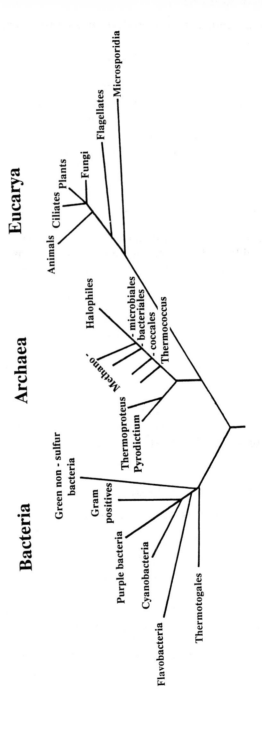

FIG. 5. Rooted universal phylogenetic tree (Wheelis et al 1992).

attempts to determine the gene products in cell extracts by immunoprecipitation, it appears likely that the polypeptides encoded by these sequences are not produced in a 1:1 stoichiometry relative to the α, β and γ subunits. In fact, a sub-stoichiometric production of the *mcrD* gene product has been demonstrated (Reeve 1992). The role of these polypeptides remains to be elucidated. Indirect evidence suggests they are involved in the activation of methyl-coenzyme-M reductase, which probably proceeds via reduction of the enzyme bound coenzyme F$_{430}$ from the Ni(II) state to the Ni(I) state. They may also play a role in the assembly of the enzyme, or could have a function not yet envisaged.

Evolutionary considerations

All methanogenic bacteria, as mentioned above, contain, besides coenzyme F$_{430}$, corrinoids, and some additionally contain haems or sirohaems or both. The methanogens all belong to the domain Archaea, considered to be more closely related phylogenetically to the first living organisms on earth, the progenotes, than the Bacteria and Eucarya (Wheelis et al 1992) (Fig. 5). This consideration suggests that biosynthesis of coenzyme F$_{430}$ and of the other macrocyclic tetrapyrroles found in methanogenic bacteria must have developed early in the evolution of life.

Of the genes involved in coenzyme F$_{430}$ biosynthesis only that encoding 5-aminolaevulinic acid dehydratase (porphobilinogen synthase, EC 4.2.1.24) has been cloned and sequenced. It shows a high degree of sequence identity (40–53%) to the genes encoding the dehydratase in *Escherichia coli*, *Bacillus subtilis*, yeast and mammals; the substrate-binding sites and the zinc-binding regions are almost perfectly conserved (Bröckl et al 1992).

Biodegradation of coenzyme F$_{430}$

Coenzyme F$_{430}$ and the other macrocyclic tetrapyrroles present in methanogenic bacteria do not accumulate in the natural habitats of these anaerobic microorganisms. These compounds must therefore somehow be degraded. The mechanism of this degradation is an interesting question; molecular oxygen is essential for the degradation of haems and chlorophylls in plants and animals, but it is not available to methanogenic bacteria in their natural habitat. There has been only one report of a macrocycle of a tetrapyrrole being cleaved under anoxic conditions. Brumm et al (1983) showed that in cell extracts of *Clostridium tetanomorphum* uroporphyrinogen I is converted to a blue bile pigment in the complete absence of molecular oxygen. How this is achieved mechanistically is not known. (For the structure of this bactobilin see Valasinas et al 1985).

Acknowledgements

This work was supported by a grant from the Deutsche Forschungsgemeinschaft and by the Fonds der Chemischen Industrie.

References

Ahn Y, Krzycki JA, Floss HG 1991 Steric course of the reduction of ethyl coenzyme M to ethane catalyzed by methyl coenzyme M reductase from *Methanosarcina barkeri*. J Am Chem Soc 113:4700–4701

Ankel-Fuchs D, Jaenchen R, Gebhardt NA, Thauer RK 1984 Functional relationship between protein-bound and free factor F430 in *Methanobacterium*. Arch Microbiol 139:332–337

Becher B, Müller V, Gottschalk G 1992 N^5-Methyl-tetrahydromethanopterin: coenzyme M methyltransferase of *Methanosarcina* strain Göl is an Na$^+$-transporting membrane protein. J Bacteriol 174:7656–7660

Berkessel A 1991 Methyl-coenzyme M reductase: model studies on pentadentate nickel complexes and a hypothetical mechanism. Bioorg Chem 19:101–115

Blaut M, Müller V, Gottschalk G 1992 Energetics of methanogenesis studied in vesicular systems. J Bioenerg Biomembr 24:529–546

Bonacker LG, Baudner S, Thauer RK 1992 Differential expression of the two methyl-coenzyme M reductases in *Methanobacterium thermoautotrophicum* as determined immunochemically via isoenzyme-specific antisera. Eur J Biochem 206:87–92

Bröckl G, Berchtold M, Behr M, König H 1992 Sequence of the 5-aminolevulinic acid dehydratase-encoding gene from the hyperthermophilic methanogen, *Methanothermus sociabilis*. Gene 119:151–152

Brumm PJ, Fried J, Friedmann HC 1983 Bactobilin: blue bile pigment isolated from *Clostridium tetanomorphum*. Proc Natl Acad Sci USA 80:3943–3947

Diekert G, Klee B, Thauer RK 1980a Nickel, a component of factor F_{430} from *Methanobacterium thermoautotrophicum*. Arch Microbiol 124:103–106

Diekert G, Jaenchen R, Thauer RK 1980b Biosynthetic evidence for a nickel tetrapyrrole structure of factor F_{430} from *Methanobacterium thermoautotrophicum*. FEBS (Fed Eur Biochem Soc) Lett 119:118–120

DiMarco AA, Bobik TA, Wolfe RS 1990 Unusual coenzymes of methanogenesis. Annu Rev Biochem 59:355–394

Ellefson WL, Whitman WB, Wolfe RS 1982 Nickel-containing factor F_{430}: chromophore of the methylreductase of *Methanobacterium*. Proc Natl Acad Sci USA 79:3707–3710

Ellermann J, Hedderich R, Böcher R, Thauer RK 1988 The final step in methane formation. Investigations with highly purified methyl-CoM reductase (component C) from *Methanobacterium thermoautotrophicum* (strain Marburg). Eur J Biochem 172:669–677

Färber G, Keller W, Kratky L et al 1991 Coenzyme F_{430} from methanogenic bacteria: complete assignment of configuration based on an X-ray analysis of 12,13-diepi-F_{430} pentamethyl ester and on NMR spectroscopy. Helv Chim Acta 74:697–716

Fässler A, Kobelt A, Pfaltz A et al 1985 Zur Kentnnis des Faktors F_{430} aus methanogenen Bakterien: absolute Konfiguration. Helv Chim Acta 68:2287–2298

Ferry JG 1992 Methane from acetate. J Bacteriol 174:5489–5495

Friedmann HC, Thauer RK 1986 Ribonuclease-sensitive d-aminolevulinic acid formation from glutamate in cell extracts of *Methanobacterium thermoautotrophicum*. FEBS (Fed Eur Biochem Soc) Lett 207:84–88

Friedmann HC, Thauer RK, Gough SP, Kannangara CG 1987 D-aminolevulinic acid formation in the archaebacterium *Methanobacterium thermoautotrophicum* requires tRNA^{Glu}. Carlsberg Res Commun 52:363–371

Friedmann HC, Klein A, Thauer RK 1991 Biochemistry of coenzyme F_{430}, a nickel porphinoid involved in methanogenesis. In: Jordan PM (ed) Biosynthesis of tetrapyrroles, new comprehensive biochemistry. Elsevier Science Publishers, Amsterdam, p 139–154

Gilles H, Thauer RK 1983 Uroporphyrinogen III, an intermediate in the biosynthesis of the nickel-containing factor F_{430} in *Methanobacterium thermoautotrophicum*. Eur J Biochem 135:109–112

Gilles H, Jaenchen R, Thauer RK 1983 Biosynthesis of 5-aminolevulinic acid in *Methanobacterium thermoautotrophicum*. Arch Microbiol 135:237–240

Gunsalus RP, Wolfe RS 1978 Chromophoric factors F_{342} and F_{430} of *Methanobacterium thermoautotrophicum*. FEMS (Fed Eur Microbiol Soc) Microbiol Lett 3:191–193

Hüster R, Gilles HH, Thauer RK 1985 Is coenzyme M bound to factor F_{430} in methanogenic bacteria? Experiments with *Methanobrevibacter ruminantium*. Eur J Biochem 148:107–111

Jaenchen R, Diekert G, Thauer RK 1981 Incorporation of methionine-derived methyl groups into factor F_{430} by *Methanobacterium thermoautotrophicum*. FEBS (Fed Eur Biochem Soc) Lett 130:133–136

Jaun B 1993 Methane formation by methanogenic bacteria: redox chemistry of coenzyme F_{430}. In: Sigel H, Sigel M (eds) Metal ions in biological systems, vol 29: Biological properties of metal alkyl derivates. Marcel Dekker, New York, p 287–337

Jones WJ, Nagle DP Jr, Whitman WB 1987 Methanogens and the diversity of archaebacteria. Microbiol Rev 51:135–177

Länge S, Fuchs G 1987 Autotrophic synthesis of activated acetic acid from CO_2 in *Methanobacterium thermoautotrophicum*. Synthesis from tetrahydromethanopterin-bound C_1 units and carbon monoxide. Eur J Biochem 163:147–154

Lin SK, Jaun B 1991 Coenzyme F_{430} from methanogenic bacteria: detection of a paramagnetic methylnickel (II) derivative of the pentamethyl ester by ^{2}H-NMR spectroscopy. Helv Chim Acta 74:1725–1738

Livingston DA, Pfaltz A, Schreiber J et al 1984 Zur Kenntnis des Faktors F_{430} aus methanogenen Bakterien: Struktur des proteinfreien Faktors. Helv Chim Acta 67: 334–351

Moura JJG, Moura I, Santos H, Xavier AV, Scandellari M, LeGall J 1982 Isolation of P_{590} from *Methanosarcina barkeri*: evidence for the presence of sulfite reductase activity. Biochem Biophys Res Commun 108:1002–1009

Mucha H, Keller E, Weber H, Lingens F, Trösch W 1985 Sirohydrochlorin, a precursor of factor F_{430} biosynthesis in *Methanobacterium thermoautotrophicum*. FEBS (Fed Eur Biochem Soc) Lett 190:169–171

Pfaltz A, Jaun B, Fässler A, Eschenmoser A et al 1982 Zur Kenntnis des Faktors F_{430} aus methanogenen Bakterien: Struktur des porphinoiden Ligandsystems. Helv Chim Acta 65:828–865

Pfaltz A, Livingston DA, Jaun B, Diekert G, Thauer RK, Eschenmoser A 1985 Zur Kenntnis des Faktors F_{430} aus methanogenen Bakterien: Über die Natur der Isolierungsartefakte, ein Beitrag zur Chemie von F_{430} und zur konformationellen Stereochemie der Ligandperipherie von hydroporphinoiden Nickel(II)-Komplexen. Helv Chim Acta 68:1338–1358

Pfaltz A, Kobelt A, Hüster R, Thauer RK 1987 Biosynthesis of coenzyme F_{430} in methanogenic bacteria. Identification of 15,17^3-seco-F_{430}-17^3-acid as an intermediate. Eur J Biochem 170:459–467

Reeve JN 1992 Molecular biology of methanogens. Annu Rev Microbiol 46:165–191

Rospert S, Voges M, Berkessel A, Albracht SPJ, Thauer RK 1992 Substrate-analogue-induced changes in the nickel-EPR spectrum of active methyl-coenzyme-M reductase from *Methanobacterium thermoautotrophicum*. Eur J Biochem 210:101–107

Stupperich E, Eisinger HJ, Schurr S 1990 Corrinoids in anaerobic bacteria. FEMS (Fed Eur Microbiol Soc) Microbiol Rev 87:355–360

Weiss DS, Thauer RK 1993 Methanogenesis and the unity of biochemistry. Cell 72:819–822

Valasinas A, Diaz L, Frydman B, Friedmann HC 1985 Total synthesis of the urobiliverdin isomers. Identification of bactobilin as urobiliverdin I. J Org Chem 50:2398–2400

Wheelis ML, Kandler O, Woese CR 1992 On the nature of global classification (domain/systematics/molecular evolution). Proc Natl Acad Sci USA 89:2930–2934

Whitman WB, Wolfe RS 1980 Presence of nickel in factor F_{430} from *Methanonbacterium bryantii*. Biochem Biophys Res Commun 92:1196–1201

Wolfe RS 1991 My kind of biology. Annu Rev Microbiol 45:1–35

DISCUSSION

Kräutler: You mentioned that the membrane-bound corrinoid enzyme which transfers a methyl group to coenzyme M, N^5-methyltetrahydromethanopterin:coenzyme-M methyltransferase, is apparently associated with the build up of a sodium gradient which, amazingly, seems not to be associated with a redox process. Of course, formally, Co(I) goes to Co(III), and vice versa, but really there is no exchange of electrons; but, on the other hand, the methyl cation is transported, so there must be some reverse transport of a cation. This may be how a gradient of cations can be built up.

Thauer: I agree that in principle you could envisage some way to make the methyl transfer a vectorial process. My argument against that is that all the straightforward mechanisms proposed for the build up of electrochemical gradients have turned out to be too simplistic. For example, the H^+-transporting F_1/F_0 ATPase (H^+-transporting ATP synthase, EC 3.6.1.34) of bacteria, mitochondria and chloroplasts was thought to transport protons along specific carboxyl groups with different pK values, but this turned out not to be the case, because, with minor mutational changes, the bacterial enzyme can also translocate sodium ions. The transfer of sodium ions coupled to the methyltransferase reaction in methanogenic bacteria must be a complicated process, because it requires an enzyme with seven different subunits; we are fairly sure that most of the subunits are part of the machinery. The genes encoding at least three of the seven subunits are arranged in a transcription unit.

Kräutler: Could this not rationally allow for the transport of a sodium ion by the enzyme complex when bound to an intact membrane? If a methyl cation is transported in one direction, something must be transported in the other to maintain electroneutrality.

Arigoni: Electroneutrality is not essential. There could be accumulation of charge on one face of the membrane to provide chemical potential.

Eschenmoser: What you suggest would be an energy-creating process.

Arigoni: That's exactly what the methyl-transferring enzyme system seems to be doing; it creates a potential which is then exploited to drive the synthesis of ATP.

Kräutler: In which direction does the methyl group go?

Arigoni: Are you suggesting that it is jumping across the membrane?

Thauer: The methyl group donor and the acceptor are on the same side of the membrane, just as NADH and O_2 in the respiratory chain react with membrane proteins on the same side. The current theory is that the free energy change associated with the methyl transfer reaction, or redox reaction, results in conformational changes of proteins in the membrane, such that the energy is released by vectorial cation translocations; an electrochemical potential is thus built up.

Arigoni: This is an important point. Many people think the chemiosmotic theory consists of the building up of a pH gradient, but charge separation was also included in the original proposal. This is something we can mimic in normal organic systems only with electrodes.

Eschenmoser: After transfer of the methyl group, electroneutrality should be restored, as Bernhard Kräutler said.

Thauer: Consider a well-known case, the hydrolysis of ATP by the membrane-associated H^+-transporting ATPase. This hydrolysis is coupled to the translocation of three, four or five protons against an electrochemical potential, the stoichiometry depending on the thermodynamic conditions. It is difficult to understand how the proton translocation works, especially if one considers that minor changes in the enzyme's primary structure can change its specificity for protons into one for sodium ions.

Scott: During the transfer of the methyl group from the cobalt, can you observe the changes in cobalt valency?

Thauer: No. You can see Co(II) in electron paramagnetic resonance spectroscopy, but this doesn't tell you that Co(II) is involved in the catalytic mechanism. The Co(II)–Co(I) redox potential of the corrinoid protein was found to be around -400 mV in the membrane fraction.

Scott: Was this a fairly persistent radical species?

Arigoni: Even if it is persistent that doesn't mean that it is kinetically competent.

Thauer: Strong reducing conditions are needed to get the methyl transfer reaction going, so it is more likely that Co(I) is taking up the methyl group.

Arigoni: Has the stereochemical course of this methyl transfer been examined?

Thauer: No.

Battersby: In the final step of methanogenesis, in which methyl-coenzyme-M reductase generates this interesting unsymmetrical disulphide, what happens to this disulphide?

Thauer: The disulphide is reduced to regenerate the 7-mercaptoheptanoyl-threonine phosphate.

Arigoni: Is molecular hydrogen used?

Thauer: Yes, but not directly. Molecular hydrogen is first activated by a nickel-containing hydrogenase and the electrons are transferred to the heterodisulphide via an electron-transport chain. The energy released in this electron transfer reaction is conserved in an electrochemical proton potential which can be used to drive either the endergonic synthesis of ATP from ADP and P_i or the first step in methanogenesis, reduction of CO_2 to formylmethanofuran.

Akhtar: Are you not using purified enzymes? I thought that membrane potential was a relevant matter only in the whole organism. You have purified the first enzyme, which catalyses the reduction of carbon dioxide, and this works in solution. This is an ordinary, classical enzyme, which, when given hydrogen and carbon dioxide, produces the formyl derivative.

Thauer: It will not work with H_2, because the redox potential of the H^+–H_2 couple at pH 7 is -420 mV whereas that of the CO_2–formylmethanofuran couple is -500 mV.

Akhtar: How do you follow that reaction?

Thauer: We test it in the opposite direction.

Akhtar: Do you mean you make molecular hydrogen?

Thauer: No, we use an artificial electron acceptor such as a viologen dye.

Akhtar: With the purified enzyme you used an artifical electron donor with a more negative redox potential than hydrogen which deputizes for the physiological donor.

Thauer: Yes. Instead of molecular hydrogen, you can use titanium(III) citrate, which has a redox potential of about -500 mV and which is a one-electron donor. In the presence of mediators such as paraquat (a viologen dye), Ti(III) will drive the reduction of CO_2 to the formyl oxidation level. We have tested the enzymic activity in both directions, but only the scalar reaction, not the coupling with vectorial sodium ion translocation. We can see cation pumping only when we use intact membrane vesicles.

Akhtar: That simplifies matters, because one can forget about the membrane potential and the proton gradient and look just at the mechanism in an artificial system. The only question remaining is how *in vivo* H_2 is used to generate the reducing power which one can't mimic in a soluble system.

Thauer: At the partial pressures of H_2 prevailing in the natural habitats of methanogens, the redox potential of the H^+–H_2 couple is only -300 mV. The electrons therefore have to be pumped uphill 200 mV for the CO_2 to be reduced to formylmethanofuran.

Akhtar: You are quite right that the first reaction is endergonic. In glycolysis, some of the early reactions are endergonic by more than 4 kcal, but are driven by highly exergonic reactions further down the pathway. You have suggested that in a classical tetrahydrofolate (H_4F), because of the carbonyl in a *para* position, the N^{10} nitrogen is more activated towards nucleophilic attack than the N^{10} nitrogen in N^5,N^{10}-tetrahydromethanopterin (H₄MPT), which is in a position *para* to a methylene group (Weiss & Thauer 1993). I would not expect that to be the case.

Arigoni: It is always possible to deprotonate the NH group of *p*-amino-benzoate to produce a better nucleophile.

Akhtar: But surely if N^{10} was vinylogous to an electron-withdrawing carbonyl group that would deactivate the N^{10} nitrogen.

Arigoni: But it would facilitate the deprotonation, and once the NH group is deprotonated, it is a better nucleophile.

Akhtar: In tetrahydromethanopterin the N^{10} has a pK of about 4.0 (similar to that in aniline), in comparison with one of 1.4 in tetrahydrofolate.

Thauer: One thing is certain. The energy which is conserved in the formamide bond in N^{10}-formyltetrahydrofolate is sufficient to drive the synthesis of ATP. The enzyme which catalyses this energy-conserving reaction is well studied. In contrast, N^{10}-formyltetrahydromethanopterin hydrolysis is not sufficiently exergonic to be coupled with ATP synthesis.

Arigoni: That agrees with what I was just saying. One is an imid-type of compound, and the other is just a formamide derivative.

Akhtar: One is an amino nitrogen, and the other is, as you say, an amide nitrogen. The ordinary amino nitrogen, once it is deprotonated, will be more reactive, a better nucleophile.

Eschenmoser: Dr Thauer, could you clarify exactly which reaction drives the formation of ATP?

Thauer: N^{10}-Formyltetrahydrofolate is converted into formate and tetrahydrofolate with the concomitant formation of ATP from ADP and phosphate. The enzyme involved is called N^{10}-formyltetrahydrofolate synthetase (formate–tetrahydrofolate ligase, EC 6.3.4.3).

Eschenmoser: Does the system require the dihydro form of a coenzyme, or does it work directly with H_2?

Thauer: There are two enzymes. The more important one goes directly with H_2. Surprisingly, the only transition metal this enzyme appears to contain is zinc.

Eschenmoser: It is most interesting that an organic molecule can react with H_2 without a redox-active metallic catalyst.

Thauer: The novel hydrogenase has some other interesting features. For example, it does not catalyse the transfer of electrons to one-electron acceptors.

Eschenmoser: Does the methene hydrogen stay there during the process?

Thauer: Yes. We have, however, observed an exchange of the methenyl hydrogen with protons from water, which is slow in comparison with the enzyme-catalysed reduction of the methenyl group to the methylene group.

Arigoni: The hydrogen in the methenyl compound (**1**) ought to be as acidic as the C–H bond in thiamine pyrophosphate (**2**).

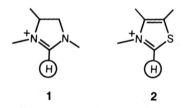

1   **2**

Akhtar: Why doesn't this exchange happen in methenyltetrahydrofolate?

Arigoni: How do you know it doesn't?

Akhtar: I thought it didn't.

Thauer: Professor Arigoni told me when I reported our exchange results that he did not know of any literature describing the exchange of the methenyl hydrogen of N^5,N^{10}-methenyltetrahydrofolate with water protons, but that he would not be surprised if this had been overlooked. Later, we found that a report of this exchange was hidden in the materials and methods section of a paper by Slieker & Benkovic (1984). We then did quantitative studies and found that with N^5,N^{10}-methenyltetrahydrofolate, the exchange is considerably slower than with N^5,N^{10}-methenyltetrahydromethanopterin.

Akhtar: Beilman, who incorporated a methenyl group via formic acid into serine, was lucky that the exchange wasn't strong enough to upset his stereochemical goal.

Arigoni: As Rolf Thauer pointed out, the non-enzymic loss is slow in comparison with the enzymic reaction.

Thauer: The half-life of the non-enzymic loss is about an hour, depending somewhat on the pH.

Beale: The enzyme catalysing the first step in methanogensis, formylmethanofuran dehydrogenase, which has a requirement for molybdenum or tungsten, is reminiscent of the alternative nitrogenases (isoenzymes) which require vanadium or molybdenum. You mentioned that extreme thermophiles tend to have the tungsten enzyme. Is there something about the relative availability of tungsten and molybdenum in high temperature environments which might explain that?

Thauer: I can only speculate. At the beginning of life, the environment was anaerobic, and molybdenum and tungsten were probably present as the sulphides MoS_2 and WS_2. Tungsten sulphide is a little more soluble in water than molybdenum sulphide, though neither are particularly soluble. If extreme thermophiles are more closely related to the first living organisms, as phylogenetic trees tend to suggest, their tungsten requirement might reflect the fact that this transition metal was more available than molybdenum in the early days of life. Later on, when the environment became more oxidized, the sulphides were converted to molybdate and tungstate, both of which are highly soluble in water. Molybdate would then have become the preferred metal because its reduction from the VI oxidation state to the V state, which is involved in the catalytic mechanism, operates at a less negative redox potential than the tungsten redox couple and also because Mo(V) is less susceptible to auto-oxidation than W(V).

Spencer: Can nickel in the environment be a limiting factor determining how much methane methanogenic bacteria can produce?

Thauer: In the natural environment nickel is probably never limiting. This is in contrast to the situation with iron, cobalt, molybdenum and vanadium, the availability of which in some habitats is growth rate-limiting. Nickel can be growth-limiting under laboratory conditions.

References

Slieker LJ, Benkovic SJ 1984 Synthesis of (6*R*,11*S*)- and (6*R*,11*R*)-5,10-methylene-[11-^{1}H,^{2}H]tetrahydrofolate. Stereochemical paths of serine hydroxymethyltransferase, 5,10-methylenetetrahydrofolate dehydrogenase, and thymidylate synthetase catalysis. J Am Chem Soc 106:1833–1838

Weiss DS, Thauer RK 1993 Methanogenesis and the unity of biochemistry. Cell 72:819–822

Haem d_1 and other haem cofactors from bacteria

C. K. Chang

Department of Chemistry, Michigan State University, East Lansing, MI, 48824, USA

Abstract. Several bacterial haem prosthetic groups whose structures deviate significantly from the ubiquitous protohaem (Fe–protoporphyrin) have been discovered recently. These newly discovered pigments contain dramatic modifications in their aromatic core and/or side chains. Examples include the dioxoisobacteriochlorin-type haem d_1 and the chlorin-type haem d as well as the haem a-like haem o. Total syntheses of these macrocycles have been accomplished. Synthetic haem d_1 and its analogues were used in reconstitution studies with nitrite reductase which revealed the importance of the oxo groups and the acrylate side chain for enzymic activity. The structural features of these porphyrinoids immediately suggest some possible, but as yet unproven, biosynthetic pathways.

1994 The biosynthesis of the tetrapyrrole pigments. Wiley, Chichester (Ciba Foundation Symposium 180) p 228–246

One of the most colourful enzyme cofactors found in Nature is the iron–porphyrin (haem) prosthetic group. Three classes of haem, differing only in porphyrin side chains, are commonly observed: the a-type haem of cytochrome oxidases; the b-type haems, which include iron–protoporphyrin IX (protohaem), found at the active site of haemoglobin, and cytochromes b and P_{450}; and the c-type haem of cytochrome c. Recently, an increasing number of haem groups isolated from bacterial sources have been shown to possess a tetrapyrrole macrocycle that is quite different from the common haems. In some of these the porphyrin ring is saturated (as in sirohaem), whereas in others oxygen is added to the pyrrole unit (as in haem d and d_1). Because of the modifications of their π-system, these haems often exhibit unique properties not shared by protohaem. Haem d_1 is widely present in denitrifying bacteria that reduce nitrate salts in soil to nitrogen or nitrous oxide. Haem d and haem o are typically associated with the terminal oxidases that mediate the reduction of oxygen to water. The chemical mechanism involved in these biologically important multi-electron reductions is intriguing but not well understood. Elucidation of the structural features and intrinsic reactivities of an enzyme's prosthetic group is a crucial step in the study of its function. Here, I summarize recent synthetic work that helped to establish the structure of haems d_1, d and o. Structural, functional and molecular properties of haem d_1 are also presented to highlight the attributes of this macrocycle.

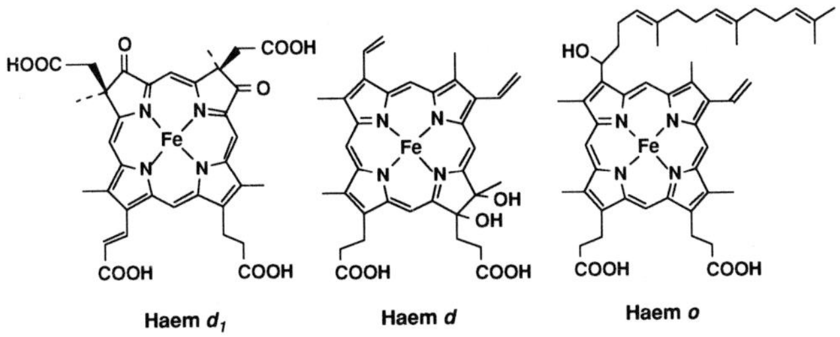

Haem d_1 Haem d Haem o

Haem d_1: structure and synthesis

Microbes such as *Pseudomonas aeruginosa* anaerobically catalyse dissimilatory reduction of nitrite using the soluble enzyme cytochrome cd_1, which also functions *in vitro* as a cytochrome oxidase (Henry & Bessières 1984, Hochstein & Tomlinson 1988). The cd_1 nitrite reductase has an M_r of 120 000 and is composed of two equivalent subunits each containing a c-type haem and a green-coloured haem d_1. During the enzymic cycle, electrons released in phosphorylation reactions pass to haem c and haem d_1 and then to the substrates that coordinate to the haem d_1 binding site. Because of its intense green colour, haem d_1 was for many years assumed to possess a chlorin core structure, and characterization of this extractable pigment was not reported until 1984. On the basis of the detailed spectroscopic data of Timkovich et al (1984a,b), Chang (1985) proposed that the d_1 macrocycle is not a chlorin but a porphyrindione (dioxoisobacteriochlorin). The proposed structure, although unprecedented in the biological world, was substantiated by a series of comparative studies with model compounds (Chang et al 1986a, Chang & Wu 1986), and has now been unambiguously proven by synthesis.

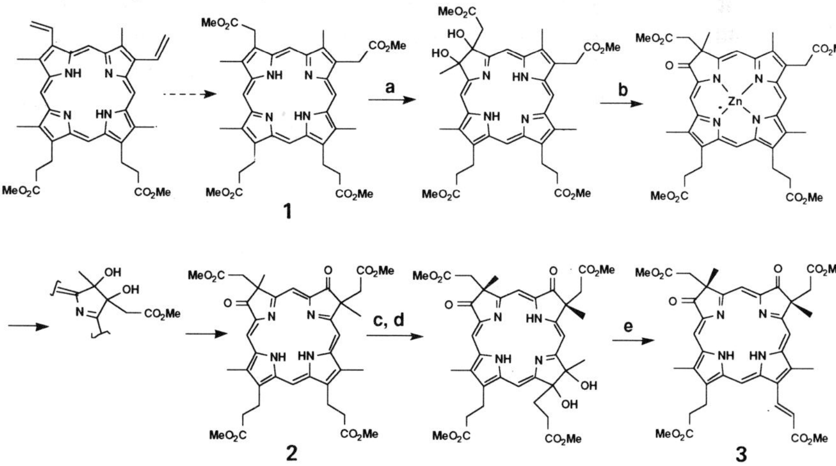

a. OsO_4
b. i) FSO_3H/H_2SO_4 ii) $ZnOAc$
c. separation of cis/trans isomers
d. OsO_4
e. i) HCl/benzene, reflux
   ii) CH_2N_2

Scheme 1

In the first synthesis of d_1 a two-stage pinacolic rearrangement was used to generate the oxo groups on the porphyrin ring (Wu & Chang 1987). The initial chlorin diol was prepared by osmium tetroxide addition to porphyrin, which is specific about the pyrrole β–β double bonds but not selective for a particular pyrrole unit. Of the four diol isomers, the differing migratory aptitudes of the side chains and electronic effects were such that only one could be used to generate the desired oxo structure (Chang & Sotiriou 1985). As illustrated in Scheme 1, the diacetate porphyrin 1 was first converted to the porphyrinone and then to the dione 2 in a step-wise fashion. Insertion of metal ion in to the porphyrinone altered the course of the second OsO_4 attack to give isobacteriochlorin (Chang et al 1986b). The diastereoisomers of 2 were separated before being converted to the acrylic form. In the final step, the acrylate double bond was selectively introduced by an unconventional bishydroxylation–elimination sequence to complete the unique d_1 aromatic system (Chang & Wu 1986). By comparing our product with the natural sample in NMR studies with shift reagents and by comparing chromatographic patterns we determined that d_1 is the *cis* isomer. The absolute stereochemistry of natural d_1, of course, could not be established from the synthetic racemate. Other keto and acrylate isomers were obtained during this synthesis, albeit in minor quantities. These compounds turned out to be very useful in other facets of our study on the structure–function relationships of haem d_1.

Porphyrinones and porphyrindiones can also be prepared from haematoporphyrin through a Claisen rearrangement followed by oxidative cleavage of a C–C double bond (Montforts et al 1989). Using this approach, Montforts et al (1992) were able to obtain the above porphyrindione 2, and showed by X-ray crystallography that the racemate known to correspond to d_1 indeed has the C-methyl groups at the A and B rings in the *syn* orientation. Battersby's group have recently accomplished a stereoselective synthesis of an isobacteriochlorin precursor which was converted to the dione 2 by SeO_2 oxidation. After the last step, acrylate formation, they obtained a compound with a circular dichroism spectrum matching that of the natural sample (Micklefield et al 1993). Thus, haem d_1 was finally established to have the absolute configuration illustrated in structure 3.

There is an obvious stereochemical relationship between haem d_1 and sirohaem and coenzyme F_{430}, suggesting that these porphyrinoids are biosynthetically related. By using [^{13}C-*methyl*]methionine in the growth medium, Timkovich (Yap-Bondoc et al 1990) showed that the two angular methyl groups of rings A and B of haem d_1 from *P. aeruginosa* cytochrome cd_1 are derived from methionine, presumably via S-adenosylmethionine, thereby establishing the link of haem d_1 to the sirohydrochlorin and vitamin B_{12} pathway (Battersby 1986). The biological mechanism by which the two propionate groups are removed from sirohydrochlorin or an earlier precursor leading to the dioxo structure is intriguing and remains to be elucidated. A second natural pigment with a dioxobacteriochlorin core has now been discovered in an alga (Prinsep et al 1992), indicating that d_1 is not an isolated case in which porphyrinoidal side chains are replaced by conjugating oxo functions.

Haem d_1: properties and enzyme reconstitution experiments

Having proven the structure of haem d_1, we must ask why this structure is needed. Does the structure have any obligatory roles in dissimilatory nitrite reduction? As a first step towards answering this question, we compared the characteristics of this macrocycle with other better known porphyrinoids. First, I shall summarize the redox potentials. A prominent feature of saturated porphyrin systems is that their redox potentials are invariably less positive than those of unsaturated porphyrins—chlorin and isobacteriochlorin are easier to oxidize but harder to reduce than the corresponding porphyrin (Chang 1982). Extended Hückel MO calculations (Chang et al 1981) explained that as the porphyrin β–β double bonds become saturated, the a_{1U} orbitals as well as the π^* orbitals are progressively raised. However, the oxo groups altered the picture (Chang et al 1986c, Barkigia et al 1992). The presence of these conjugated keto groups pulls down the π stacks so that the redox potentials are in fact closer to the parent porphyrin than to their chlorin and isobacteriochlorin cousins. As a result, cyclic voltammograms indicate that the redox potential of the Fe(III)/Fe(II) couple of the dionehaem is even more positive than those of the porphyrin haems.

Next, we may consider the molecular structure. X-ray crystallographic structures of Cu(II), Fe(III)Cl and free base octaethylporphyrin dione have been obtained (Chang et al 1986c, Barkigia et al 1992). The crystal structures revealed an expansion of the core size resulting from the lengthening of the (metal) centre to pyrroline-N distance. This elongation seems to be a general phenomenon, also occurring in other saturated porphyrinoid structures (Eschenmoser 1986). Thus, the cores of haem d_1 and isobacteriochlorin are about 4% larger than a porphyrin core. Haem d_1's larger core and electronegative nature ought to stabilize low-valent metal complexes. Ligand coordination with the ferrous dionehaems appears to differ very little from the porphyrin haems. We have measured binding equilibria and kinetic rates of CO coordination in model compounds and in reconstituted myoglobin. The variations in $P_{1/2}$ (half-saturation pressure in O_2 binding) in comparison with the corresponding mesohaem system are within a factor of two (C. Chang, unpublished work 1992). There is no immediate clue about the special intrinsic characteristics that make haem d_1 uniquely suitable for reactions with nitrite. The uncertainty surrounding the end product of cytochrome cd_1 nitrite reductases—whether it is NO or N_2O—further complicates the issue.

To explore the relationship between the structure of the prosthetic group and the reductase's functioning we have reconstituted the apoprotein of a cytochrome cd_1 nitrite reductase isolated from *Pseudomonas stutzeri*, using native and

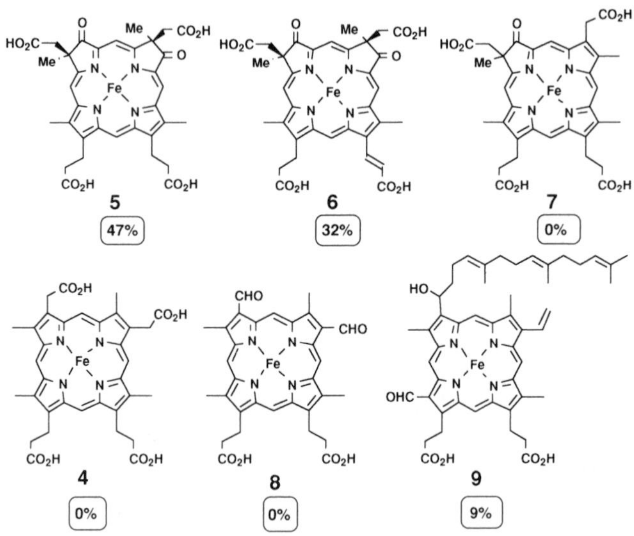

FIG. 1.   Structures of haems reconstituted with cytochrome cd_1 nitrite reductase from *Pseudomonas stutzeri*. The recovery of enzyme activity is indicated under each haem as a percentage of the activity (NO and N_2O evolution) shown by the native enzyme. Another haem not shown here, the *trans*-isomer of **3**, had activity on reconstitution (77%) essentially indistinguishable from that of the *cis*-isomer.

synthetic haem d_1 (Weeg-Aerssens et al 1991). The nitrite reductase activity, as measured by NO and N_2O gas evolution, can be restored to 82% of the original enzyme activity when the protein is reconstituted with native haem d_1, and to 77% of the activity when reconstituted with the synthetic (racemic) haem d_1. Using the same experimental protocol, we have replaced the native haem with the synthetic *trans*-d_1 and six additional Fe–porphyrinoids (**4–9**) (W. Wu & C. Chang, unpublished work 1991). These homologous haems were chosen to probe the importance of the oxo group and the acrylic side chain; also, the electron-withdrawing haem **8** has a redox potential similar to that of haem d_1. Reconstitution with all of these haems was successful, as shown by the well-defined narrow protein band produced in the final DEAE column. In contrast, non-specifically bound haems (protohaem, for example) always resulted in a very diffuse band. The enzymic activity of the reconstituted protein was measured by monitoring gas evolution on a gas chromatograph equipped with a ^{63}Ni-electron capture detector. In all our experiments, the yield of N_2O was quite low, typically $\leqslant 10\%$ of the NO production. The recovered activities summarized in Fig. 1 were based on the initial rate (first 20 min) of combined NO and N_2O evolution. These results indicate that the dioxo functionality is important for nitrite reductase activity; removal of just one oxo group results in a total loss of activity. The acrylate side chain also appears to be critical, as shown by the diminished activity of the enzyme reconstituted with haem **5** or **6**. Among the porphyrin haems, only haem *a* **(9)** produced an appreciable yield of NO, which is interesting in view of Hill & Wharton's (1978) observation that haem *a* is the only porphyrin-type haem capable of restoring a small degree (5%) of *oxidase* activity on reconstitution with *P. aeruginosa* cytochrome cd_1. The interactions between the protein and the peripheral groups of the haem moiety are clearly very important for enzyme functioning.

Haem *d* and haem *o*

We now turn to the oxidase enzymes which reduce O_2 to water. Aerobically grown *Escherichia coli* expresses two membrane-bound, multi-haem respiratory proteins: cytochrome-*o* oxidase at high O_2 concentration and cytochrome-*d* oxidase at low O_2 concentration. Both are linked to dehydrogenases via a ubiquinone pool.

Haem *d*

Cytochrome-*d* oxidase has two *b*-type haems and an oxygen-binding haem *d* which catalyses reduction of O_2 to H_2O. The structure of the green-coloured haem *d* prosthetic group originally deduced from spectral characterizations (Timkovich et al 1985) has been tested by synthesis (Sotiriou & Chang 1988). This ring structure formally belongs to the chlorin class but the saturation of the pyrrole double bond is brought by the addition of two hydroxy groups. Our synthesis of this chlorin was based on dihydroxylation of the porphyrin ring by OsO_4. Separation of the four possible isomers by chromatography was necessary to obtain the desired diol. The resultant chlorin diol, bearing a geminal propionate ester side chain, could easily form a γ-spirolactone (Andersson et al 1987), the stereochemistry of which depends, interestingly, on the chromatography conditions. A variety of reagents, including pyridine and NaOAc, are effective in bringing about lactone formation without altering the *cis*-configuration of the diol. On the other hand, prolonged contact of the *cis*-diol with silica gel during chromatography not only results in lactone formation but also in epimerization to a *trans* configuration. As shown in Scheme 2, all four diastereoisomers of the diol and lactone can be produced.

Timkovich et al (1985) showed that the extracted haem *d* pigment, after the iron had been removed and the propionate groups methylated, was in the form of the *trans*-lactone **10**. This is not the true form of haem *d* (Vavra et al 1986) and probably arises as a consequence of the isolation procedures which inevitably involved chromatography. In a related case, however, a *d*-type haem from *Escherichia coli* catalase HPII, extracted and purified under similar conditions, was reported to have an ^1H NMR spectrum closely matching that of the *cis* isomer of **10** (Chiu et al 1989). Further work is necessary to establish the true

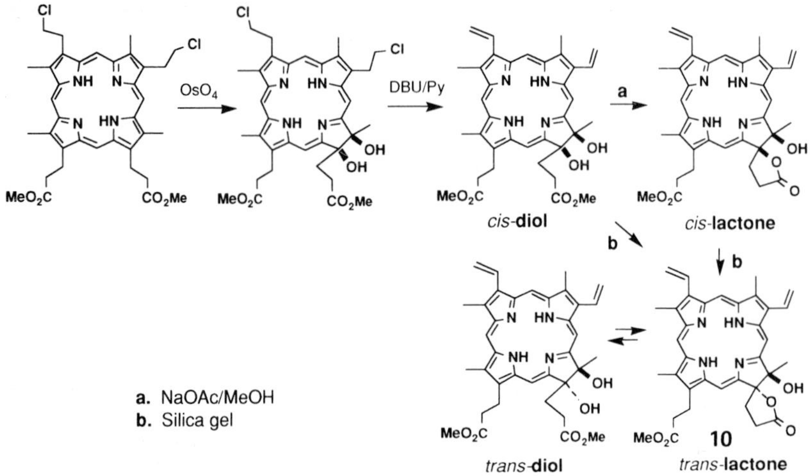

Scheme 2

structure of the nascent haem *d* present in these proteins. The presence of the hydroxyl group on the haem moiety promotes hydrogen bonding, as revealed by X-ray diffraction (Barkigia et al 1991); such interactions with protein residues may modulate the haem's reactivity.

Haem *o*

The cytochrome-*o* oxidase of *E. coli* has two haems and one Cu ion; one of the haems and the copper centre form a binuclear metal catalytic site for binding and reduction of O_2. This enzyme bears a strong resemblance to the mitochondrial aa_3 oxidase system. Until recently, because of apparent spectral similarities, the structure of haem *o* had been assumed to be that of protohaem. We recently proved that the O_2-binding haem *o* is an unusual haem containing a C_{17} hydroxyethylfarnesyl side chain, similar to haem *a* (**9**) but with a methyl group at pyrrole ring D (Wu et al 1992). This structure has been verified by synthesis (C. Chang & W. Wu, unpublished work 1992). As shown in Scheme 3, the synthesis of the parent porphyrin was achieved by a standard [2+2] dipyrrylmethene condensation; however, the electron-withdrawing carboxyl group kept the yield low. The farnesyl side chain was introduced to the ring by reacting acid chloride with the magnesium chelate of farnesyl malonate, as demonstrated by Clezy & Fookes (1981).

The close structural similarity between haem *o* and haem *a* raises interesting possibilities. Little is known about the biosynthesis of haem *a*, but protoporphyrin IX is believed to be a precursor. Haem *o* logically, could be an intermediate between protoporphyrin and haem a, with the oxidation of the C-8 methyl group occurring at a later stage than the addition of the C-3 farnesyl chain.

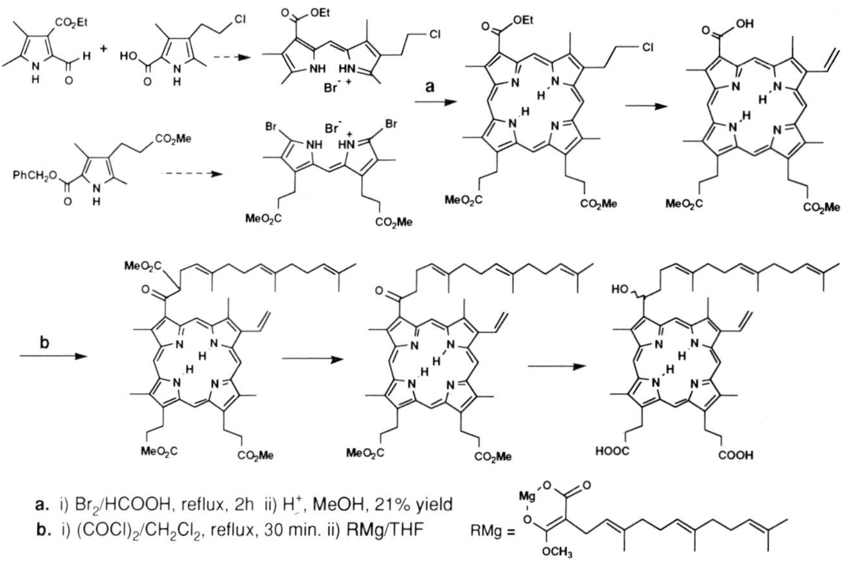

a. i) Br_2/HCOOH, reflux, 2h ii) H^+, MeOH, 21% yield

b. i) $(COCl)_2$/CH_2Cl_2, reflux, 30 min. ii) RMg/THF

Scheme 3

The haem compounds described here function as redox catalysts in the major known types of terminal oxidases. Figure 2 provides an overall picture of the enzyme catalytic unit and the probable course of O_2 binding and reduction. The mechanism for O–O bond cleavage has been studied extensively (Babcock & Wikström 1992). Among these oxidases, only the a and o systems contain a copper ion near the oxygen-binding haem which would allow the formation of binuclear Fe–OO–Cu intermediates. The d system contains no metal ions near the haem d binding site, and in the cd_1 enzyme, haem c is quite distant from haem d_1. Therefore, the Fe–OO complex in these enzymes must be activated by means other than formation of a ferric-cupric peroxide. Endogenous acid functions near the oxygen-binding site may provide conditions that facilitate the O–O bond cleavage (Yoshikawa & Caughey 1990). It remains to be seen whether such proton- or Lewis acid-assisted scission is important in bacterial cytochromes that reduce O_2 to H_2O.

In summary, the successful syntheses of the newly discovered cofactors has not only firmly established their novel structures but also expanded the horizon of porphyrin research by providing impetus for an exploration of the new chemistry related to these systems.

236

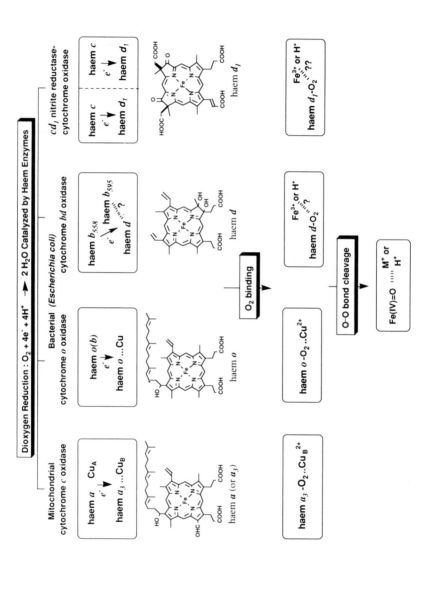

FIG. 2. Haem-containing enzymes that reduce O_2 to H_2O utilize the haem cofactors to coordinate and activate the O–O bond.

Acknowledgements

This work was supported by NIH grants GM 34468 and 36520. I would like to thank particularly two former graduate students, Weishi Wu and Lia Sotiroiu, who carried out most of the studies described here. This research would not have been possible without the collaboration of Professors Jerry Babcock, Jack Fajer, Thomas Loehr, James Tiedje and Russ Timkovich. I also thank Professor Sir Alan Battersby for communicating results prior to publication.

References

Andersson LA, Sotiriou C, Chang CK, Loehr TM 1987 Facile lactonization and inversion of vicinal diols in heme d-type chlorins: a spectroscopic study. J Am Chem Soc 109:258–264

Babcock GT, Wikström 1992 Oxygen activation and the conservation of energy in cell respiration. Nature 356:301–309

Barkigia KM, Chang CK, Fajer J 1991 Molecular structure of a dihydroxychlorin. J Am Chem Soc 113:7445–7447

Barkigia KM, Chang CK, Fajer J, Renner MW 1992 Models of heme d_1. Molecular structure and NMR characterization of an iron(III) dioxoisobacteriochlorin (porphyrindione). J Am Chem Soc 114:1701–1707

Battersby AR 1986 Biosynthesis of vitamin B_{12}. Acc Chem Res 19:147–153

Chang CK 1982 Hemes of hydroporphyrins. In: Dunford HB, Dolphin D, Raymond KN, Sieker L (eds) The biological chemistry of iron. D Reidel, Dordrecht, p 313–334

Chang CK 1985 On the structure of heme d_1. J Biol Chem 260:9520–9522

Chang CK, Sotiriou C 1985 Migratory aptitudes in pinacol rearrangement of vic-dihydroxy-chlorins. J Heterocyclic Chem 22:1739–1741

Chang CK, Wu W 1986 The porphinedione structure of heme d_1. J Biol Chem 261:8593–8596

Chang CK, Hanson LK, Richardson PF, Young R, Fajer J 1981 π-cation radicals of ferrous and free base isobacteriochlorins: models for siroheme and sirohydrochlorin. Proc Natl Acad Sci USA 78:2652–2656

Chang CK, Timkovich R, Wu W 1986a Evidence that heme d_1 is a 1,3-porphyrindione. Biochemistry 25:8447–8453

Chang CK, Sotiriou C, Wu W 1986b Differentiation of bacteriochlorin and isobacteriochlorin formation by metallation. J Chem Soc Chem Commun, p 1213–1215

Chang CK, Barkigia KM, Hanson LK, Fajer J 1986c Models of heme d_1. Structure and redox chemistry of dioxoisobacteriochlorin. J Am Chem Soc 108:1352–1354

Chiu JT, Loewen PC, Switala J, Gennis RB, Timkovich R 1989 Proposed structure for the prosthetic group of the catalase HPII from *Escherichia coli*. J Am Chem Soc 111:7046–7050

Clezy PS, Fookes CJR 1981 The synthesis of a possible biogenetic precursor of haem *a* and its relationship to the prosthetic group of myeloperoxidase and to cryptoporphyrin a. Aust J Chem 34:871–883

Eschenmoser A 1986 Chemistry of corphinoids. Ann NY Acad Sci 471:108–129

Henry Y, Bessières P 1984 Denitrification and nitrite reduction: *Pseudomonas aeruginosa* nitrite-reductase. Biochimie 66:259–289

Hill KE, Wharton DC 1978 Reconstitution of the apoenzyme of cytochrome oxidase from *Pseudomonas aeruginosa* with heme d_1 and other heme groups. J Biol Chem 253:489–495

Hochstein LI, Tomlinson GA 1988 The enzymes associated with denitrification. Annu Rev Microbiol 42:231–261

Micklefield J, Mackman RL, Aucken CJ et al 1993 A novel stereoselective synthesis of the macrocycle of heme d_1 that establishes its absolute configuration as 2R, 7R. J Chem Soc Chem Commun, p 275–277

Montforts FP, Romanowski F, Bats JW 1989 A simple synthesis of dioxoisobacteriochlorins from hematoporphyrin. Angew Chem Int Ed Engl 28:480–483

Montforts FP, Mai G, Romanowski F, Bats JW 1992 A convenient synthesis of the nitrite reducing cofactor heme d_1 from hematoporphyrin. Tetrahedron Lett 33:765–768

Prinsep MR, Caplan FR, Moore RE, Patterson GML, Smith CD 1992 Tolyporphin, a novel multidrug resistance reversing agent from the blue–green alga *Tolypothrix nodosa*. J Am Chem Soc 114:385–387

Sotiriou C, Chang CK 1988 Synthesis of the heme d prosthetic group of bacterial terminal oxidase. J Am Chem Soc 110:2264–2270

Timkovich R, Cork MS, Taylor PV 1984a Proposed structure for the noncovalently associated heme prosthetic group of dissimilatory nitrite reductases. J Biol Chem 259:1577–1585

Timkovich R, Cork MS, Taylor PV 1984b Proposed structure for the noncovalently associated heme prosthetic group of dissimilatory nitrite reductases. J Biol Chem 259:15089–15093

Timkovich R, Cork MS, Gennis RB, Johnson PY 1985 Proposed structure of heme d, a prosthetic group of bacterial terminal oxidases. J Am Chem Soc 107:6069–6075

Vavra MR, Timkovich R, Yap F, Gennis RB 1986 Spectroscopic studies on heme d in the visible and infrared. Arch Biochem Biophys 250:461–468

Weeg-Aerssens E, Wu W, Ye RW, Tiedje JM, Chang CK 1991 Purification of cytochrome cd_1 nitrite reductase from *Pseudomonas stutzeri* JM 300 and reconstitution with native and synthetic heme d_1. J Biol Chem 266:7496–7502

Wu W, Chang CK 1987 Structure of dioneheme. Total synthesis of the green heme prosthetic group in cytochrome cd_1 dissimilatory nitrite reductase. J Am Chem Soc 109:3149–3150

Wu W, Chang CK, Varotsis C, Babcock GT, Puustinen A, Wikström M 1992 Structure of the heme o prosthetic group from the terminal quinol oxidase of *Escherichia coli*. J Am Chem Soc 114:1182–1187

Yap-Bondoc F, Bondoc LL, Timkovich R, Baker DC, Hebbler A 1990 C-methylation occurs during the biosynthesis of heme d_1. J Biol Chem 265:13498–13500

Yoshikawa S, Caughey WS 1990 Infrared evidence of cyanide binding to iron and copper sites in bovine heart cytochrome c oxidase. J Biol Chem 265:7945–7958

DISCUSSION

Timkovich: I would like to share with you some recent studies that reflect on the biosynthesis of haem d_1. The most striking characteristic of haem d_1 is its oxo groups, or the carbonyls at C-3 and C-8. Haem d_1 is formed in *Pseudomonas* only when it is grown anaerobically, yet, with its oxo groups, it is an oxygenated tetrapyrrole. A second major question concerns the biogenesis of the double bond in the unusual acrylate. We have gained some insight into these issues by searching for precursors to haem d_1. Doss et al (1971) and Jacobs et al (1972) originally showed that anaerobic cultures of *Pseudomonas* fed high levels of 5-aminolaevulinic acid biosynthesize excess tetrapyrroles to the point that they are actually secreted into the culture medium. They characterized some of the secreted metabolites, but we have re-investigated the products by HPLC, mass spectrometry and NMR spectroscopy. The pattern of secreted metabolites is complicated and not all the products have yet been purified and characterized. Many expected tetrapyrroles have been found in the mixture, including uroporphyrin, decarboxylated porphyrins and coproporphyrin, but it has been possible to fish out some novel and interesting tetrapyrroles from this metabolic soup.

Included in this set are porphyrins with the following substituents: compound **a**, four propionates, three acetates and a $-CH_2OH$ group; compound **b**, four propionates, one acetate, two methyls and one $-CH_2OH$; compound **c**, four propionates, three acetates and $-CH(OH)COOH$; compound **d**, four propionates, three acetates and $-(C=O)COOH$; compound **e**, three propionates, four acetates and $-CH(OH)CH_2COOH$; compound **f**, three propionates, four acetates and $-CH=CHCOOH$. Type I compounds predominate.

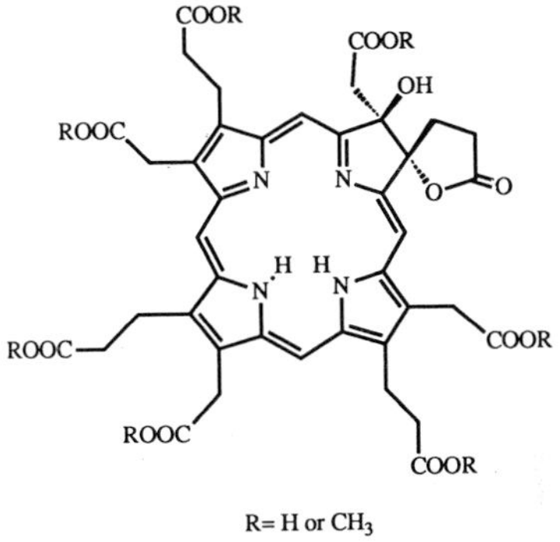

R= H or CH$_3$

FIG. 1. (*Timkovich*) Structure of the tetrapyrrole **g** isolated from aerobic cultures of *Pseudomonas* fed high concentrations of 5-aminolaevulinic acid.

Perhaps the most interesting animal in the zoo is a green pigment (**g**) which has a typical chlorin-type spectrum with a prominent band for the methyl ester derivative at 641 nm. It has the structure shown in Fig. 1. Its most striking feature is the spirolactone substituent derived from a former propionate. In this regard, it is an analogue of haem d, found in *E. coli*. On the basis of the known chemistry of haem d, we would surmise that the spirolactone in **g** arose spontaneously from a diol precursor, and that the diol may well have come from an earlier precursor that contained an epoxide across a pyrrolic $\beta-\beta$ double bond.

With the probable origin of the spirolactone in mind, it seems to us that compounds **a**, **b**, **c**, **e** and **g** could have been produced by insertion of oxygen either into CH bonds or across a carbon double bond. This chemistry is

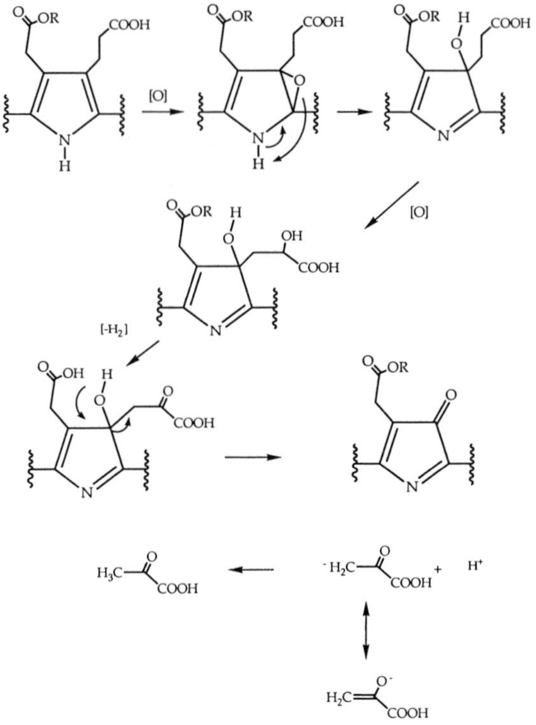

FIG. 2. (*Timkovich*)   Proposed scheme for the biosynthesis of the oxo groups of haem d_1 based on an oxygen-inserting enzymic activity.

reminiscent of the action of monooxygenases, a group of enzymes among which cytochrome P_{450} is the most familiar. However, monooxygenases require molecular oxygen as their source of added oxygen, and our compounds are formed under anaerobic conditions.

These are the facts: how might the compounds relate to the biosynthesis of haem d_1? First, the compounds are not necessarily all direct precursors of haem d_1. In the presence of excess 5-aminolaevulinic acid (ALA) abnormally high, even massive, amounts of uroporphyrin are formed (Harris et al 1993), and we believe that this uroporphyrin is being accepted by the biosynthetic enzymes present as a poor but competitive substrate; these dead-end products then accumulate.

The acrylate found in **f** and the alcohol in **e** suggest that the acrylate in haem d_1 may be formed by addition of oxygen to generate the hydroxy, followed by dehydration to form the double bond. Because the hydroxyl in **e** has benzylic character, this may be an especially easy dehydration.

The oxygenated porphyrins, especially **g**, suggest a speculative route to the oxo groups in d_1, as shown in Fig. 2. Instead of oxygen being added β–β to start production of **g**, it is possible that the physiological reaction of correct specificity adds an oxygen across an $\alpha-\beta$ pyrrole double bond to produce a transient $\alpha\beta$ epoxide. This is an amine epoxide which could spontaneously open to generate a β pyrrole hydroxyl group. Oxygen addition α to the propionate carbonyl would generate an alcohol isomer to **e**. The existence of **d** indicates that there is a type of tetrapyrrole alcohol dehydrogenase in *Pseudomonas* that could convert the α-alcohol into an α-keto carbonyl compound. This now sets up an absolutely wonderful leaving group: retroaldo elimination would generate the oxo group directly, and free the well-known biochemical, pyruvic acid.

The key to this hypothesis and the production of the oxygenated tetrapyrroles is the presence of a monooxygenase activity that does not require O_2. We know of no precedents for such an enzyme, but there may not be any fundamental reasons why O_2 should always be required by oxygenases. The critical transition state in P_{450} chemistry is an activated oxygen bonded to heam iron. This is achieved in a series of steps whereby P_{450} first binds O_2, then uses reducing equivalents delivered by coupled electron transport proteins to reduce an atom to water and liberate it, while generating the highly activated oxygen which will subsequently be inserted into the substrate. It may be that in *Pseudomonas* a redox enzyme system can generate the same type of activated oxygen from some other source such as water. An enzyme-bound water molecule could be stripped of two protons and two electrons to generate such a highly reactive atom. P_{450} chemistry is thermodynamically downhill, whereas with our hypothetical enzyme the reactions would not be, and would clearly require the input of biochemical energy. This is not impossible. Electron flow in mitochondrial respiration can be reversed under suitable conditions by the addition of excess ATP.

Eschenmoser: Your formulae are type I.

Timkovich: The dominant fractions isolated are type I. When you feed the microorganisms ALA, you by-pass the bottle-neck of ALA synthesis, so the cosynthase, which produces uroporphyrinogen III, becomes limiting. The massive amount of porphobilinogen that's being generated and acted on by the deaminase is flowing mostly into type I skeletons. There are type III compounds, but type I predominates. These type I skeletons are poor substrates for the enzymes that are present, but they are present in such massive amounts that the enzymes accept them.

K. Smith: How do you know they are type I?

Timkovich: From NOE studies.

Castelfranco: What reductant could oxidize water to superoxide under anaerobic conditions?

Timkovich: I don't think there is a single reductant. I would speculate that an elaborate process, driven by an elaborate electron transport chain, produces an activated oxygen equivalent. These bacteria have very complicated, branching electron transport chains.

Thauer: The reduction of N_2O to N_2, which is the last step in denitrification, has an $E^{\circ\prime}$ of $+1335$ mV, almost 400 mV more positive than that of the oxygen electrode. Thermodynamically, N_2O, in principle, could do it.

Rebeiz: Did I understand correctly that one of the compounds excreted had a methylhydroxy group? How might this be formed?

Timkovich: Yes. Again, I think an as yet uncharacterized oxygenase enzyme inserts an oxygen across the C–H bond. Because O_2 is not available, I proposed the reverse P_{450} reaction.

Rebeiz: Could that play a role in the formation of the formyl group in chlorophyll *b* biosynthesis?

Timkovich: These are not photosynthetic organisms. I don't know.

Battersby: Dr Chang, in your reconstitution experiments using racemic synthetic haem d_1, presumably the apoprotein is specific for one of the two enantiomers. Was there anything left over?

Chang: We used a large excess of the haem, so couldn't tell which isomer the protein is specific for. However, the activity of the *trans* isomer is still high, suggesting that the stereochemistry is not a critical factor.

K. Smith: Because both the protein and the porphyrin (haem) are asymmetric, and because the two propionates face outwards from the protein pocket, there are two ways in which reconstitution can occur. This is called haem disorder (La Mar et al 1975), because two orientations which differ by a 180° rotation about the α,γ (Fischer nomenclature) axis are possible. Is there any evidence of two rotational isomers in your reconstitution experiments?

Chang: We did not do an NMR study. We monitored the reconstitution using visible absorption, the common method, and the spectrum looks like the original native haem.

Arigoni: There is an intriguing biosynthetic problem associated with the structure of haem d_1. It could be derived from a sirohaem precursor through elimination of two propionic acid side chains. Alternatively, the two angular methyl groups might stem from the decarboxylation of the acetic acid side chains followed by rearrangement of a truncated version of the two propionic acid chains. Is it known which of those two pathways is operative?

Chang: If you follow the pinacol rearrangement route, as we did in the laboratory synthesis, what you suggest is true. I think the sirohydrochlorin pathway is the more likely, because the source of the methyl group is *S*-adenosylmethionine.

Beale: Does knowing which organisms have haem *o* and which haem *a* as a terminal acceptor tell you anything? If *E. coli* can get by with haem *o*, without the formyl group, why do other organisms need it?

Chang: I don't know. All the known sources of haem *o* cytochrome are bacteria.

Eschenmoser: Is there any evidence for the existence of pseudoacid forms (**11**)?

11

Chang: The evidence against the formation of pseudoacid forms of d_1 is as follows. The green dioxoisobacteriochlorin structure of d_1 (3) has electronic, resonance Raman and NMR spectra very different from those of a pink isobacteriochlorin which the cyclized lactol forms of d_1 must become. We have not observed any spectral characteristics to indicate that either the free base d_1 or its metal complexes undergo such cyclizations under common acetal-forming conditions. Reducing agents such as L-selectride (lithium tri-*sec*-butylborohydride) can reduce the oxo groups of d_1 to yield directly the lactonized form with typical isobacteriochlorin spectral features.

Kräutler: The Fe(III) to Fe(II) redox potential of haem d_1 is more positive than that of other iron porphyrins, iron chlorins and isobacteriochlorins. Could that not be a consequence of the particular ligand structure of the porphyrin dione?

Chang: Yes. The dione is like a quinone of a porphyrin; the conjugating oxo group is electron-withdrawing.

Kräutler: You raised the question of the presumed biological relevance of the particular constitution of the ligand. Wouldn't you consider the redox properties of the metal to be an important consequence of the structure of the ligand?

Chang: Yes. The d_1 ring, being slightly electron-withdrawing and larger in core size, would be expected to stabilize a low valency metal ion. The iron d_1 systems have not shown any significant differences from other haems in terms of coordination, binding constants or kinetics. Differences in redox chemistry are evident in model chloro–iron complexes, but we still do not know the implication for catalytic function.

Timkovich: The redox potential of the d_1 haem iron is about $+300\,mV$ in the enzyme. There are *b*-type cytochromes and *c*-type cytochromes ranging in redox potential from $-150\,mV$ up to $+400\,mV$. The surrounding tetrapyrrole structure is always the same, *c*-type or *b*-type. It is therefore the protein that determines the final observed redox potential.

Scott: Does the structure of the unusual marine oxoporphinoid tolyporphin (Prinsep et al 1992) give any clue as to how the propionate might have disappeared from d_1? Are there one or two ketones in tolyporphin?

Chang: There are two ketones at the diagonal position, so it is a dioxobacteriochlorin. You can draw a pathway similar to d_1 by clipping off the propionate groups from rings A and C in uroporphyrinogen III and replacing them with conjugating oxo groups.

Rebeiz: In cytochrome-c oxidase, if you replaced the haem prosthetic group with zinc–protoporphyrin, how would that modify the function?

Chang: I think it would lose its redox function, because the haem is the primary binding site for oxygen. The Cu_B copper is close to the a_3 haem-binding site, so there's always a possibility that a μ-peroxo Fe–O_2–Cu complex is involved, but there is no spectroscopic evidence that this copper can form a bond to the Fe–O_2.

Arigoni: I understand that some of the enzymes that use cofactors such as haem d_1 can reduce N_2O to nitrogen.

Chang: More than one enzyme is needed. The cd_1 enzyme by itself cannot catalyse the entire conversion.

Arigoni: What I really want to know is how many other biochemical systems are known to evolve nitrogen. We are all familiar with nitrogen fixation, but which reactions are responsible for returning nitrogen back to the atmosphere?

Chang: That's what these denitrifying bacteria do. They convert all the oxidized nitric oxides and nitrate and nitrite back to N_2.

Arigoni: Is the amount of nitrogen so released sufficient to compensate for the fixation?

Chang: Yes. It has been suggested, in fact, that all the nitrogen in our atmosphere is derived from the denitrification process.

Thauer: Like nitrogen fixation, denitrification is a widespread process. The halobacteria, for example, which belong to the archaebacteria, can denitrify. One interesting aspect is that N_2O reductase (nitrous-oxide reductase, EC 1.7.99.6) shows significant sequence identity to cytochrome-c oxidase (EC 1.9.3.1).

Chang: This seems to suggest that the modification of the macrocycle is not significant catalytically in comparison with external interactions, that the presence of another metal ion and other endogenous functional groups may be more important than modification of the side chains.

Scott: Is it known what happens to the NO produced in this system? It isn't used as a messenger, is it?

Chang: A lot of haem enzymes, particularly if there is a proton around, will react with nitrite and split off the NO automatically. NO release is not necessarily a function of nitrite reductase.

Scott: I wondered if there might have been hidden message in it.

Chang: I wouldn't like to speculate on that!

Castelfranco: In cytochrome oxidase there are two haems and two coppers. That stands to reason, because to reduce one oxygen molecule, four electrons are needed. I would expect there to be some mechanism to conserve these four electrons until a full O_2 molecule is reduced, otherwise destructive oxygen radicals would tend to be released. Does anyone know how this works? In the opposite process in photosynthesis tremendous efforts have been made to find out how a full molecule of oxygen is released.

Chang: A mechanistic picture of oxygen reduction by cytochrome-*c* oxidase is now emerging. First comes oxyhaem formation, followed by several electron transfer steps. The reduced oxygen species, whatever it is, never leaves the iron, but the O–O bond is cleaved to produce an $Fe(IV)=O$ species; further reduction of this complex produces a second molecule of water. It's critical to trap these half-reduced oxygen species at the active site.

Castelfranco: What role does copper play in this?

Chang: The copper, according to Yoshikawa & Caughey (1990), acts like a metal anchor, as a Lewis acid, binding another endogenous acid function, such as a carboxylate, to provide hydrogen bonding towards the haem–oxygen complex. The copper also functions as an electron reservoir, even though the inner sphere electron transfers may be absent.

Akhtar: Paul Castelfranco raised an important question. There are four redox centres, two coppers and two haems, in the conventional cytochrome oxidase. Under artificial conditions you can reduce all four centres with four molecules of cytochrome *c*, and the fully reduced enzyme will then reduce one molecule of oxygen. There are, therefore, four electrons sitting in the four redox centres.

Castelfranco: That's what I was speculating.

Chang: You can short-circuit catalysis by reacting, for example, hydrogen peroxide with oxidase.

Rebeiz: The intermediates must be tightly bound until a full molecule is ready for release. One could infer from that that if anything happens to prevent the free radical intermediates from being held tightly to the prosthetic group, so that they are released into the stroma, catastrophe would ensue because the released free radicals would oxidize everything in their path. That's why I raised the issue of the putative Zn–protoporphyrin prosthetic group (p 244). If, by some unknown mechanism, cytochrome-*c* oxidase were to be adulterated so that some of the haem prosthetic groups became Zn–protoporphyrin groups, this would be a disaster. With some of our photodynamic insecticides and herbicides a lot of Zn–protoporphyrin is formed by the treated plants and this appears to be photodynamically toxic. It is possible here that Zn–protoporphyrin prosthetic groups have replaced the haem prosthetic group in cytochrome-*c* oxidase.

Chang: The phototoxicity arising from the zinc system is due to singlet oxygen. I don't believe the Zn–porphyrin system would bind oxygen.

References

Doss M, Philipp-Dormston WK 1971 Porphyrin and heme biosynthesis from endogenous and exogenous δ-aminolevulinic acid in *Escherichia coli*, *Pseudomonas aeruginosa*, and *Achromobacter metalcaligenes*. Hoppe-Seyler's Z Physiol Chem 352:725–733

Harris WF III, Burkhalter RS, Lin W, Timkovich R 1993 Enhancement of bacterial porphyrin biosynthesis by exogenous aminolevulinic acid and isomer specificity of the products. Bioorg Chem 21:209–220

Jacobs NJ, Jacobs JM, Morgan HE Jr 1972 Comparative effect of oxygen and nitrate on protoporphyrin and heme synthesis from δ-amino levulinic acid in bacterial cultures. J Bacteriol 112:1444–1445

La Mar GN, Budd DL, Viscio DB, Smith KM, Langry KC 1975 Proton nuclear magnetic resonance characterization of heme disorder in hemoproteins. Proc Natl Acad Sci USA 75:5755–5759

Prinsep MR, Caplan FR, Moore RE, Patterson GML, Smith CD 1992 Tolyporphin, a novel multidrug resistance reversing agent from the blue-green alga *Tolypothrix nodosa*. J Am Chem Soc 114:385–387

Yoshikawa S, Caughey WS 1990 Infrared evidence of cyanide binding to iron and copper sites in bovine heart cytochrome c oxidase. J Biol Chem 265:7945–7958

Genetics and enzymology of the B_{12} pathway

N. P. J. Stamford

University Chemical Laboratory, Lensfield Road, Cambridge CB2 1EW, UK

Abstract. The chemical complexity of vitamin B_{12} suggests that its formation may involve a large number of enzymic steps. However, until recently, little was known of the number, mechanism and stereochemical course of the many enzymic interconversions that are essential to vitamin B_{12} biosynthesis. In response to this the French groups led by Francis Blanche and Joel Crouzet have carried out extensive investigations into the genetic and biochemical organization of this remarkable biosynthetic pathway. Through heterologous complementation studies with cobalamin-producing mutants they were able to clone and identify a total of 22 unique *cob* genes from four genomic regions (A–D) of the *Pseudomonas denitrificans* chromosome. This was the first report of a genetic analysis of *cob* genes at the molecular level and provided a suitable genetic model from which biosynthetic investigations could be initiated. The metabolic roles of most of the products of these genes have now been defined and in light of this progress current research concentrates on the development and use of a variety of techniques to investigate the chemistry involved in these individual enzymic steps. Here, my focus is on the recent efforts and successes of the French groups that have led to the elucidation of almost the entire enzymic sequence of events in vitamin B_{12} biosynthesis. From this perspective, recent developments at Cambridge (UK) regarding the utilization of reconstituted enzymic systems to manufacture substrates as probes for this biosynthetic pathway are illustrated.

1994 The biosynthesis of the tetrapyrrole pigments. Wiley, Chichester (Ciba Foundation Symposium 180) p 247–266

Vitamin B_{12}, in its coenzyme forms, mediates an important and interesting series of rearrangements and group transfers. However, it is the complex structure of this biological molecule which has made its biosynthesis one of the most challenging biosynthetic problems in Nature. Vitamin B_{12} itself is just one member of a much larger family of structurally similar cyclic tetrapyrrole compounds that, at least in part, share a common biosynthetic pathway. Owing to their diverse and fundamental roles in biosynthetic and catabolic processes in living systems these molecules have been affectionately dubbed the 'Pigments of Life'.

The chemical complexity of vitamin B_{12} suggests that its biosynthesis may involve many enzymic steps, the phases of which can be divided into three main sequences: (i) the biosynthesis of the cyclic tetrapyrrole porphyrin nucleus, uroporphyrinogen III (uro'gen III), from 5-aminolaevulinic acid (ALA); (ii) the formation of the first true corrin macrocycle, cobyrinic acid, from uro'gen III; and (iii) the ultimate conversions that take cobyrinic acid to vitamin B_{12} (cyanocobalamin) and its coenzyme forms (Leeper 1989) (Scheme 1).

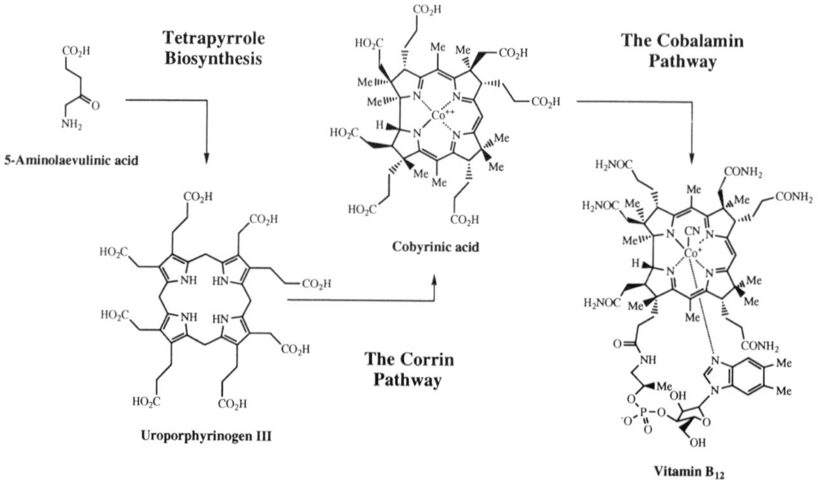

Scheme 1

Investigations into the biosynthesis of vitamin B_{12} established over a decade ago that ALA, porphobilinogen (PBG), uro'gen III and three C-methylated uroporphyrins, precorrin-1, 2 and 3, were essential early intermediates in the formation of the corrin nucleus (Battersby 1993). However, until recently, little was known of the number, mechanism and stereochemical course of other enzymic interconversions along the biosynthetic pathway from these intermediates to vitamin B_{12}. These investigations remained restricted by the lack of a genetic base from which to examine vitamin B_{12} biosynthesis systematically. In response to this the groups led by Francis Blanche and Joel Crouzet at Rhône–Poulenc Rorer, France have carried out extensive investigations into the molecular biology and biochemistry of vitamin B_{12} biosynthesis in the bacterium *Pseudomonas denitrificans*. The ultimate aim of their studies was to improve industrial vitamin B_{12}-producing strains, but they have provided a much broader understanding of the genetic and biochemical organization of the vitamin B_{12} biosynthetic pathway and will continue to assist in the identification and characterization of the enzymes and intermediates involved.

Other chapters in this volume on the enzymes ALA dehydratase (porphobilinogen synthase, EC 4.2.1.24), hydroxymethylbilane synthase (EC 4.3.1.8) and uro'gen III synthase (EC 4.2.1.75) (Leeper 1994, Jordan 1994, Spencer & Jordan 1994) have already made it evident that the synthesis of uro'gen III occurs in three distinct and well-characterized enzymic steps from ALA, the first committed intermediate in porphyrin biosynthesis. I therefore shall not reiterate this work, but shall focus on the French groups' recent efforts and successes with the second and third phases of vitamin B_{12} biosynthesis outlined above. Then, from this perspective, I shall illustrate recent developments at Cambridge in the utilization of cell-free enzymic systems to manufacture substrates as probes for this remarkable biosynthetic pathway.

Isolation and characterization of *cob* genes

The French researchers identified genes involved in vitamin B_{12} biosynthesis in *Ps. denitrificans* by heterologous complementation of mutants unable to synthesize cobalamin (*cob* mutants) (Cameron et al 1989). Initially they constructed mutants deficient in cobalamin synthesis from *Pseudomonas putida* and *Agrobacterium tumifaciens* using chemical and transposon mutagenesis strategies. Mutants of *Ps. putida* were selected and identified on the basis of their inability to use ethanolamine as a source of nitrogen in the absence of added cobalamin (deamination of ethanolamine requires coenzyme B_{12}), whereas *cob* mutants of *A. tumifaciens* were selected simply by their reduced cobalamin synthesis. After the construction of a genomic library of *Ps. denitrificans* in a wide-host-range vector, each clone of the genomic library was mobilized individually into the *Ps. putida* and *A. tumifaciens cob* mutants and screened for its ability to restore the lost phenotype. Clones from the genomic library identified as capable of complementing at least one *cob* mutation were then also tested to determine whether they might carry enough genetic information from the genome of *Ps. denitrificans* for complementation of another *cob* mutation. This ultimately led to the isolation of plasmids from the genomic library that displayed heterogeneous complementation patterns. On the basis of this complementation data, four complementation groups were defined, each identified by a set of plasmids that complemented a specific set of mutants not complemented by plasmids from other groups (Fig. 1).

This work complements similar systematic genetic investigations of cobalamin biosynthesis in *Bacillus megaterium* and *Salmonella typhimurium*. In these studies the isolation and genetic characterization of mutants of these organisms blocked in cobalamin biosynthesis were described and in each case clustering of *cob* genes was reported (Brey et al 1986, Wolf & Brey 1986, Jeter et al 1984, Jeter & Roth 1987). In *B. megaterium* all the *cob* genes identified were located in the same region of the chromosome, as they also were in *S. typhimurium*, with the exception of the *cysG* gene. In *Ps. denitrificans* such clustering of the four identified complementation groups has not yet been established. However, restriction endonuclease maps of the inserts of the plasmids from these complementation groups revealed no correlation between the groups, indicating that there is no overlap between the four different fragments (Cameron et al

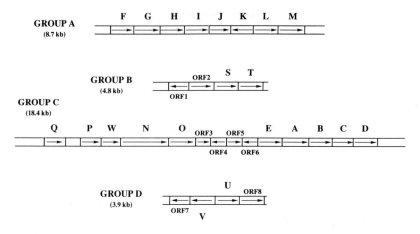

FIG. 1.   Arrangement of the genes encoding the enzymes of vitamin B_{12} biosynthesis in the four *Pseudomonas denitrificans* complementation groups (A–D). The *cob* genes are shown by their corresponding letters and open reading frames (ORFs) of unknown function are also represented. The amount of genomic sequence determined is listed for each complementation group in kilobase pairs (kb) DNA and arrows indicate the direction of the coding strand for each identified ORF.

1989). This suggests that at least four genetic loci are involved in cobalamin biosynthesis in *Ps. denitrificans*.

After heterologous complementation of *cob* mutants with the genomic library from *Ps. denitrificans*, a detailed genetic analysis of the complementation groups was also undertaken in order to localize individual *cob* genes (Cameron et al 1991a,b, Crouzet et al 1990a,b, Crouzet et al 1991) (Fig. 1). Each complementing fragment was sequenced and potential open reading frames (ORFs) were identified. Inactivating insertion mutagenesis was then used to determine which of these potential ORFs were *cob* genes; from subsequent complementation analysis with these insertion mutants a total of 22 unique *cob* genes were identified within the four original complementation groups. This was the first report of a genetic analysis of *cob* genes at the molecular level.

The cloning and identification of the genes involved in cobalamin biosynthesis has provided a genetic model from which considerable efforts have been launched towards the systematic elucidation of the enzymic sequence of events along this pathway. The metabolic roles of the products of nearly all the genes identified in complementation groups A–D have now been characterized. By combining knowledge of complementation ability and biochemical activities it is possible to draw a parallel between the genomic arrangement of *cob* genes and the metabolic roles associated with each gene product (Fig. 2). It is not known whether this correlation reflects a common regulation of the expression of *cob* genes belonging to the same cluster in *Ps. denitrificans*, but similar

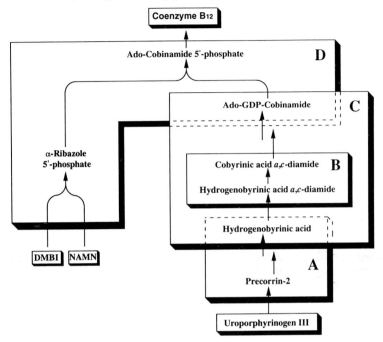

FIG. 2. The pathway for coenzyme B$_{12}$ biosynthesis from uroporphyrinogen III in *Ps. denitrificans* indicating the genomic arrangement of *cob* genes and the metabolic roles associated with each identified gene product. The complementation group responsible for the outlined biosynthetic transformations is indicated by the corresponding letters, A, B, C or D. DMBI, dimethylbenzimidazole; NAMN, nicotinic acid mononucleotide.

arrangements of *cob* genes have also been reported in *B. megaterium* (Brey et al 1986, Wolf & Brey 1986) and *S. typhimurium* (Jeter et al 1984, Jeter & Roth 1987).

Identification and analysis of enzymes of the corrin pathway

The first enzyme on the pathway from uro'gen III to cobalamin in *Ps. denitrificans*, *S*-adenosyl-L-methionine:uroporphyrinogen-III *C*-methyltransferase (SUMT, EC 2.1.1.107), is a key enzyme because it acts at the branch-point of two divergent pathways. SUMT is responsible for catalysing the first two methylations of uro'gen III, at positions C-2 and C-7, to yield precorrin-2 (Blanche et al 1989), and was identified as the product of the *cobA* gene (Crouzet et al 1990b). In addition to these *S*-adenosylmethionine (SAM)-dependent *C*-methylations, the biotransformation of uro'gen III to hydrogenobyrinic acid (the cobalt-free analogue of cobyrinic acid) requires at least six further SAM-dependent methylations, a ring scission leading to the extrusion of C-20 and

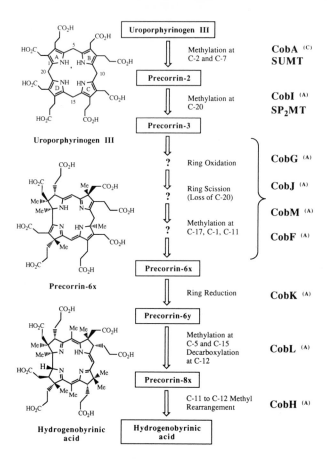

FIG. 3. The pathway of corrin biosynthesis from uro'gen III and reactions catalysed by the products of the *Ps. denitrificans cob* genes. Structures of uro'gen III, precorrin-6x and hydrogenobyrinic acid are shown. Other established corrin pathway intermediates are indicated in outline. Reactions occurring between the indicated intermediates are presented in sequence and represent current knowledge. Superscript letters after Cob proteins refer to the complementation group (A–D) to which the corresponding structural gene belongs.

its attached methyl group, and a decarboxylation of the acetic acid side chain at C-12 (Battersby 1993, Leeper 1989) (Fig. 3). With the exception of SUMT, all the enzymes required for these transformations are the products of the eight genes identified in complementation group A. These genes, *cobF→cobM*, were initially implicated by their ability to complement those *cob* mutants that did not accumulate corrinoid intermediates (Crouzet et al 1990a).

A single gene from complementation group A, *cobI*, was identified as the gene encoding the enzyme *S*-adenosyl-L-methionine:precorrin-2 methyltransferase

(SP$_2$MT). This enzyme is responsible for the SAM-dependent methylation of precorrin-2 at position C-20 to form the well-established biosynthetic intermediate precorrin-3 (Thibaut et al 1990a). However, the next biosynthetic intermediate towards the corrin nucleus isolated was not the product of the next enzymic reaction, but a much later intermediate in corrin biosynthesis, precorrin-6x (Thibaut et al 1990b,c) (Fig. 3). The three enzymic reactions involved in the conversion of precorrin-6x to hydrogenobyrinic acid have since been investigated and are now well characterized; for a detailed account of the chemistry of these transformations see Battersby (1994, this volume).

The first and last enzymic reactions in these transformations are catalysed by the *cobK* and *cobH* gene products respectively. CobK is responsible for the NADPH-dependent reduction of precorrin-6x to a dihydro derivative called precorrin-6y (Blanche et al 1992a). The final transformation takes an octamethylated intermediate, precorrin-8x, on to hydrogenobyrinic acid via a C-11 to C-12 methyl migration (Thibaut et al 1992). The transformation that takes precorrin-6y forward to the substrate for the CobH-mediated reaction is catalysed by a unique, bifunctional SAM-dependent methyltransferase encoded by the *cobL* gene (Blanche et al 1992b). This enzyme not only sequentially methylates precorrin-6y at the C-5 and C-15 positions but also catalyses decarboxylation of the acetic acid side chain at C-12.

The N-terminal half of CobL shows amino acid sequence similarity to other SAM-dependent methyltransferases of the corrin pathway (SUMT and SP$_2$MT) and the *cysG* gene product of *Escherichia coli*, which has been shown to catalyse SUMT activity in this organism (Crouzet et al 1990a). CobL is a considerably larger protein than either SUMT or SP$_2$MT and the additional C-terminal domain may encode structure and functionality necessary for the decarboxylation reaction. The sequence similarity these SAM-dependent methyltransferases show may reflect possible similarities in structural domains shared by all SAM-dependent methyltransferases. A search for putative SAM-methyltransferase sequences within the four uncharacterized proteins whose genes are clustered within complementation group A revealed that the proteins encoded by the *cobF*, *cobJ* and *cobM* genes may also be responsible for catalysing SAM-dependent methylations during corrin biosynthesis (Crouzet et al 1990a). Indeed, purified CobF showed significant SAM-binding affinity, and it is proposed that CobF, CobJ and CobM are the SAM-dependent methyltransferases responsible for catalysing the remaining *C*-methylations at positions C-17, C-1 and C-11.

The only other enzymic processes in the formation of the corrin macrocycle are an oxidation and a ring contraction. Like the three uncharacterized methylation reactions, these two steps must occur during the transformation of precorrin-3 to precorrin-6x and might well be initiated or even catalysed by the product of the *cobG* gene, the only other *cob* gene identified in complementation group A.

Identification and analysis of enzymes of the cobalamin pathway

The final phase of vitamin B_{12} biosynthesis, involving the interaction of several enzymic processes along the cobalamin pathway (Scheme 1), follows the formation of hydrogenobyrinic acid by enzymes of the corrin pathway. The necessary steps are: (i) insertion of cobalt into the corrin nucleus; (ii) step-wise amidation of six peripheral carboxyl groups of the chromophore to form cobyric acid; (iii) attachment of 1-amino-2-propanol to the ring D propanol side chain, to produce cobinamide; and (iv) the step-wise attachment of ribose and dimethylbenzimidazole (DMBI) to form a 'nucleotide loop' which is covalently linked to cobalt as the axial lower ligand (Fig. 4). Apart from these key elements the covalent linkage of an adenosyl group through C-5 of the ribose moiety as the axial upper ligand of cobalt is also required to transform the corrinoids to their coenzyme forms (Crouzet et al 1991, Huennekens et al 1982, Leeper 1989).

It is evident from this outline that the unique cobalt ion of vitamin B_{12} ultimately forms a hexa-co-ordinated complex with the four equatorial nitrogen atoms of the corrin nucleus and the two axial ligands in adenosylcobalamin. In *Ps. denitrificans*, however, the enzyme-mediated insertion of this metal ion occurs at a relatively early stage along the cobalamin pathway following amidation of hydrogenobyrinic acid (Debussche et al 1992) (Fig. 4). Insertion of cobalt into the corrin ring system at this point means that cobyrinic acid is probably not a true intermediate of vitamin B_{12} biosynthesis in this organism. This is in contrast to the process of cobalt insertion in cobyrinic acid-producing bacteria, such as *Propionibacterium shermanii*, where cobalt chelation may occur at a very early stage in the corrin pathway (Müller et al 1991, R. Vishwakarma, S. Balachandran, N. P. J. Stamford, S. Monaghan, A. Prelle, F. J. Leeper & A. R. Battersby, unpublished work 1987–1992).

Through enzymic studies the amidated natural substrate of the *Ps. denitrificans* cobalt-chelating enzyme was identified as the first intermediate beyond the corrin pathway, hydrogenobyrinic acid *a,c*-diamide (Debussche et al 1992). This intermediate is produced by step-wise amidation of two of the peripheral carboxyl groups, at positions *c* then *a*, by a glutamine amidotransferase enzyme specified by the *cobB* gene, hydrogenobyrinic acid *a,c*-diamide synthase (Debussche et al 1990). Characterization of the enzyme that catalyses the subsequent cobalt insertion reaction into this amidated intermediate showed it to be a complex ATP-dependent enzyme consisting of two separable components (Debussche et al 1992), the first component encoded by the *cobN* gene and the second by the *cobS* and *cobT* genes. These features make the cobaltochelatase from *Ps. denitrificans* similar in many ways to the complex two-component ATP-dependent magnesium-chelating enzyme from pea chloroplasts, which catalyses the first committed step in chlorophyll synthesis via insertion of magnesium into protoporphyrin IX (Walker & Weinstein 1991; Castelfranco et al 1994, this volume).

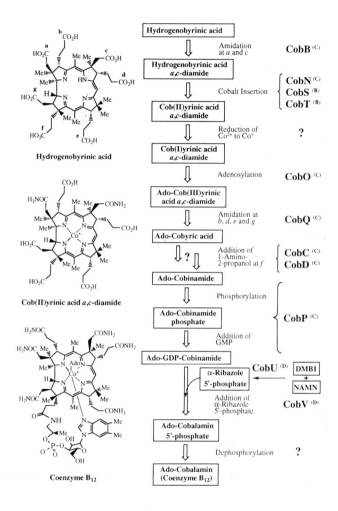

FIG. 4. Pathway of coenzyme B$_{12}$ (5′-deoxyadenosylcobalamin) biosynthesis from hydrogenobyrinic acid and reactions catalysed by the products of the *Ps. denitrificans cob* genes. Structures of hydrogenobyrinic acid, cob(II)yrinic acid *a,c*-diamide and adenosylcobalamin (Ado-cobalamin, coenzyme B$_{12}$) are shown. Other established cobalamin pathway intermediates are indicated in outline. Reactions occurring between the indicated intermediates are presented in sequence and represent current knowledge. The adenosylation reaction is thought to occur at the stage of cob(I)yrinic acid *a,c*-diamide, therefore all following intermediates are shown as adenosylated and at the oxidation state of Co(III). The superscript letter after each Cob protein refers to the complementation group (A–D) to which the corresponding structural gene belongs. Ado, 5′-deoxy-5′-adenosyl; DMBI, dimethylbenzimidazole; NAMN, nicotinic acid mononucleotide.

Following transformation of the first two carboxyl groups of hydro-genobyrinic acid at *a* and *c* and insertion of cobalt only four further amidations of peripheral carboxyl groups are required to complete the transformation of hydrogenobyrinic acid to cobyric acid (Fig. 4). All four of these remaining amidations are catalysed by a single enzyme, cobyric acid synthase (Blanche et al 1991a), which is encoded by the *cobQ* gene. As does hydrogenobyrinic acid *a,c*-diamide synthase, this enzyme utilizes glutamine as the amide donor and amidation follows a strict reaction sequence. However, cobyrinic acid *a,c*-diamide is not a substrate for this enzymic reaction and CobQ is specific only for its coenzyme form.

The major corrinoids occurring in Nature apparently exist in the coenzyme form. During coenzyme B_{12} biosynthesis the addition of the adenosyl group as the axial upper ligand is thought to occur soon after formation of cobyrinic acid (Escalante-Semerena et al 1990, Huennekens et al 1982). Adenosylation of corrinoids in *Ps. denitrificans* is catalysed by the enzyme cobalamin adenosyltransferase encoded by the *cobO* gene (Debussche et al 1991). This enzyme is capable of adenosylating all corrinoids isolated from this organism, although its specificities suggest its physiological substrate is cob(I)yrinic acid *a,c*-diamide. It is therefore appropriate to refer to all intermediates that follow formation of adenosylcobyrinic acid *a,c*-diamide as their adenosylated (Ado) forms. Because cob(II)yrinic acid *a,c*-diamide is the product of cobaltochelatase activity, by necessity, there must be a cob(II)yrinic acid *a,c*-diamide reductase to provide substrate for the adenosylation reaction. Such an enzyme was purified from cell-free extracts of *Ps. denitrificans* (Blanche et al 1992c) and shown to be an NADH-dependent flavoprotein similar to cob(II)alamin reductase from *Clostridium tetanomorphum* (Huennekens et al 1982). The N-terminal amino acid sequence of this protein does not correspond to any protein predicted from genes within the four complementation group fragments reported to encode cobalamin biosynthetic enzymes in *Ps. denitrificans* (Cameron et al 1989).

The final steps leading from Ado-cobyric acid in the biosynthesis of vitamin B_{12} in *Ps. denitrificans* involve the attachment of 1-amino-2-propanol to the ring D propanol side chain to produce Ado-cobinamide and elaboration of the 'nucleotide loop' to form Ado-cobalamin (Fig. 4). The enzymes responsible for the first of these steps have not yet been isolated, but the *cobD* and *cobC* genes code for proteins that are implicated in this transformation. Their involvement probably occurs at the point of specific amidation of the carboxyl group at position *f* of the corrin nucleus with (*R*)-1-amino-2-propanol (Crouzet et al 1990b). In contrast, conversion of Ado-cobinamide into Ado-cobalamin in *Ps. denitrificans* is now well characterized (Blanche et al 1991b, Cameron et al 1991a); as in *Pr. shermanii* (Leeper 1989), a total of four biosynthetic steps are required: (i) phosphorylation of the hydroxyl group of the (*R*)-1-amino-2-propanol residue of Ado-cobinamide to yield Ado-cobinamide phosphate; (ii) transformation of Ado-cobinamide phosphate into Ado-GDP–

cobinamide by the addition of the GMP moiety of a molecule of GTP; (iii) exchange of GMP from Ado-GDP–cobinamide with 1-α-ribofuranoside-5,6-dimethylbenzimidazole (α-ribazole 5′-phosphate) to yield Ado-cobalamin 5′-phosphate; and (iv) dephosphorylation of Ado-cobalamin 5′-phosphate to cobalamin.

The first two enzymic steps during biosynthesis of Ado-cobalamin from Ado-cobinamide are catalysed by a single bifunctional enzyme (Blanche et al 1991a). This protein, encoded by the *cobP* gene, catalyses both the conversion of Ado-cobinamide to Ado-cobinamide phosphate (cobinamide kinase) and the conversion of Ado-cobinamide phosphate to Ado-GDP–cobinamide (cobinamide-phosphate guanylyltransferase). The final enzymic steps leading to formation of Ado-cobalamin, the formation and addition of α-ribazole 5′-phosphate to complete the nucleotide loop, are carried out by two proteins encoded by the only two *cob* genes (*cobU* and *cobV*) identified in complementation group D (Cameron et al 1991a) (Fig. 1). CobU is the nicotinate-nucleotide:dimethylbenzimidazole phosphoribosyltransferase catalysing the synthesis of α-ribazole 5′-phosphate from DMBI and β-nicotinic acid mononucleotide. CobV (cobalamin-5′-phosphate synthase) then utilizes the product of the reaction mediated by CobU (α-ribazole 5′-phosphate) and Ado-GDP–cobinamide to catalyse the synthesis of Ado-cobalamin 5′-phosphate. Interestingly, CobP is also able to synthesize Ado-cobalamin from Ado-GDP–cobinamide and α-ribazole directly, making this enzyme both a cobalamin-5′-phosphate synthase and a cobalamin synthase enzyme. This only leaves the final step in the synthesis of cobalamin, the dephosphorylation of Ado-cobalamin 5′-phosphate or α-ribazole 5′-phosphate, which is catalysed by an as yet unidentified endogenous 5′-phosphatase. The metabolic roles of the *cobE* and *cobW* gene products have yet to be defined.

New developments in vitamin B$_{12}$ research

It is clear from this outline of the corrin and cobalamin pathways constituting the final two phases of vitamin B$_{12}$ biosynthesis in *Ps. denitrificans* that elucidation of the entire enzymic sequence of events along this pathway is almost complete. In light of this progress, research is now concentrating on the development and use of a variety of techniques that utilize this knowledge of the genetic and biochemical organization of the pathway to define the chemistry involved in individual enzymic steps.

In the past, efforts at Cambridge (UK) have successfully approached the problems faced in investigating the biosynthesis of vitamin B$_{12}$ through the use of various labelling techniques (Battersby 1993). With this in mind, research at Cambridge has now focused on the development of engineered multi-enzyme *in vitro* systems that allow, for the first time, the *biosynthetic* preparation of early intermediates of the corrin pathway. Through these systems it is now

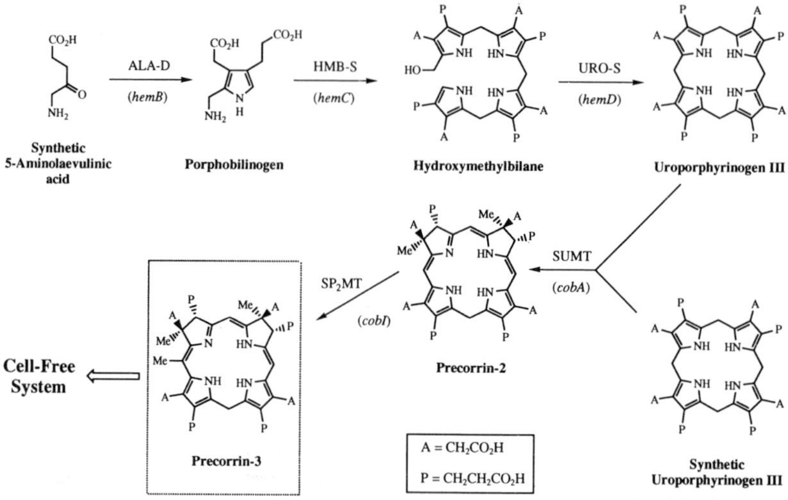

FIG. 5. The target transformations for the engineered biosynthesis of early corrin pathway intermediates. The structures of intermediates 5-aminolaevulinic acid (ALA), porphobilinogen (PBG) and hydroxymethylbilane (HMB) involved in tetrapyrrole biosynthesis and precorrins-2 and 3 from the corrin pathway are shown. The enzymes involved in these interconversions are also given (D, dehydratase; S, synthase), together with their corresponding structural genes from which genetically engineered enriched sources of these enzymes could be manufactured. The intermediates ALA and uro'gen III are the favoured points of entry for a reconstituted enzyme system using synthetically manufactured substrates.

possible to manufacture specific tailor-labelled substrates to use as probes in feeding experiments with cell-free systems designed to elucidate aspects of enzymic interconversions that may take place several steps further along the biosynthetic pathway to vitamin B_{12}.

In the development of these coupled-enzyme systems the target transformation was the manufacture of the trimethylated corrin pathway intermediate, precorrin-3, from earlier more accessible substrates (Fig. 5). This intermediate was the target molecule of choice because it is not only committed to the biosynthetic pathway to vitamin B_{12} but is also efficiently utilized by cell-free systems of *Ps. denitrificans* and is currently the most advanced intermediate for which manufacture is possible given current knowledge of the enzymic sequence of events in this pathway. However, the overriding requirement for efficient production of usable quantities of precorrin-3 from a reconstituted enzymic system is the manufacture of very large quantities of each of the enzymes involved in its biosynthesis. If readily available substrates such as ALA are to be used, both the first and second SAM-dependent methyltransferases of the corrin pathway (SUMT and SP₂MT) and the enzymes involved in uro'gen III formation are required.

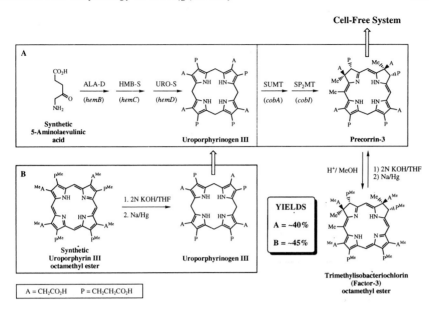

FIG. 6.   Two reconstituted multi-enzyme systems (A and B) for the biosynthesis of early corrin pathway intermediates. Enzymes extracted and partially purified from genetically engineered sources are recombined in a one-pot *in vitro* synthesis of the target molecule. Systems for the transformation of both ALA (A) and uro'gen III (B) to precorrin-3 have been developed. Incubations are routinely carried out at 37 °C in degassed buffers, under an argon stream and in the dark. Precorrin-3 is isolated by a combination of ion-exchange and thin-layer chromatography and converted by air oxidation and acidic esterification to its didehydro-octamethyl ester. Precorrin-3 can be regenerated from tri-methylisobacteriochlorin octamethyl ester by base-catalysed ester hydrolysis and reduction on NaHg. Synthetically manufactured uroporphyrin III octamethyl ester can be similarly treated for use as a substrate in the coupled enzyme system (B). Enzymes required for each set of transformations are given and the corresponding structural gene from which enriched sources of each protein was engineered is indicated.

The *Bacillus subtilis hem* genes encoding ALA dehydratase, hydroxymethyl-bilane synthase and uro'gen-III synthase (Hansson et al 1991), and the *Ps. dentrificans cobA* and *cobI* genes encoding SUMT (Crouzet et al 1990b) and SP$_2$MT (Thibaut et al 1990a) respectively, have been used to construct enriched sources of each of the appropriate enzymes. Two coupled-enzyme systems were developed with the enzymes isolated from these enriched sources; these systems can use either synthetically manufactured ALA or uro'gen III as a substrate to produce precorrin-3 on the multi-milligram scale (Fig. 6). In each case, overall yields from the target transformation were in excess of 40%, including not only the enzymic conversion of substrate to precorrin-3 but also subsequent chemical modification and isolation of the target intermediate.

In collaboration with Francis Blanche and Joel Crouzet we are now using blocked mutants of vitamin B_{12}-producing organisms for feeding experiments with our biosynthetically produced substrates. Through enzymic interconversion of the tailor-labelled substrates in these cell-free systems it will be possible to do experiments that will allow rigorous structural determinations of newly isolated intermediates. With knowledge gained from these investigations, further specifically labelled precursors will be designed for similar feeding experiments to provide insights into the chemical mechanism and stereochemical course of enzymic interconversions between intermediates.

Acknowledgements

The work described in this short review is largely the achievement of Francis Blanche and Joel Crouzet (Rhône–Poulenc Rorer, France) and their colleagues, whose names will be found in the references. I am therefore extremely grateful to them for allowing me to recount their outstanding success in vitamin B_{12} biosynthesis. I should also like to acknowledge the dedication and skills of my colleagues in Cambridge, Dr A. I. Alanine and Dr A. R. Pitt, who developed our coupled enzyme systems. Finally, it has been my immense good fortune to be associated with Professor Sir Alan Battersby, and I thank him sincerely for his knowledgeable advice, ceaseless enthusiasm and generous financial support.

References

Battersby AR 1993 Biosynthesis of vitamin B_{12}. Acc Chem Res 26:15–21
Battersby AR 1994 New intermediates in the B_{12} pathway. In: The biosynthesis of the tetrapyrrole pigments. Wiley, Chichester (Ciba Found Symp 180) p 267–284
Blanche F, Debussche L, Thibaut D, Crouzet J, Cameron B 1989 Purification and characterization of S-adenosyl-L-methionine:uropophyrinogen III methyltransferase from *Pseudomonas denitrificans*. J Bacteriol 171:4222–4231
Blanche F, Couder M, Debussche L, Thibaut D, Cameron B, Crouzet J 1991a Biosynthesis of vitamin B_{12}: stepwise amidation of carboxyl groups *b*, *d*, *e* and *g* of cobyrinic acid *a*,*c*-diamide is catalyzed by one enzyme in *Pseudomonas denitrificans*. J Bacteriol 173:6046–6051
Blanche F, Debussche L, Famechon A, Thibaut D, Cameron B, Crouzet J 1991b A bifunctional protein from *Pseudomonas denitrificans* carries cobinamide kinase and cobinamide phosphate guanylyltransferase activities. J Bacteriol 173:6052–6057
Blanche F, Thibaut D, Famechon A, Debussche L, Cameron B, Crouzet J 1992a Precorrin-6x reductase from *Pseudomonas denitrificans*: purification and characterization of the enzyme and identification of the structural gene. J Bacteriol 174:1036–1042
Blanche F, Famechon A, Thibaut D, Debussche L, Cameron B, Crouzet J 1992b Biosynthesis of vitamin B_{12} in *Pseudomonas denitrificans*: the biosynthetic sequence from precorrin-6y to precorrin-8x is catalyzed by the *cobL* gene product. J Bacteriol 174:1050–1052
Blanche F, Maton L, Debussche L, Thibaut D 1992c Purification and characterization of cob(II)yrinic acid *a*,*c*-diamide reductase from *Pseudomonas denitrificans*. J Bacteriol 174:7452–7454

Brey RN, Banner CDB, Wolf JB 1986 Cloning of multiple genes involved with cobalamin (vitamin B$_{12}$) biosynthesis in *Bacillus megaterium*. J Bacteriol 167:623–630

Cameron B, Briggs K, Pridmore S, Brefort G, Crouzet J 1989 Cloning and analysis of genes involved in coenzyme B$_{12}$ biosynthesis in *Pseudomonas denitrificans*. J Bacteriol 171:547–557

Cameron B, Blanche F, Rouyez M-C et al 1991a Genetic analysis, nucleotide sequence, and products of two *Pseudomonas denitrificans cob* genes encoding nicotinate-nucleotide:dimethylbenzimidazole phosphoribosyltransferase and cobalamin (5′-phosphate) synthase. J Bacteriol 173:6066–6073

Cameron B, Guilhot C, Blanche F et al 1991b Genetic and sequence analysis of a *Pseudomonas denitrificans* DNA fragment containing two *cob* genes. J Bacteriol 173:6058–6065

Castelfranco PA, Walker CJ, Weinstein JD 1994 Biosynthetic studies on chlorophylls: from protoporphyrin IX to protochlorophyllide. In: The biosynthesis of the tetrapyrrole pigments. Wiley, Chichester (Ciba Found Symp 180) p 194–209

Crouzet J, Cameron B, Cauchois L et al 1990a Genetic and sequence analysis of an 8.7-kilobase *Pseudomonas denitrificans* fragment carrying eight genes involved in transformation of precorrin-2 to cobyrinic acid. J Bacteriol 172:5980–5990

Crouzet J, Cauchois L, Blanche F et al 1990b Nucleotide sequence of a *Pseudomonas denitrificans* 5.4 kilobase DNA fragment containing five *cob* genes and identification of structural genes encoding S-adenosyl-L-methionine:uroporphyrinogen III methyltransferase and cobyrinic acid *a,c*-diamide synthase. J Bacteriol 172:5968–5979

Crouzet J, Levy-Schil S, Cameron B et al 1991 Nucleotide sequence and genetic analysis of a 13.1-kilobase-pair *Pseudomonas denitrificans* DNA fragment containing five *cob* genes and identification of structural genes encoding Cob(I)alamin adenosyltransferase, cobyric acid synthase, and bifunctional cobinamide kinase-cobinamide phosphate guanylyltransferase. J Bacteriol 173:6074–6087

Debussche L, Thibaut D, Cameron B, Crouzet J, Blanche F 1990 Purification and characterization of cobyrinic acid *a,c*-diamide synthase from *Pseudomonas denitrificans*. J Bacteriol 172:6239–6244

Debussche L, Couder M, Thibaut D, Cameron B, Crouzet J, Blanche F 1991 Purification and partial characterization of cob(I)alamin adenosyltransferase from *Pseudomonas denitrificans*. J Bacteriol 173:6300–6302

Debussche L, Couder M, Thibaut D, Cameron B, Crouzet J, Blanche F 1992 Assay, purification and characterisation of cobaltochelatase, a unique complex enzyme catalyzing cobalt insertion in hydrogenobyrinic acid *a,c*-diamide during coenzyme B$_{12}$ biosynthesis in *Pseudomonas denitrificans*. J Bacteriol 174:7445–7451

Escalante-Semerena JC, Suh S-J, Roth JR 1990 *cobA* function is required for both de novo cobalamin biosynthesis and assimilation of exogenous corrinoids in *Salmonella typhimurium*. J Bacteriol 172:273–280

Hansson M, Rutberg L, Schröder I, Hederstedt L 1991 The *Bacillus subtilis hemAXCDBL* gene cluster, which encodes enzymes of the biosynthetic pathway from glutamate to uroporphyrinogen III. J Bacteriol 173:2590–2599

Huennekens FM, Vitols KS, Fujii K, Jacobsen DW 1982 Biosynthesis of cobalamin coenzymes. In: Dolphin D (ed) B$_{12}$. Wiley, Chichester, vol 1:146–167

Jeter RM, Roth JR 1987 Cobalamin (vitamin B$_{12}$) biosynthetic genes of *Salmonella typhimurium*. J Bacteriol 169:3187–3189

Jeter RM, Olivera BM, Roth JR 1984 *Salmonella typhimurium* synthesizes cobalamin (vitamin B$_{12}$) de novo under anaerobic growth conditions. J Bacteriol 159:206–213

Jordan PM 1994 The biosynthesis of uroporphyrinogen III: mechanism of action of porphobilinogen deaminase. In: The biosynthesis of the tetrapyrrole pigments. Wiley, Chichester (Ciba Found Symp 180) p 70–96

Leeper FJ 1989 The biosynthesis of porphyrins, chlorophylls and vitamin B_{12}. Nat Prod Rep 6:171–203

Leeper FJ 1994 The evidence for a spirocyclic intermediate in the formation of uroporphyrinogen III by cosynthase. In: The biosynthesis of the tetrapyrrole pigments. Wiley, Chichester (Ciba Found Symp 180) p 111–130

Müller G, Zipfel F, Hlineny K et al 1991 Timing of cobalt insertion in vitamin B_{12} biosynthesis. J Am Chem Soc 113:9893–9895

Spencer P, Jordan PM 1994 5-Aminolaevulinic acid dehydratase: characterization of the α and β metal-binding sites of the *Escherichia coli* enzyme. In: The biosynthesis of the tetrapyrrole pigments. Wiley, Chichester (Ciba Found Symp 180) p 50–69

Thibaut D, Couder M, Crouzet J, Debussche L, Cameron B, Blanche F 1990a Assay and purification of *S*-adenosyl-L-methionine:precorrin-2 methyltransferase from *Pseudomonas denitrificans*. J Bacteriol 172:6245–6251

Thibaut D, Blanche F, Debussche L, Leeper FJ, Battersby AR 1990b Biosynthesis of vitamin B_{12}: structure of precorrin-6x octamethyl ester. Proc Natl Acad Sci USA 87:8800–8804

Thibaut D, Debussche L, Blanche F 1990c Biosynthesis of vitamin B_{12}: isolation of precorrin-6x, a metal-free precursor of the corrin macrocycle retaining five *S*-adenosylmethionine-derived peripheral methyl groups. Proc Natl Acad Sci USA 87:8795–8799

Thibaut D, Couder M, Famechon A et al 1992 The final step in the biosynthesis of hydrogenobyrinic acid is catalyzed by the *cobH* gene product with precorrin-8x as the substrate. J Bacteriol 174:1043–1049

Walker CJ, Weinstein JD 1991 *In vitro* assay of the chlorophyll biosynthetic enzyme Mg-chelatase: resolution of the activity into soluble and membrane-bound fractions. Proc Natl Acad Sci USA 88:5789–5793

Wolf JB, Brey RN 1986 Isolation and genetic characterization of *Bacillus megaterium* cobalamin biosynthesis-deficient mutants. J Bacteriol 166:51–58

DISCUSSION

Arigoni: Perhaps I should point out, on behalf of the organizers, that Blanche and Crouzet were of course invited here, but were unfortunately unable to accept.

Battersby: Could I just echo that? They asked me to act as their spokesman and I shall do my best on their behalf.

Arigoni: What is known about the criteria which led the French group to select this specific organism, *Ps. denitrificans*?

Stamford: Their point of view is not one of fundamental basic research but rather one of developing an industrial strain which produces large quantities of cobalamin.

Arigoni: Was *Ps. denitrificans* the best strain to begin with?

Stamford: It is a good producer of cobalamin.

Hädener: Why were two different organisms used to make these *cob* mutants, rather than *Ps. denitrificans*?

Stamford: This was because molecular genetic techniques were far better established in *Ps. putida* and *A. tumifaciens* than in *Ps. denitrificans*, all of which are cobalamin-synthesizing Gram-negative aerobic rods.

Arigoni: There are 22 genes in the complete set. For how many of these is a specific function known?

Stamford: The specific functions of the products of all of these genes are known except those of CobE and CobW, and the proteins encoded by the *cobC* and *cobD* genes, which are implicated in the transformation of cobyric acid to cobinamide.

Warren: If I remember correctly, the majority of the *cob* genes are transcribed in one direction, but one gene is actually transcribed in the opposite direction. Is this important physiologically?

Stamford: The genes identified in complementation group D (which includes the *cobU* and *cobV* genes whose products are involved in the last steps of cobalamin biosynthesis) are divergently transcribed. This arrangement, although not unique, is interesting from the point of view of regulation. Such an arrangement may allow the transcription and regulation of expression of the two *cob* genes from a single control region. Also, the *cobK* gene is transcribed in the opposite direction from all the other genes in complementation group A, but it is less easy to speculate on the significance of this.

Scott: A couple of *cob* genes, *cobE* and *cobW*, have been referred to sporadically in the Rhône-Poulenc groups' publications (Cameron et al 1989). At one time it appeared that they played a special role, but at other times they're not mentioned.

Stamford: We know nothing at all about *cobE*. Its predicted amino acid sequence gives us no indication as to its role. The product of *cobW*, in contrast, is an interesting protein (Crouzet et al 1991). CobW contains an NAD(H)-binding site, which suggests it is involved in an NAD(H)-dependent oxido-reduction step in the pathway. However, the cob(II)yrinic acid *a,c*-diamide reductase from *Ps. denitrificans* that leads to the cobalt(I) corrinoids has already been identified as a separate protein (Blanche et al 1992). CobW also contains an ATP-binding site consensus sequence, which might mean that it is responsible for a reaction completely different from that catalysed by the cob(II)yrinic acid reductase. Whatever their roles, CobE and CobW are obviously essential; strains of *Ps. denitrificans* with mutations in these particular genes are unable to synthesize cobalamin.

Scott: As far as I remember, there's no sequence similarity between *cobE* or *cobW* and any of the *S. typhimurium cbi* genes. We are left with some orphan genes in both systems, genes that we know are necessary but not why.

Jordan: If cobalt is inserted in one of the B$_{12}$ pathways early on and in the other later, I would have thought the conformations of the two sets of substrates would differ radically. Consequently, I would expect there to be major differences in the protein sequences of the two sets of enzymes for this non-common part of the pathway.

Stamford: We know nothing about the genetics of cobalamin biosynthesis in cobyrinic acid-producing bacteria such as *Pr. shermanii* which insert cobalt early in the corrin pathway. Professor Scott may be able to enlighten us with regard to the *Salmonella* genes.

Scott: The identities between the *S. typhimurium cbi* genes with methyltransferase activity and the methyltransferases encoded by *Ps. denitrificans cob* genes are not great, typically 30% or less.

Stamford: Is there any greater similarity between the other cobalamin synthetic genes?

Scott: No. There is little or no similarity between the orphan genes in *S. typhimurium* and the unassigned *cob* gene sequences of *Ps. denitrificans*. Metal insertion in *S. typhimurium* (just as in *Pr. shermanii*) has to has to be early. We have shown that diamide formation can be achieved with cobyrinic acid as substrate, rather than hydrogenobyrinic acid, by the product of the *cbiA* gene, the *Salmonella* gene corresponding to *cobB* of *Ps. denitrificans*.

Warren: The 5,15-methylase from *Ps. denitrificans* (CobL) is fused onto the decarboxylase, giving a multi-enzyme complex. In *Salmonella* those genes are separate. Is this significant?

Scott: The homologous genes in *Salmonella*, *cbiT* and *cbiE*, encode separate decarboxylase and 5,15-methylase activities, respectively. The proteins have been expressed, but their activities haven't been demonstrated because the substrate is unavailable. There's every chance that in *Salmonella* we could pause at the 5,15-bismethylation step with CbiE and then use CbiT to do the ring C decarboxylation.

Stamford: There are significant regions of sequence similarity shared by all the methyltransferases identified in both *Ps. denitrificans* and *E. coli* (Crouzet et al 1990). In *E. coli* CysG these regions are towards the C-terminus whereas in *Ps. denitrificans* CobL they are at the N-terminus. Martin Warren has shown that the N-terminus of CysG might be involved in the other reactions the protein is known to catalyse. The C-terminal domain of CobL may be involved in the decarboxylation of C-12 of the corrin ring. It's interesting to speculate that these methyltransferases may well be the fusion products of enzymes that once carried out these reactions independently.

Scott: Have the French groups expressed the *cobG* gene, whose product, by default, ought to catalyse the oxidative ring contraction? John Roth has been looking for the CobG homologue in *Salmonella* but has found similarity to only *Salmonella* CysI, sulphite reductase, and to spinach nitrite reductase (personal communication). Sequence comparisons are, of course, fraught with danger.

Stamford: I do not know whether they have expressed *cobG*.

Jordan: To which substrate, hydrogenobyrinic acid or cobyrinic acid, does CobB add the two amide groups?

Stamford: The cobyrinic acid *a,c*-diamide synthase from *Ps. denitrificans* can amidate either of those two substrates but shows 100-fold higher affinity for hydrogenobyrinic acid.

Beale: You said that the amidations follow a strict order. The amidation sequence was previously reported to be random (Bernhauer et al 1968); is this work no longer held to be correct, or was it done in a different species?

Stamford: The sequence of amidations by the product of the *cobB* gene (cobyrinic acid *a,c*-diamide synthase) is well characterized; group *c* is the first amidated carboxyl group. Similarly, a strict sequence of amidation is also observed for amidations carried out by the product of the *cobQ* gene, cobyric acid synthase; the order of these amidations is not known, but the French groups have shown by HPLC analysis that following each amidation step only a single intermediate is isolated.

Scott: Bernhauer et al's (1968) work was in *Pr. shermanii*. In those days the analytical protocols necessary to address this question properly were not available.

Kräutler: It has not so far been possible to remove cobalt from an intact cobyrinic acid derivative. One would therefore assume that insertion of cobalt into the intact core of a natural corrin would be difficult. How well established is it that hydrogenobyrinic acid *a,c*-diamide is the substrate for cobalt insertion? Could not the corresponding 6,7-lactam be an intermediate, the actual substrate for the enzymic metal insertion?

Stamford: It is well established.

Pitt: The metallation presumably requires energy. You said that the cobaltochelatase is an ATP-dependent enzyme. Is it an ATP hydrolase?

Stamford: This is not known. Non-hydrolysable analogues of ATP inhibit the reaction at the same concentration as the K_m value for ATP itself. This suggests that ATP hydrolysis is required for cobalt insertion.

Arigoni: One can solve part of this puzzle by assuming that in the cobalt-free compounds another metal is operating at the core of the ligand and is then washed out during isolation.

Scott: We don't know about surrogate metals in *Ps. denitrificans*. The only metal that is ever added to incubations is magnesium. Would magnesium be a good central metal for directing *C*-methylation?

Eschenmoser: Magnesium is one of the most hydrolytically sensitive metal ions in the corrin ring.

Scott: That might be why the *Pseudomonas* intermediates are isolated without any co-ordinating metal.

I like zinc, intuitively, because of Gerhard Müller's experiments on cobalt-deficient *Pr. shermanii* (Müller et al 1987). Although he didn't isolate any post-precorrin-3 intermediates *per se*, he isolated a number of interesting 'diverted' products, Factors S_1-S_4, all of which contained zinc.

References

Bernhauer K, Wagner F, Michna H, Rapp P, Vogelmann H 1968 Amidation of cobyrinic acid. Hoppe-Seyler's Z Physiol Chem 349:1297–1306

Blanche F, Maton L, Debussche L, Thibaut D 1992 Purification and characterization of cob(II)yrinic acid a,c-diamide reductase from *Pseudomonas denitrificans*. J Bacteriol 174:7452–7454

Cameron B, Briggs K, Pridmore S, Brefort G, Crouzet J 1989 Cloning and analysis of genes involved in coenzyme B_{12} biosynthesis in *Pseudomonas denitrificans*. J Bacteriol 171:547–557

Crouzet J, Cameron B, Cauchois L et al 1990 Genetic and sequence analysis of an 8.7-kilobase *Pseudomonas denitrificans* fragment carrying eight genes involved in transformation of precorrin-2 to cobyrinic acid. J Bacteriol 172:5980–5990

Crouzet J, Levy-Schil S, Cameron B et al 1991 Nucleotide sequence and genetic analysis of a 13.1-kilobase-pair *Pseudomonas denitrificans* DNA fragment containing five *cob* genes and identification of structural genes encoding Cob(I)alamin adenosyltransferase, cobyric acid synthase, and bifunctional cobinamide kinase-cobinamide phosphate guanylyltransferase. J Bacteriol 173:6074–6087

Müller G, Schmiedl J, Schneider E et al 1987 Factor S_1, a natural corphin from *Propionibacterium shermanii*. J Am Chem Soc 109:6902–6904

New intermediates in the B_{12} pathway

Alan R. Battersby

University Chemical Laboratory, Lensfield Road, Cambridge, CB2 1EW, UK

Abstract. Vitamin B_{12} has a complex structure which represents one of the most challenging biosynthetic problems in Nature. Exciting progress has been made by combining the techniques, approaches and strengths of chemistry, spectroscopy and biology. Most of the advances until recently came from experiments based either on labelling simpler precursors with radioactive isotopes followed by controlled degradation of the labelled products, or on the use of stable isotopes, ^{13}C in particular, because it can be detected and its environment can be studied by NMR spectroscopy. These experiments imposed heavy demands on synthesis which provided the specifically labelled starting materials. More recently, the powerful methods of genetics and molecular biology have been added to the armoury, leading to another massive surge forward by allowing the preparation, through gene overexpression, of large quantities of the enzymes of the biosynthetic pathway. Equally important has been the generation of mutant forms of B_{12}-producing organisms in which the biosynthetic pathway is blocked at specific points. Here I focus on the latest advances. The structures of the newly discovered intermediates are described and some of the chemistry involved is explored. In conclusion, the presently known pathway to vitamin B_{12} is reviewed.

1994 The biosynthesis of the tetrapyrrole pigments. Wiley, Chichester (Ciba Foundation Symposium 180) p 267–284

The question of how vitamin B_{12} (**1**) is constructed in Nature is one of the great biosynthetic problems, possibly the greatest. It will take a lifetime of scientific effort to work out even the broad form of the biosynthetic pathway; though this first phase has not yet been completed, you will see later that the work has, if I can adapt Churchill's words, passed the end of the beginning and reached the beginning of the end. My task is to continue this marvellous B_{12} story from the point at which Pat Stamford left off (Stamford 1994, this volume), and I am glad he has laid down the foundations of the genetics, molecular biology and some of the enzymology that will carry us along the B_{12} pathway as far as the trimethylated intermediate, precorrin-3 (**3**).

The structure of vitamin B_{12} (**1**) is rather complex, but because it is known to be biosynthesized from cobyrinic acid (**4**) in one organism we shall consider, and in another organism from the corresponding metal-free form, **5**, called hydrogenobyrinic acid, our aim is to understand how these somewhat simpler molecules are built.

Already the message coming across is that the biosynthesis of B_{12} is a problem involving stepwise *C*-methylation with some other transformations thrown in for good measure. Inspection of the overall change from precorrin-3 (**3**) to cobyrinic acid (**4**) shows that five more methyl groups have been introduced, and that ring contraction, a decarboxylation and cobalt insertion have occurred, together with possible redox changes. Simple permutation shows that these operations could be carried out in a huge number of ways, and our problem is (a) to discover the naturally selected sequence, which may have minor variations in different living systems, and (b) to determine the structures of all the intermediates.

| 1 R = CN |
| |

| 2 R = H, ● = ^{12}C |
| 3 R = Me, ● = ^{12}C |
| 3a R = Me, ● = ^{13}C |

| 4 M = Co^{III}, Cobyrinic acid |
| 5 M = H, Hydrogenobyrinic acid |

I am spokesman for two teams of scientists, one in France at Rhône–Poulenc Rorer, and the other at Cambridge. Some of the work was done independently in France, some independently in Cambridge, and some has been carried out jointly by the two teams. I want to acknowledge at the outset the marvellous collaboration we have enjoyed with our French colleagues led by Francis Blanche and Joel Crouzet. The French groups have done great things on B_{12}. You have already heard of their critically important work on the genetics and molecular biology of *Pseudomonas denitrificans*, an organism that biosynthesizes the metal-free corrin **5** for subsequent cobalt insertion and transformation into vitamin B_{12} (Debussche et al 1992).

We should focus first on the French team's development of a genetically engineered strain of *Ps. denitrificans* in which eight of the genes involved in the biosynthesis of hydrogenobyrinic acid (**5**) are overexpressed (Crouzet et al 1990). The result is that the eight corresponding enzymes are overproduced by this strain. It must be emphasized that this strain is in no way odd or aberrant; it simply produces more of the normal enzymes than the wild strain, and so permits greater turnover of material. The amounts produced are sufficient for structural work by NMR spectroscopy on the newly discovered intermediates we shall consider. A cell-free protein extract from this strain was able to convert

6a   X = Y = ^{2}H
7a   2'-Phosphate of 6a
6b   X = H, Y = ^{2}H
7b   2'-Phosphate of 6b
6c   X = ^{2}H, Y = H
7c   2'-Phosphate of 6c
6d   X = H, Y = ^{3}H
7d   2'-Phosphate of 6d
6e   X = ^{3}H, Y = H
7e   2'-Phosphate of 6e

precorrin-3 (3) into hydrogenobyrinic acid (5) with high yield provided various cofactors were included in the incubation mixture, one being reduced nicotinamide adenine dinucleotide phosphate, NADPH (part structure 7). When NADPH was omitted, a previously unobserved yellow pigment was formed instead of hydrogenobyrinic acid (5). That a new biosynthetic intermediate had been isolated was clear when the yellow pigment was converted in high yield into 5 by incubation with the enzyme system and all the cofactors including NADPH (Thibaut et al 1990a). This new intermediate had as profound an effect on the field of B$_{12}$ biosynthesis as did the earlier isolation and structure determination of derivatives of precorrin-1, precorrin-2 and precorrin-3 (3) (Stamford 1994, this volume). It was named precorrin-6x.

Structure of precorrin-6x

The first exciting results came from the preparation of precorrin-6x from ^{14}C-labelled precorrin-3 using [*methyl*-^{3}H] S-adenosylmethionine, [*methyl*-^{3}H] SAM. The ^{3}H:^{14}C ratio in the precorrin-6x produced indicated (a) that three new methyl groups had been added to the three already present in precorrin-3, making six in all added to uroporphyrinogen (uro'gen) III (8), (hence the name precorrin-6x) and (b) that, surprisingly, ring-contraction had already occurred with loss from precorrin-3 (3) of C-20 and its attached methyl group. ^{13}C-Labelling experiments in combination with NMR then established that the *C*-methyl groups which appear in hydrogenobyrinic acid (5) at C-17, C-12α and C-1 are present in precorrin-6x (Thibaut et al 1990a).

More surprises were to come. Precorrin-6x was found to be an octacarboxylic acid with the 12-acetate group still intact. Yet if the methyl group at 12α in the final corrin (5) is at the same position in its precursor, precorrin-6x, then subsequent decarboxylation of the 12-acetate would surely be blocked. Also, precorrin-6x was found to have seven double bonds, whereas a corrin has six. A reductive step is therefore needed before 5 is reached, which fits in with the production of precorrin-6x when the reducing cofactor NADPH is omitted.

All the experiments I have described so far on precorrin-6x were carried out entirely by the French groups (Thibaut et al 1990a). From these experiments it was evident that precorrin-6x was a most important substance; at this point, a joint effort was launched involving both the Paris and the Cambridge teams. Our plan was to use biosynthetic methods to ^{13}C-label every carbon in the macrocycle, and also the three new methyl groups; this work provided three differently labelled samples of precorrin-6x. The full panoply of modern NMR techniques was then applied to study the octamethyl esters of these three multiply labelled samples. The assembled mass of data led us to the quite startling structure **9** for precorrin-6x octamethyl ester (Thibaut et al 1990b).

8   ● = ◆ = ^{12}C
8a ● = ^{12}C. ◆ = ^{13}C
8b ● = ^{13}C. ◆ = ^{12}C

9   ◆ = O = ^{12}C, R = Me
9a ◆ = O = ^{12}C, R = H
9b ◆ = O = ^{13}C, R = Me

10

11 R = Me
11a R = H

These NMR studies confirmed that in precorrin-6x ester (**9**) ring contraction had already occurred, showed that the additional double bond was between C-18 and C-19, and that the C-12 acetate was still present—all surprising findings—and, most striking of all, proved that *C*-methylation had occurred not at C-12α but at C-11. This last feature was confirmed (Blanche et al 1992a) by the biosynthesis of precorrin-6x from [*methyl*-^{13}C]SAM and [11-^{13}C]uro'gen III (**8a**); the latter was unambiguously synthesized. With only one ^{13}C-labelled carbon in the macrocycle, the ^{13}C-NMR spectrum of the isolated ester **9b** provided unimpeachable evidence for C-11 methylation. Two separate short chromophores are generated by methylation at C-11, accounting for the pale yellow colour of this intermediate.

Precorrin-6y and the reductase which catalyses its formation

Under normal conditions, precorrin-6x (**9a**) is carried forward along the biosynthetic pathway by reduction. The reductase responsible for this step has been isolated from *Ps. denitrificans*. It is encoded by the *cobK* gene and is specific in its requirement for NADPH (Blanche et al 1992b).

The reductase was then used to reduce precorrin-6x (**9a**) or a tautomer, with NADPH as the only cofactor present, so allowing isolation of the next biosynthetic intermediate, precorrin-6y, which, in the absence of SAM, was not transformed further (Blanche et al 1992b). When precorrin-6y was incubated with the full enzyme system and all cofactors, it was converted into hydrogenobyrinic acid (**5**) in high yield.

Precorrin-6y octamethyl ester was proven to have structure **11** by a joint Paris–Cambridge effort (Thibaut et al 1992a) using the same ^{13}C multiple-labelling approach combined with NMR spectroscopy as outlined above for precorrin-6x. It was then established by experiments based on [4-^{2}H$_2$]NADPH (**7a**) that C-19 of precorrin-6x (**9a**) is the site to which hydride transfer from NADPH occurs (Weaver et al 1991). Presumably it is the C-18-protonated form of precorrin-6x, **10**, which is the substrate for the reductase enzyme.

Finally, the preparation of [4R-^{2}H]NADPH (**7b**) and [4S-^{2}H]NADPH (**7c**) provided proof that the reductase catalyses transfer of H$_R$ from the 4-position of NADPH (**7**) (Kiuchi et al 1992).

In summary, the foregoing reduction step was studied by focusing in turn on the reductase, the substrate and the cofactor; it was given this much attention because reduction was unexpected. This was so because comparison of the oxidation states of the early precursor, uro'gen III (**8**), with that of hydrogenobyrinic acid (**5**), and the fact that the ring-contraction process in another B$_{12}$-producing organism (*Propionibacterium shermanii*) releases acetic acid (Mombelli et al 1981, Battersby et al 1981), had led to the view that external redox reagents would not be required. The evidence above shows that there must be an oxidative step prior to precorrin-6x (**9a**) and that reduction is indeed needed for B$_{12}$ biosynthesis in *Ps. denitrificans* which is an *aerobic* organism. However, it is possible that this redox chemistry evolved only after oxygen became available in the atmosphere, so it is conceivable that an analogous reduction step is not involved in B$_{12}$ biosynthesis in anaerobic systems. The organism chosen to test this point was the anaerobe *Pr. shermanii*. This biosynthesizes vitamin B$_{12}$ by way of cobyrinic acid (**4**), which is simply the cobalt complex of hydrogenobyrinic acid (**5**) produced by *Ps. denitrificans*; the organic macrocycles are identical. I shall consider shortly this difference in the timing of cobalt insertion between these two organisms.

The reductase of *Propionibacterium shermanii*

The methods used above in the study of the reduction step were not suitable for *Pr. shermanii*. Whereas overproduced and substantially purified reductase was available from *Ps. denitrificans*, only a mixture of enzymes at the natural unenhanced level is available from *Pr. shermanii*. We therefore had to call on the greater sensitivity of tritium labelling.

If a reductase is required for the biosynthesis of vitamin B_{12} in *Pr. shermanii*, its cofactor could be NADPH (7) or NADH (6). [$4R$-^{3}H] NADPH (7d) and [$4R$-^{3}H] NADH (6d) together with the corresponding 4S-isomers 7e and 6e were prepared, all in stereospecifically labelled form (Ichinose et al 1993a). A mixture of the two 4R-cofactors 7d and 6d was then incubated with [^{14}C] precorrin-2, as 2, (generated from its aromatized form which undergoes reduction *in situ*) and the enzyme system from *Pr. shermanii*. The cobyrinic acid (4) which was biosynthesized was found to be ^{3}H-labelled, and appropriate degradation showed that the tritium was located entirely at C-19. In contrast, cobyrinic acid carrying negligible tritium was formed in a parallel run identical except that the two 4R-cofactors were replaced by the two 4S-isomers 7e and 6e. These results (Ichinose et al 1993b) demonstrated that reduction is a necessary step for B_{12} biosynthesis even in an anaerobic organism. Interestingly, the reductases involved in both *Pr. shermanii* and *Ps. denitrificans* are specific for H_R at position 4 of the cofactor.

The isolation and structure of precorrin-8x

We can now return to the experiments using *Ps. denitrificans* which led to the isolation of the next biosynthetic intermediate beyond precorrin-6y (11a) on the B_{12} pathway. Comparison of the structures of precorrin-6y (11a) and hydrogenobyrinic acid (5) shows that transformations requiring three different types of reaction are required (not necessarily in this order): (i) methylation at C-5 and C-15; (ii) decarboxylation of the C-12 acetate group; and (iii) rearrangement of the C-11 methyl group to C-12. Any of these transformations could in principle be used for an assay appropriate for screening protein fractions from the engineered strain of *Ps. denitrificans*. Success came by screening for methyltransferase activity. A single protein was isolated (Blanche et al 1992c) which was responsible not only for the methylations at C-5 and C-15 but also the decarboxylation of the acetate group at C-12. This enzyme was purified to homogeneity and was found to be encoded by *cobL* (Blanche et al 1992c). When the pure enzyme was incubated with [^{14}C] precorrin-6y, as 11a, and SAM, the next intermediate on the B_{12} pathway was generated. It was called precorrin-8x because, as will be confirmed below, two methyl groups are transferred from SAM to precorrin-6y (11a).

It is remarkable that a single protein catalyses reactions so different as methyl transfer and decarboxylation. The *cobL* gene was probably derived by a fusion of two ancestral genes, one originally encoding just the C-5/C-15 methylase activity and the other the decarboxylase activity; there are indications from the amino acid sequences to support this view (Blanche et al 1992c).

Though precorrin-8x could be produced using the purified enzyme encoded by *cobL*, a larger scale preparative method was developed on the basis of the observation that the final step of corrin biosynthesis in this organism, precorrin-8x → hydrogenobyrinic acid (5), is strongly inhibited by added 5, that is, that there is product inhibition (Thibaut et al 1992b). An incubation of precorrin-3 (3) with the enzymes of Ps. *denitrificans* which would normally yield 5 therefore produced precorrin-8x when 5 was added at the outset (Thibaut et al 1992b). Enzyme preparations from Pr. *shermanii* behave similarly in that added cobyrinic acid (4) strongly inhibits its own formation (S. M. Monaghan, H. C. Uzar & A. R. Battersby, unpublished work 1984–1987).

With the Ps. *denitrificans* system, suitable labelling experiments showed that the formation of precorrin-8x involved the addition of five methyl groups to the three already present in precorrin-3 (3). Thus the new intermediate was an octamethylated derivative of uro'gen III (8), hence its name, precorrin-8x. Finally, when doubly labelled [^{3}H:^{14}C] precorrin-8x was found to be converted very efficiently into hydrogenobyrinic acid (5) by the enzyme preparation from Ps. *denitrificans* without significant change in the labelling ratio, its standing as a new intermediate on the B$_{12}$ pathway was beyond doubt (Thibaut et al 1992b).

Though precorrin-8x was now available, the determination of its structure presented a major challenge. The reason was that unlike precorrin-6x and precorrin-6y, which yielded reasonably stable esters, the heptamethyl ester of precorrin-8x was extremely unstable (Thibaut et al 1992b). Also, although the hepta-acid, precorrin-8x itself, was more stable than its ester, it too changed in aqueous solution during NMR spectroscopy runs to give a mixture of at least five closely related forms, which of course yielded highly complex spectra. Eventually it was observed that these different forms all slowly yielded the same final stable form (Thibaut et al 1992c). The frustrations were over and the approach involving multiple ^{13}C-labelling in combination with NMR, as described earlier, allowed structure 12 to be established for the stable form of precorrin-8x.

But what is the true form of precorrin-8x on the biosynthetic pathway to B$_{12}$? The true form must be a close relative of the stable system, 12. The speed and efficiency with which each of the five closely related forms of precorrin-8x were enzymically converted into hydrogenobyrinic acid (5) was tested; one component far outstripped the others and could be stabilized at high pH for study in labelled form by NMR spectroscopy. The conclusion from these experiments was that structure **13** for precorrin-8x gave the best fit for all the ^{13}C and ^{1}H NMR data derived not only from ^{13}C-labels around the macrocycle but also from labelling of the side chain carbon attached to C-12 (Thibaut et al 1992c). The latter carbon was thus shown to be a methyl group attached to an sp^2 carbon and not the methylene group which is presumably formed as an intermediate (**15**) in the decarboxylation process **14** → **15** → **16**, see Scheme 1.

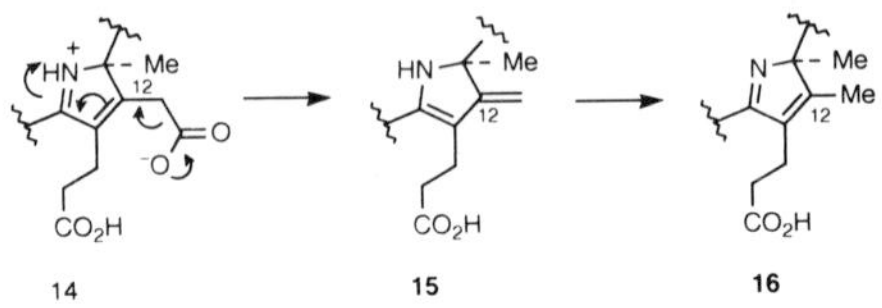

Scheme 1

Structure **13** still lacks some stereochemical detail (e.g., at C-15), but all the main features of precorrin-8x, the final intermediate for the corrin system **5**, are now clear. It is no surprise that in this structure tautomeric changes in the two separated chromophores happen readily and there could be epimerization at sp^3 centres. Various combinations of these changes would generate different structural isomers which would all finally fall into the thermodynamic sink represented by the stable form, **12**.

Precorrin-8x (**13**) has all the *C*-methyl groups of **4** and **5**, and vitamin B$_{12}$ (**1**) attached to the macrocycle, such that rearrangement of the 11-Me to C-12 yields **5**, this step being catalysed in *Ps. denitrificans* by the next enzyme to be considered.

The C-11 to C-12 methyl rearrangement leading to the corrin macrocycle

Protein fractions were tested for their ability to use precorrin-8x as a substrate to produce **5**. This led to the isolation of an enzyme which smoothly converted precorrin-8x (**13**) into hydrogenobyrinic acid (**5**); the enzyme is encoded by the *cobH* gene (Thibaut et al 1992b). This 'rearrangase' is a small protein (M_r 22 000) and it will be fascinating to learn how it catalyses the rearrangement.

Structural studies on the tetramethylated intermediate, precorrin-4x

You have now seen that the B$_{12}$ biosynthetic pathway has recently provided a feast of unexpected chemistry. One could be forgiven for thinking that perhaps the surprises are over and that steps closer to expectations might be used for the remaining unknown sections of the path; it is not turning out that way. A cell-free enzyme preparation for a strain of *Ps. denitrificans* with a mutation in the *cobM* gene was found by the French groups to convert precorrin-3 (**3**) efficiently into the previously undetected tetramethylated intermediate precorrin-4x. This readily lost two hydrogens, perhaps by oxidation in air, to give a reasonably stable conjugated form, named Factor IV, which could be isolated (Debussche et al 1993). As earlier, every carbon in the macrocycle was labelled with ^{13}C and tentative signal assignments were made. All this evidence, and the fact that precorrin-4x (the source of the isolated product) arises from **3** and is transformed enzymically into precorrin-6x (**9a**), led to structure **17** for Factor IV (Thibaut et al 1993). This structure shows three unexpected features: (i) the ring

17

contraction has already occurred; (ii) the extruded acetyl group is attached at C-1, not at C-19, the location previously considered most likely, and (iii) the composition of **17** indicates that there has been an oxidation step (which adds one oxygen overall) at some point during the formation of precorrin-4x from precorrin-3 (**3**).

A position had now been reached for **17** like that reached earlier for precorrin-6x. All the data fitted snugly together, but because so much hangs for future work on structure **17** it was sensible to confirm its key features. (i) Precorrin-3 was prepared from 5-amino[4-^{13}C]laevulinic acid (ALA) and [*methyl*-^{13}C]SAM using the necessary set of five overexpressed enzymes (A. I. D. Alanine, K. Ichinose, R. A. Vishwakarma, S. Balachandran, A. R. Pitt, N. P. J. Stamford, F. J. Leeper, B. Cameron, J. Crouzet & A. R. Battersby, unpublished work 1992–1993) as described by Stamford (1994, this volume). The intermediate was then further transformed by the enzyme system from the *Ps. denitrificans cobM* mutant and the product isolated as Factor IV. The NMR spectrum unambiguously confirmed attachment of the fourth methyl group to C-17. (ii) By work in progress, [1,10,20-^{13}C$_3$]uro'gen III (**8b**) was synthesized and *C*-methylated enzymically to yield precorrin-3 (**3a**). This will be used as the precursor of Factor IV to confirm, by ^{13}C NMR, attachment of the ketonic carbonyl group to C-1 (Alanine et al 1994).

What a remarkable structure **17** is. It raises a host of fascinating questions. For example, how is the ring contraction initiated? How is the C-1 methyl group introduced subsequently? The answers will come in time, but at present remain points of speculation.

Today's knowledge of the B$_{12}$ biosynthetic pathway

Stamford (1994, this volume) covered the pathway from ALA as far as precorrin-3 (**3**), so Fig. 5 in Stamford (1994, this volume) should be added to

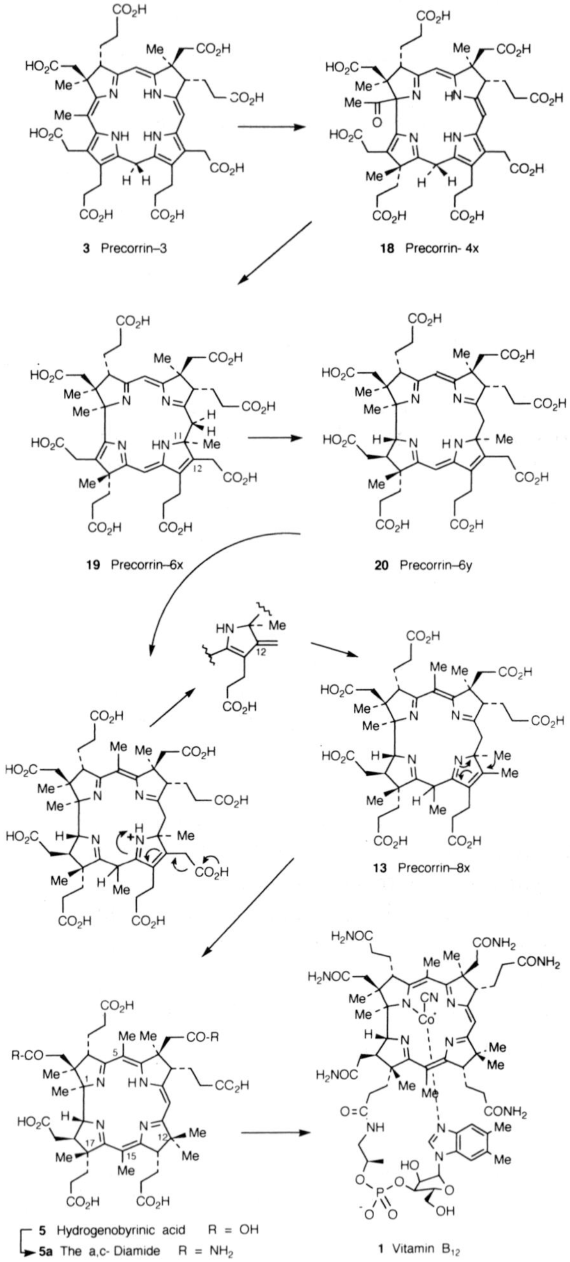

Scheme 2

Scheme 2 here to give the full sequence. The tautomeric forms of the structures in Scheme 2 have been drawn to match each other. Precorrin-6x, for example, is shown as **19**, with one double bond changed from its established position in the ester **9**. This is reasonable because tautomerism and also epimerization of *sp*^{3} centres are extremely easy processes in this series and the isolation steps do cause such changes.

Precorrin-3 (**3**) is converted into precorrin-4x via several steps including C-17 methylation, and one can give **18** as a sensible structure for this intermediate by drawing on structure **17** above. The next known intermediate is precorrin-6x, **19**, which is reduced to afford precorrin-6y, **20**, again with one double bond moved from its position in its ester. Then the remarkable double methylation at C-5 and C-15 occurs, together with decarboxylation, by a process shown as the equivalent of proton-catalysed decarboxylation of pyrrole-3-acetic acids to yield precorrin-8x (**13**). [1,5]Sigmatropic suprafacial rearrangement of the methyl group at C-11 over to C-12 forms hydrogenobyrinic acid, **5**, where the conjugated system of the corrin macrocycle is established.

In *Ps. denitrificans*, cobalt insertion then occurs into the *a,c*-diamide of **5a** (Debussche et al 1992) which is then ready for building B$_{12}$ (**1**). In *Pr. shermanii*, however, cobalt is inserted early, at or close to precorrin-3 (**3**), as indicated by two very different approaches (Müller et al 1991; R. A. Vishwakarma, S. Balachandran, N. P. J. Stamford, S. M. Monaghan, A. Prelle, F. J. Leeper & A. R. Battersby, unpublished work 1987–1992). In this organism, the later enzymes must be operating on cobalt-containing macrocycles.

My focus here (see also Battersby 1993) has been on the biosynthesis of the key corrins **4** and **5** *en route* to B$_{12}$ (**1**). However, the steps of amidation and attachment of the nucleotide loop have also been extensively studied in *Ps. denitrificans* (Crouzet et al 1991) and in other systems (Leeper 1989 and references therein); some of this work has been referred to earlier (Stamford 1994, this volume). As regards the corrins **4** and **5**, the broad picture of the biosynthetic pathway is now substantially painted; the gaps, now small, are between precorrin-3 and precorrin-4x and from precorrin-4x forward to precorrin-6x. It is evident that this phase of the research has certainly reached the beginning of the end, and the aim of the Paris and Cambridge teams, sometimes jointly, sometimes in cooperative independence, is to push on to that end.

References

Alanine AID, Ichinose K, Thibaut D et al 1994 Biosynthesis of vitamin B$_{12}$: use of specific ^{13}C-labelling for structural studies on Factor IV. J Chem Soc Chem Commun, p 193–196

Battersby AR 1993 Biosynthesis of vitamin B$_{12}$. Acc Chem Res 26:15–21

Battersby AR, Bushell MJ, Jones C, Lewis NG, Pfenninger A 1981 Biosynthesis of vitamin B$_{12}$: identity of fragment extruded during ring contraction to the corrin macrocycle. Proc Natl Acad Sci USA 78:13–15

Blanche F, Kodera M, Couder M, Leeper FJ, Thibaut D, Battersby AR 1992a Biosynthesis of vitamin B_{12}: use of a single ^{13}C label in the macrocycle to confirm C-11 methylation in precorrin-6x. J Chem Soc Chem Commun, p 138–139

Blanche F, Thibaut D, Famechon A, Debussche L, Cameron B, Crouzet J 1992b Precorrin-6x reductase from *Pseudomonas denitrificans*: purification and characterization of the enzyme and identification of the structural gene. J Bacteriol 174:1036–1042

Blanche F, Famechon A, Thibaut D, Debussche L, Cameron B, Crouzet J 1992c Biosynthesis of vitamin B_{12} in *Pseudomonas denitrificans*: the biosynthetic sequence from precorrin-6y to precorrin-8x is catalyzed by the *cobL* gene product. J Bacteriol 174:1050–1052

Crouzet J, Cameron B, Cauchois L et al 1990 Genetic and sequence analysis of an 8.7-kilobase *Pseudomonas denitrificans* fragment carrying eight genes involved in transformation of precorrin-2 to cobyrinic acid. J Bacteriol 172:5980–5990

Crouzet J, Levy-Schil S, Cameron B et al 1991 Nucleotide sequence and genetic analysis of a 13.1-kilobase-pair *Pseudomonas denitrificans* DNA fragment containing five *cob* genes and identification of structural genes encoding Cob(I)alamin adenosyltransferase, cobyric acid synthase, and bifunctional cobinamide kinase–cobinamide phosphate guanylyltransferase. J Bacteriol 173:6074–6087

Debussche L, Couder M, Thibaut D, Cameron B, Crouzet J, Blanche F 1992 Assay, purification and characterisation of cobaltochelatase, a unique complex enzyme catalysing cobalt insertion in hydrogenobyrinic acid *a,c*-diamide during coenzyme B_{12} biosynthesis in *Pseudomonas denitrificans*. J Bacteriol 174:7445–7451

Ichinose K, Leeper FJ, Battersby AR 1993a Preparation of $[4R^{-3}H]NADH$, $[4R^{-3}H]NADPH$ and the corresponding 4S-isomers all with substantial specific activities. J Chem Soc Perkin Trans I, p 1213–1216

Ichinose K, Leeper FJ, Battersby AR 1993b Proof that biosynthesis of vitamin B_{12} involves a reduction step in an anaerobic as well as an aerobic organism. J Chem Soc Chem Commun, p 515–517

Kiuchi F, Thibaut D, Debussche L, Leeper FJ, Blanche F, Battersby AR 1992 Biosynthesis of vitamin B_{12}: stereochemistry of transfer of a hydride equivalent from NADPH by precorrin-6x reductase. J Chem Soc Chem Commun, p 306–308

Leeper FJ 1989 The biosynthesis of porphyrins, chlorophylls and vitamin B_{12}. Nat Prod Rep 6:171–203

Mombelli L, Nussbaumer C, Weber H, Müller G, Arigoni D 1981 Biosynthesis of vitamin B_{12}: nature of the volatile fragment generated during formation of the corrin ring system. Proc Natl Acad Sci USA 78:11–12

Müller G, Zipfel F, Hlineny K et al 1991 Timing of cobalt insertion in vitamin B_{12} biosynthesis. J Am Chem Soc 113:9893–9895

Stamford NPJ 1994 Genetics and enzymology of the B_{12} pathway. In: The biosynthesis of the tetrapyrrole pigments. Wiley, Chichester (Ciba Found Symp 180) p 247–266

Thibaut D, Debussche L, Blanche F 1990a Biosynthesis of vitamin B_{12}: isolation of precorrin-6x, a metal-free precursor of the corrin macrocycle retaining five S-adenosylmethionine-derived peripheral methyl groups. Proc Natl Acad Sci USA 87:8795–8799

Thibaut D, Blanche F, Debussche L, Leeper FJ, Battersby AR 1990b Biosynthesis of vitamin B_{12}: the structure of precorrin-6x octamethyl ester. Proc Natl Acad Sci USA 87:8800–8804

Thibaut D, Kiuchi F, Debussche L, Leeper FJ, Blanche F, Battersby AR 1992a Biosynthesis of vitamin B_{12}: structure of the ester of a new biosynthetic intermediate precorrin-6y. J Chem Soc Chem Commun, p 139–141

Thibaut D, Couder M, Famechon A et al 1992b The final step in the biosynthesis of hydrogenobyrinic acid is catalyzed by the *cobH* gene product with precorrin-8x as the substrate. J Bacteriol 174:1043–1049

Thibaut D, Kiuchi F, Debussche L et al 1992c Biosynthesis of vitamin B$_{12}$: structural studies on precorrin-8x, an octamethylated intermediate and the structure of its stable tautomer. J Chem Soc Chem Commun, p 982–985

Thibaut D, Debussche L, Fréchet D, Herman F, Vuilhorgne M, Blanche F 1993 Biosynthesis of vitamin B$_{12}$: the structure of factor IV, the oxidized form of precorrin-4x. J Chem Soc Chem Commun, p 513–515

Weaver GW, Leeper FJ, Battersby AR, Blanche F, Thibaut D, Debussche L 1991 Biosynthesis of vitamin B$_{12}$: the site of reduction of precorrin-6x. J Chem Soc Chem Commun, p 976–979

End note: As described in the foregoing paper, the structures of the following precorrins were known at the time of the symposium, starting from the trimethylated stage: 3A, 4, 6x, 6y and 8x. New developments starting within weeks of the symposium filled the two remaining gaps. The structure of precorrin-3B, which follows precorrin-3A, has been elucidated (Debussche et al 1993a) and supported (Scott et al 1993); it is formed from precorrin-3A by CobG. Also, CobM converts precorrin-4 into precorrin-5; the structure of the latter has been both partly (Debussche et al 1993b) and fully (Min et al 1993) determined. In each case, the true intermediate may be a close double bond tautomer of the illustrated structure.

Precorrin-3B   Precorrin-4   Precorrin-5

Debussche L, Thibaut D, Danzer M et al 1993a Biosynthesis of vitamin B$_{12}$: structure of precorrin-3B, the trimethylated substrate of the enzyme catalysing ring contraction. J Chem Soc Chem Commun, p 1100–1103

Debussche L, Thibaut D, Cameron B, Crouzet J, Blanche F 1993b Biosynthesis of the corrin macrocyle of coenzyme B$_{12}$ in *Pseudomonas denitrificans*. J Bacteriol 175:7430–7440

Min C, Atshaves BP, Roessner CA, Stolowich NJ, Spencer JB, Scott AI 1993 Isolation, structure and genetically engineered synthesis of precorrin-5, the pentamethylated intermediate of vitamin B$_{12}$ biosynthesis. J Am Chem Soc 115:10380–10381

Scott AI, Roessner CA, Stolowich NJ, Spencer JB, Min C, Osaki S-I 1993 Biosynthesis of vitamin B$_{12}$: discovery of the enzymes for oxidative ring contraction and insertion of the fourth methyl group. FEBS (Fed Eur Biochem Soc) Lett 331:105–108

DISCUSSION

Hädener: You said the order of methylation from precorrin-3 to precorrin-6x is 17, 12 and 1 (these being the positions these methyl groups finally hold in cobyrinic acid and hydrogenobyrinic acid).

Battersby: That is the order, based on pulse labelling. The same order of methylation is also clear from the structures of all the new intermediates I described.

Hädener: In which organism was the pulse labelling done?

Battersby: For our pulse labelling we used *Clostridium tetanomorphum* (Uzar & Battersby 1982, 1985, Uzar et al 1987); Scott et al (1984) did it in *Pr. shermanii*, and Blanche et al (1990) worked with *Ps. denitrificans*.

Jordan: Is precorrin-6x actually an intermediate, or is 6y the real intermediate?

Battersby: Precorrin-6x is an intermediate as well. A specific reductase converts precorrin-6x into precorrin-6y (Blanche et al 1992).

Warren: Factor IV was isolated from a *cobM* mutant. Professor Scott has shown that the equivalent protein (on the basis of sequence similarity) in *Salmonella*, CbiF, methylates precorrin-3 at C-11. Does this study confirm this?

Battersby: In *Ps. denitrificans*, the protein encoded by *cobM* is quite clearly the C-11 methylase; mutation of this gene eliminates the C-11 methylase, so allowing production of precorrin-4x, which is isolated as Factor IV (Thibaut et al 1993).

Warren: You might therefore assume that precorrin-5 would still have the acetate in the C-1 position. Where would the next methylation be?

Battersby: There are a number of possibilities for the steps between precorrin-3 and precorrin-4x.*

Arigoni: I would like to present some experimental information concerning the possible participation of the ring A acetic acid side chain in the stages beyond precorrin-3 (S. Broers, A. Berry & D. Arigoni, unpublished work 1993). Such a participation was first suggested by Albert Eschenmoser (1988) in his mechanistic interpretation of the reaction responsible for the contraction of the macrocyclic ring, and subsequent work (Kurumaya et al 1989, Scott et al 1991) has shown that loss of oxygen from this side chain in the set of reactions leading to the formation of the ring A acetamido group during vitamin B_{12} biosynthesis in *Pr. shermanii* is in excess of the statistically expected amount.

In an attempt to shed light on the fate of the lost oxygen we incubated an enzyme preparation from *Pr. shermanii* with a sample of 5-aminolaevulinic acid labelled with ^{13}C at the C-5 position and with ^{18}O in its three oxygen atoms. There is quite a lot of endogenous acetic acid in this preparation and it is essential to get rid of as much of it as one can by prolonged dialysis. Scheme 1 indicates the distibution of label expected for the precorrin-3 derived from such a

---

*For recent information on stages after precorrin-3 arising after the symposium, see Debussche et al (1993) and *End note* to paper (p 279).

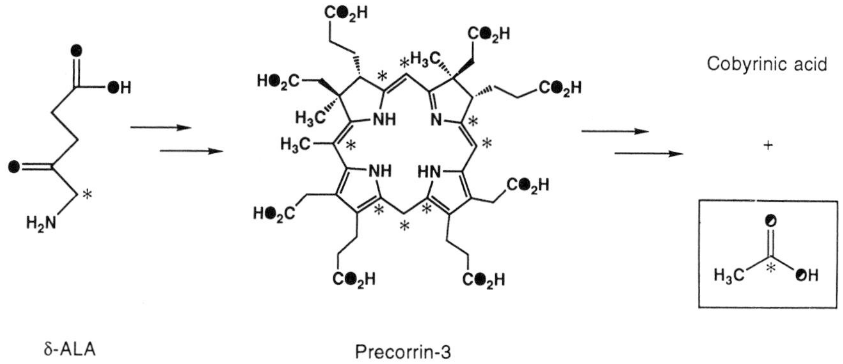

Scheme 1 (*Arigoni*)

precursor. During the conversion of this intermediate to cobyrinic acid an equivalent amount of acetic acid is produced, which is known to be derived from C-20 and the methyl group attached to it. We isolated this acetic acid as its *p*-phenyl phenacyl ester, which was then analysed by ^{13}C NMR spectroscopy. As indicated in Fig. 1, different signals appear in the region around 170 p.p.m. The main peak (A + A$'$) is due to the ester carbonyl of molecules free of ^{18}O. Using the keto carbonyl group of the compound as an internal reference (= 100%) one can show that the intensity of the main peak (A + A$'$) is increased by 10.8%. This excess corresponds to the amount of acetic acid generated from the labelled precursor with complete loss of the oxygen label. Comparison with the spectra of authentic specimens confirms that the B and C peaks correspond to molecules with an ^{18}O label in the carbonyl group (C) or in the ester bridge (B). It is rewarding to see that the relative intensities of these two peaks meet the statistical requirement. A small protruberance indicated in the spectrum as D is barely discernable at the position expected for molecules containing two ^{18}O labels linked to the ^{13}C atom. The percentage distribution of labelled species generated during the experiment is indicated in Fig. 1. In control experiments using authentic specimens of appropriately labelled acetic acid we have established that no oxygen loss is occurring during the reisolation procedure, whereas a loss of about 50% can be detected under the conditions of the incubation experiment. Our experimental figures can, therefore, be taken as proof that at least one oxygen from the ring A acetic acid side chain of precorrin-3 is transferred at some stage of the subsequent transformations into the acetate released as a fragment after the ring contraction reaction.

Battersby: I would like to continue this theme, to outline the experiments we are doing to determine at which stage of the biosynthesis the loss of oxygen occurs from the ring A acetate. This can be done by studying earlier intermediates and we are making these backward jumps in two stages, firstly to cobyrinic acid, and secondly to precorrin-4x. The first study is complete (Vishwakarma

282

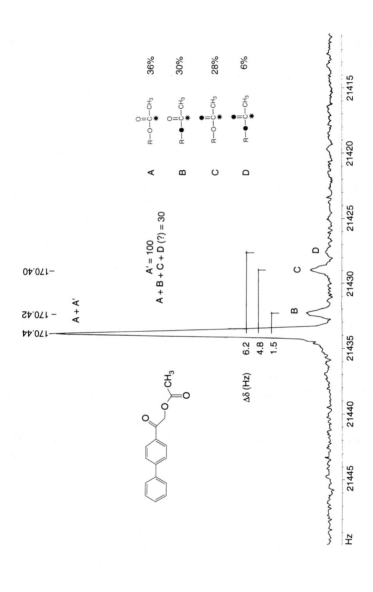

FIG. 1. (*Arigoni*) ^{13}C NMR spectrum of acetic acid (isolated as the *p*-phenyl ester) produced during the conversion of precorrin-3 to cobyrinic acid in a preparation of *Pr. shermanii* given 5-aminolaevulinic acid labelled with ^{13}C at C-5 and ^{18}O at its three oxygens. The labelled species corresponding to the peaks A, B, C and D are shown at the right together with their relative abundancies.

et al 1993) and the second far advanced (R. A. Vishwakarma, S. Balachandran, A. I. D. Alanine, L. Debussche, D. Thibaut, F. Blanche, N. P. J. Stamford, F. J. Leeper & A. R. Battersby, unpublished work 1993).

Sirohydrochlorin was prepared enzymically as described elsewhere in this volume (Stamford 1994) from 5-amino[1-^{13}C,1-^{18}O$_2$]laevulinic acid such that the ^{18}O level was high but not 100%. This allows the small unshifted ^{13}C signal from the eight carbomethoxy groups of the corresponding ocatmethyl ester (prepared under mild *basic* conditions) to be seen as well as the isotopically shifted ones in the ^{13}C NMR spectrum. The spectrum showed that the major signal from each of the eight carbomethoxy groups of sirohydrochlorin ester corresponded to C^{18}O$_2$Me. Thus, all the carboxyl groups had retained both ^{18}O labels. A second sample of the same multiply labelled sirohydrochlorin was incubated with the enzyme system from *Pr. shermanii* and the cobyrinic acid so formed was isolated as cobester (again with esterification under basic conditions). We had previously assigned unambiguously the ^{13}C signals from the seven CO$_2$Me groups of cobester; they are well separated. The ^{13}C spectrum of the labelled sample showed very clearly that the acetate on ring A no longer contained any C^{18}O$_2$Me species, but was almost entirely composed of $-C{\overset{\text{^{18}O}}{\underset{\text{^{16}OMe}}{<}}}$ and $-C{\overset{\text{^{16}O}}{\underset{\text{^{18}OMe}}{<}}}$. The other six ester groups *retained* the C^{18}O$_2$Me species. It follows that at some stage on the pathway before cobyrinic acid one oxygen is removed from the carboxyl group of the ring A acetate and is replaced by one from the medium.

The results from a similar study of precorrin-4x (as Factor IV) will be fascinating. We are also investigating whether transfer of ^{18}O occurs from the ring A acetate during the formation of precorrin-4x (cf., the experiment outlined by Professor Arigoni, above).

Arigoni: I must stress again the fact that *Pr. shermanii* preparations have a fairly high acetate content which must be removed by dialysis if the experiment is to work.

Scott: A. J. Irwin did several experiments with tritiated NADH and NADPH and worried about the exchange with the medium (unpublished work). The overall rate of incorporation of ^{3}H from NADP^{3}H into cobester was rather low. What was it like in your experiments?

Battersby: The strength of the experiments on the reduction step in *Pr. shermanii* lies in our use of a stereochemical test superimposed on the ^{3}H labelling test. This was necessary because NADH and NADPH undergo exchange with the medium at C-4. The results we obtained make it certain that transfer of a hydride equivalent *does* occur from one of these cofactors to a biosynthetic intermediate *en route* to cobyrinic acid. The labelling of the biosynthesized cobyrinic acid was at least 500 times higher than could have resulted from exchange of *all* the tritium from both cofactors into the medium followed by derivation of the cobyrinic acid from the labelled medium. The specific molar activity of the cobyrinic acid was about 2% of the intial specific molar activity of the cofactors. This is a good value; when we designed the experiments we thought the level might be lower. Even with the considerably purified reductase used for the joint work on *Ps. denitrificans* I described there was about 70% exchange at C-4 of the cofactor. For the experiments with *Pr. shermanii* a crude enzyme system had to be used, so substantially higher exchange was expected, 90% at least. Superimpose on that an adverse isotope effect of 6–8 and the specific activity of the product falls below 2% of the starting value. We were then very pleased with a stereospecific and regiospecific transfer giving at least 2% of the initial specific molar activity and at a level more than 500 times that achievable by exchange.

Scott: In *Pr. shermanii* if we wipe out all the NADH and NADPH in a cell-free extract with snake venom phosphodiesterase it doesn't make cobyrinic acid. Of course, we might have removed one or two other vital cofactors. We don't yet know if precorrin-6x accumulates.

References

Blanche F, Thibaut D, Frechet D et al 1990 Hydrogenobyrinic acid: isolation, biosynthesis and function. Angew Chem Int Ed Engl 29:884–886

Blanche F, Thibaut D, Famechon A, Debussche L, Cameron B, Crouzet J 1992 Precorrin-6x reductase from *Pseudomonas denitrificans*: purification and characterization of the enzyme and identification of the structural gene. J Bacteriol 174:1036–1042

Debussche L, Thibaut D, Danzer M et al 1993 Biosynthesis of vitamin B_{12}: structure of precorrin-3B, the trimethylated substrate of the enzyme catalysing ring contraction. J Chem Soc Chem Commun, p 1100–1103

Eschenmoser A 1988 Vitamin B_{12}: experiments concerning the origin of its molecular structure. Angew Chem Int Ed Engl 27:5–39

Kurumaya K, Okazaki T, Kajiwara M 1989 Studies on the biosynthesis of corrinoids and porphyrinoids. I. The labeling of oxygen of vitamin B_{12}. Chem & Pharm Bull (Tokyo) 37:1151–1154

Scott AI, Mackenzie NE, Santander PJ et al 1984 Biosynthesis of vitamin B_{12}. The timing of the methylation steps between uro'gen III and cobyrinic acid. Bioorg Chem 12:356–362

Scott AI, Stolowich NJ, Atshaves BP 1991 Timing and mechanistic implications of regiospecific carbonyl oxygen isotope exchange during vitamin B_{12} biosynthesis. J Am Chem Soc 113:9891–9893

Stamford NPJ 1994 Genetics and enzymology of the B_{12} pathway. In: The biosynthesis of the tetrapyrrole pigments. Wiley, Chichester (Ciba Found Symp 180) p 247–266

Thibaut D, Debussche L, Fréchet D, Herman F, Vuilhorgne M, Blanche F 1993 Biosynthesis of vitamin B_{12}: the structure of factor IV, the oxidized form of precorrin-4x. J Chem Soc Chem Commun, p 513–515

Uzar HC, Battersby AR 1982 Biosynthesis of vitamin B_{12}: pulse labelling experiments to locate the fourth methylation site. J Chem Soc Chem Commun, p 1204–1206

Uzar HC, Battersby AR 1985 Biosynthesis of vitamin B_{12}: order of the later C-methylation steps. J Chem Soc Chem Commun, p 585–588

Uzar HC, Battersby AR, Carpenter TA, Leeper FJ 1987 Biosynthesis of porphyrins and related macrocycles. Part 28. Development of a pulse labelling method to determine the C-methylation sequence for vitamin B_{12}. J Chem Soc Perkin Trans I, p 1689–1696

Vishwakarma RA, Balachandran S, Alanine AID 1993 Biosynthesis of porphyrins and related macrocycles. Part 41, Fate of oxygen atoms as precorrin-2 carrying eight labelled carboxyl groups ($^{13}C^{18}O_2H$) is enzymatically converted into cobyrinic acid. J Chem Soc Perkin Trans I, p 2893–2899

Recent studies of enzymically controlled steps in B_{12} biosynthesis

A. I. Scott

Center for Biological NMR, Department of Chemistry, Texas A&M University, College Station, TX 77843-3255, USA

Abstract. The acquisition and sequencing of the genes encoding the enzymes for vitamin B_{12} biosynthesis in *Salmonella typhimurium* and *Pseudomonas denitrificans* has dramatically altered the direction of research on the pathway from uroporphyrinogen III to the corrinoids. Through a combination of molecular biology, organic chemistry and NMR spectroscopy, logical progression along the sequence is being made. Recent work from our laboratory is focused on the discovery and specificities of the methyltransferases connecting uroporphyrinogen III with cobyrinic acid, the temporal resolution of cobalt insertion and a comparison of the anaerobic pathway in *S. typhimurium* and the aerobic pathway in *Ps. denitrificans.* The implication of two parallel routes to corrins in these bacteria is discussed.

1994 The biosynthesis of the tetrapyrrole pigments. Wiley, Chichester (Ciba Foundation Symposium 180) p 285–308

The complex mechanisms underlying the biosynthesis of the tetrapyrrolic cofactors (Fig. 1), which mediate such diverse processes as respiration (haem), photosynthesis (chlorophyll), methanogenesis (coenzyme F_{430}), six-electron sulphite and nitrite reduction (sirohaem) and vitamin function (B_{12} and its coenzyme—the corrinoids), have intrigued biochemists since the 1950s when Shemin and Neuberger began their classic studies of haem biosynthesis. Our own work in this field began 25 years ago when, by whole-cell feeding of ^{13}C-enriched precursors to *Propionibacterium shermanii*, we established by NMR spectroscopy that the pathway to B_{12}, beginning with 5-aminolaevulinic acid (ALA), continues via porphobilinogen (PBG) to the unsymmetrical uroporphyrinogen (uro'gen) III (a route common to all the cofactors in Fig. 1), before entering the chiral world of corrins, whose stereochemical complexity is largely a result of *C*-methylation of the pyrrolic rings by *S*-adenosylmethionine (SAM) (Scott 1978). The next 10 years witnessed the isolation (with L. Siegel & G. Müller) of two partially methylated intermediates (Factors II and III) on the way to corrins from cell-free systems of *Pr. shermanii* and *Clostridium tetanomorphum.* It is now clear that these oxidized pigments or 'factors', are

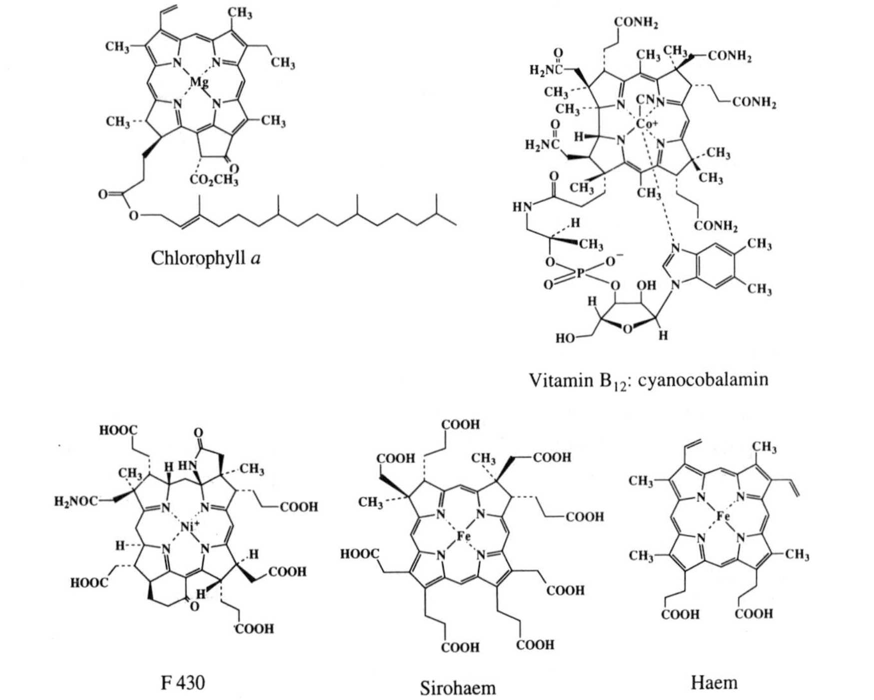

Chlorophyll *a*

Vitamin B$_{12}$: cyanocobalamin

F 430   Sirohaem   Haem

FIG. 1.   The cofactors derived from uroporphyrinogen (uro'gen) III.

carried through the biosynthetic machinery at the level of hexahydroporphyrin, and although the stable isobacteriochlorin structures of Factors II and III were conveniently elucidated by ^{13}C NMR spectroscopy of sub-milligram amounts (Scott 1978), the true substrates for the biosynthetic methyltransferases are now known to be the reduced (and much more labile) versions of these Factors, precorrins 1–3 (Leeper 1985, 1987, 1989). Here, I shall concentrate on what is now known about the biosynthetic enzymes which catalyse the conversion of uro'gen III to cobyrinic acid, a conversion involving as many as 12 steps along a pathway rich in mechanistic complexity.

There is no apparent formal change in oxidation level throughout the pathway from ALA to the corrinoid structure of B$_{12}$. This fact, taken together with the phylogeny of the B$_{12}$-producing organisms (Georgopapadakou & Scott 1977) and Eschenmoser's remarkable prebiotic simulation of uro'gen III synthesis from glutamine dinitrile, suggests that a prebiotic type III corrin template, bearing protons at the positions associated with biological *C*-methylation, was quite feasible. The primitive anaerobes (methanogenic bacteria) which make B$_{12}$ (but not haem) can be dated back to around 3.79×10^9 years (Eigen et al 1989, Mason 1991, Schopf 1983), whereas the modulation of oxygen via the respiratory (haem) pigments and oxygenic photosynthesis (chlorophyll) was not required until sufficient O$_2$ entered the earth's atmosphere about 1.7×10^9 years ago. The ancient organisms had learned to synthesize corrins via type III uro'gen or to produce methane from coenzyme F$_{430}$ long before the oxidized haems and cytochromes were required. The first chiral centres in the B$_{12}$ pathway are introduced as a result of C-methylation of uro'gen III, which involves the sequential formation of Factors I and II by the enzyme S-adenosylmethionine uro'gen-III methyltransferase (SUMT) (uroporphyrinogen-III C-methyltrans-ferase, EC 2.1.1.107), overexpressed in *Pseudomonas denitrificans* by Cameron et al (1989, Blanche et al 1989). In *Escherichia coli* it was found that the *cysG* gene encodes uro'gen-III methylase (M-1) as part of the synthetic pathway to sirohaem, the cofactor for sulphite reductase, and overproduction (30 mg/l) was achieved by the appropriate genetic engineering (Warren et al 1990a). Although SUMT and M-1 appear to perform the same task, they differ in M_r and substrate specificity (Spencer et al 1993). In fact, M-1 is a trifunctional enzyme which not only inserts CH$_3$ at C-2 and C-7 but also catalyses the NAD-dependent dehydrogenation of precorrin-2 (dihydrosirohydrochlorin) to Factor II (sirohydrochlorin) and insertion of iron to give sirohaem. M-1 should therefore be renamed sirohaem synthase. It has been possible to study in detail the reaction catalysed by M-1 directly by NMR spectroscopy and to provide rigorous proof that the structure of precorrin-2 is that of the dipyrrocorphin tautomer of dihydro-Factor II (dihydrosirohydrochlorin). Uro'gen III (enriched from [5-^{13}C] ALA at the positions shown in Fig. 2) was prepared by mixing ALA dehydratase (PBG synthase), PBG deaminase (hydroxymethylbilane synthase) and uro'gen III synthase (cosynthase) and then incubating with M-1 and [^{13}CH$_3$] SAM. The resultant spectrum of precorrin-2 revealed only one sp^3-enriched carbon, assigned to C-15, thereby locating the reduced centre. Use of a different set of ^{13}C labels, ● from [3-^{13}C] ALA and [^{13}CH$_3$] SAM, located the sp^2 carbons at C$_{12}$ and C$_{18}$ as well as the sp^3 centres coupled to the pendent ^{13}CH$_3$ groups at (*) C-2 and C-7. This result confirms an earlier NMR analysis of precorrin-2 isolated by careful anaerobic purification of the methyl ester (Battersby et al 1982), and shows that no further tautomerization takes place during the anaerobic purification. The two sets of experiments mutually reinforce the postulate that precorrins-1, 2 and 3 all exist as hexahydroporphinoids, and more recent labelling experiments have provided good evidence that precorrin-1 is discharged from the methylating enzyme (SUMT) at this oxidiation level (Brunt et al 1989).

However, prolonged incubation (two hours) of uro'gen III with M-1 had a surprising result, for the UV and NMR spectra changed dramatically from those of precorrin-2 (a dipyrrocorphin) to those of the chromophore of a pyrrocorphin, previously known only as a synthetic tautomer of hexahydroporphyrin (Waditschatka 1985), and when [^{13}CH$_3$] SAM was added to the incubation a third methyl group signal appeared in the 19–21 p.p.m. region of the NMR

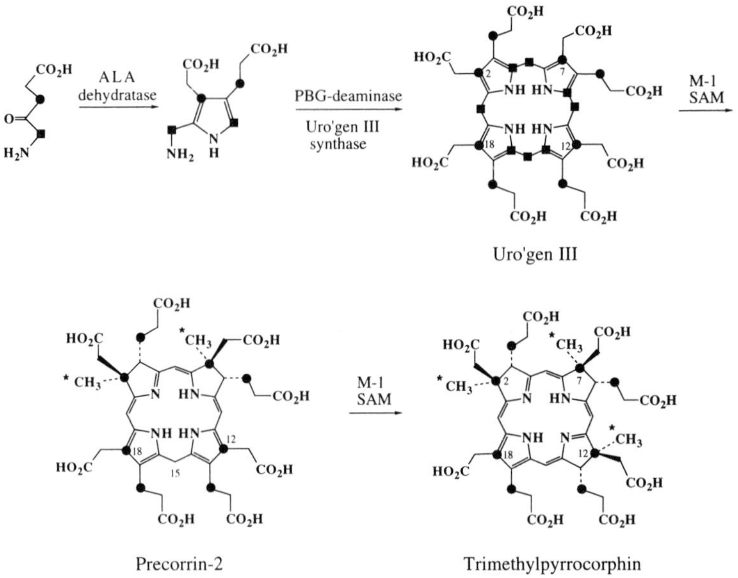

FIG. 2. Synthesis of precorrin-2, *C*-methylation of uro'gen III by the *Escherichia coli* enzyme M-1 (uroporphyrinogen-III methylase) and 'overmethylation' of the product in a clockwise direction. The ^{13}C-enriched centres used to assign the carbon centres and C–C coupling were inserted from: [3-^{13}C]ALA, ●; [5-^{13}C]ALA, ■; [^{13}CH$_3$]SAM, *.

spectrum. The necessary labelling experiments led to the proposal (Scott et al 1990) for the structure of this novel trimethylpyrrocorphin shown in Fig. 2. Thus, M-1 has been recruited to insert a ring C methyl, to synthesize the long-sought 'natural' chromophore corresponding to that of the postulated precorrin-4, although in this case the regiospecificity is altered from ring D to ring C. This lack of specificity on the part of M-1 was further exploited to synthesize a range of 'unnatural' isobacteriochlorins and pyrrocorphins based on isomers of uro'gen III. Thus, uro'gen I can be used to produce the methylated products corresponding to precorrin-2 and the type I pyrrocorphin (Fig. 3). These compounds are reminiscent of a series of tetramethyl type I corphinoids, Factors S$_1$–S$_4$, isolated from *Pr. shermanii* (Müller et al 1986, 1987), which occur naturally as their zinc complexes, as discussed below. The totally synthetic isomers uro'gens II and IV in which the acetate (A) and propionate (P) side chains are positioned in the orders APPAAPPA and APAPPAPA, respectively, can also serve as substrates for M-1, remarkably producing the corresponding 2,7-dimethylisobacteriochlorins in both cases (Warren et al 1990b). After incubation of uro'gen I with SUMT (Scott et al 1989), isolation of Factor II of the type I (sirohydrochlorin I) family revealed a lack of specificity

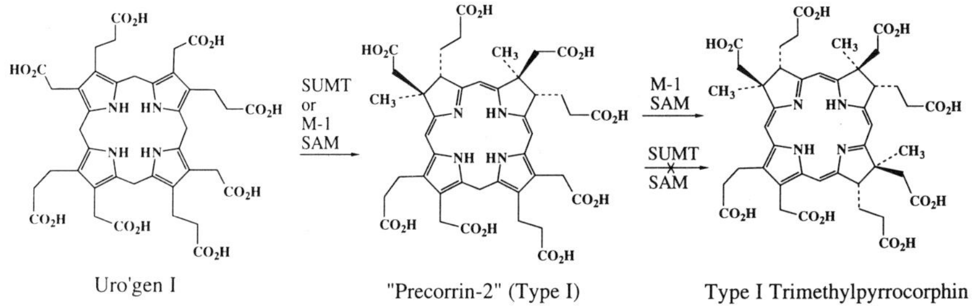

Uro'gen I   "Precorrin-2" (Type I)   Type I Trimethylpyrrocorphin

FIG. 3.   Overmethylation of the type I uro'gen by M-1 (but not by SUMT!).

on the part of this methyltransferase also, although in these studies no pyrrocorphins were observed. This may reflect a control mechanism in the *Ps. denitrificans* enzyme (SUMT), which does not 'overmethylate' precorrin-2, unlike the *E. coli* enzyme (M-1), whose physiological function is to manufacture sirohydrochlorin (Factor II) and sirohaem (Fig. 1) for sulphite reductase, an essential enzyme in cysteine synthesis. *E. coli* does not synthesize vitamin B$_{12}$; although the *C*-methylation machinery has been retained throughout the bacterium's evolution, it is required only to insert the C-2 and C-7 methyl groups.

The sites of *C*-methylation in both the type I and the type III series are also reminiscent of the biomimetic *C*-methylation of the hexahydroporphyrins discovered by Eschenmoser, and the regiospecificity is in accord with the principles he outlined (Eschenmoser 1988) for the stabilizing effect of a vinylogous ketimine system (Fig. 4). The chemical conversion of dipyrrocorphin to pyrrocorphin not only reaches the favoured ketimine system of the pyrrocorphin but also, when a C-20 methyl group is present, becomes controllable through the presence of coordinating metal. The fascinating result of these *in vitro* models is that ring D (C-17) methylation can be achieved in the presence of magnesium, which corresponds to the methylation sequence found in Nature.

On the road to corrins: a change of horses (from *E. coli* to *Salmonella*)

Our search for the biosynthetic enzymes leading to precorrin-2 up to this point had relied on the 'haem gene box' (Fig. 5) in *E. coli* responsible for encoding the proteins involved in the synthesis of precorrin-2, sirohydrochlorin and sirohaem (Scott 1978). Fortunately, the haem and corrin pathways intersect in *E. coli* and *Salmonella typhimurium*. Although the former organism does not make B$_{12}$, the discovery by J. Roth (Jeter & Roth 1987) that anaerobic fermentation of *S. typhimurium* produces vitamin B$_{12}$ has opened the way for us to enhance our knowledge of vitamin B$_{12}$ synthesis using the vast array of genetic and cloning techniques available with this organism. Three loci at minutes 14, 34 and 42 have been identified by mutation and complementation studies

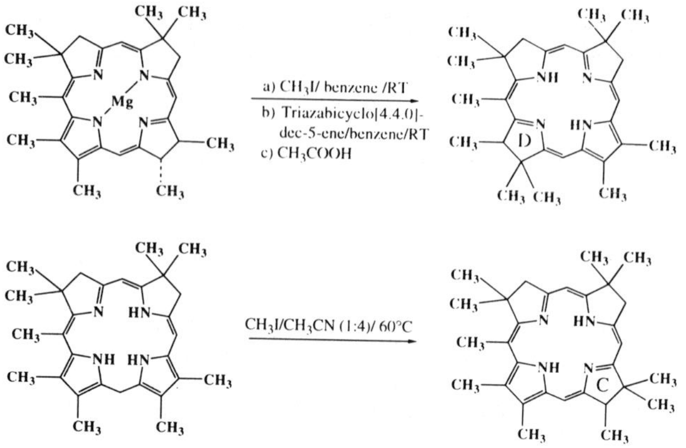

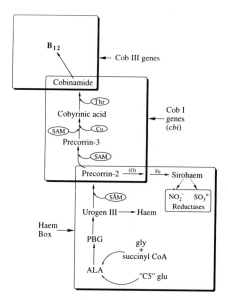

FIG. 4. 'Metallic' control of regiospecificity in Eschenmoser's biomimetic *C*-methylation of pyrrocorphins. Note that in the presence of Mg^{2+}, methylation occurs in ring D rather than ring C.

FIG. 5. The haem (*E. coli*) and Cob (*Salmonella*) gene boxes of B_{12} biosynthesis.

(Jeter & Roth 1987). The main gene cluster at 42 min contains the machinery (Cob I) necessary for the synthesis of cobinamide from precorrin-2 (Fig. 5), a process involving six *C*-methylations (at C-1, C-5, C-12, C-15, C-17 and C-20), decarboxylation (of the acetate residue at C-12), ring contraction, loss of acetic acid (from C-20 and its attached methyl), amidation and cobalt insertion under control of the *cbi* genes.

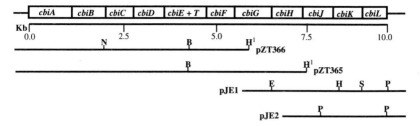

FIG. 6.   Subcloning of the *S. typhimurium cbi* genes for expression. The *cbi* genes in plasmid pBR329 (pZT365, pZT366, pJE1 and pJE2) were a generous gift from Dr John Roth, University of Utah. Restriction enzymes: B, *Bam*HI; H, *Hind*III; N, *Nae*I; P, *Pst*I; S, *sal*I. [1]H indicates the *Hind*III site in pBR329. See Table 1.

TABLE 1   Subcloning of the *Salmonella typhimurium cbi* genes

Genes	Plasmid	Construction
cbiC & D	pCAR345	pZT366(N-B) into pHN1 +
cbiE	pCAR340	ecpcr product into pHN1 +
cbiT & F	pCAR276	pZT366(B-H) into pUC18
cbiT, F, G, H[a]	pCAR293	pZT365(B-H) into pUC18
cbiH & J[a]	pCAR309	pJE1(E-H) into pUC18
cbiK & L	pCAR292	pJE1(P-P) into pUC18
cbiL	pCAR311	pJE2(S-P) into pUC18
cobF[b]	pCAR332	ecpcr product into pHN1 +
cobI[b]	pCAR333	ecpcr product into pHN1 +

[a]Form insoluble inclusion bodies.
[b]*Pseudomonas denitrificans* gene.
ecpcr, expression cassette polymerase chain reaction.
See Fig. 6 for further details.

Ten of the *cbi* genes found in the *S. typhimurium cob* operon were subcloned for expression from the four different plasmids (Fig. 6, Table 1). All of these genes except *cbiD* were expressed in *E. coli* (as determined by SDS–PAGE), and seven of the gene products were purified from the soluble fraction of cell lysates of the appropriate strain. Two further gene products, those of *cbiG* and *cbiJ*, were found in the insoluble fraction. The amino terminal sequence of each of the purified proteins was determined and used to confirm the location of the open reading frames predicted from the nucleotide sequence (J. Roth, personal communication). The gene products of *cbiE*, *cbiF*, *cbiH* and *cbiL* were shown to be SAM-binding proteins, and, on the basis of their similarity to other methyltransferases (Crouzet et al 1990, Thibaut et al 1990a and references therein), are considered to be the most likely candidates for methyltransferase

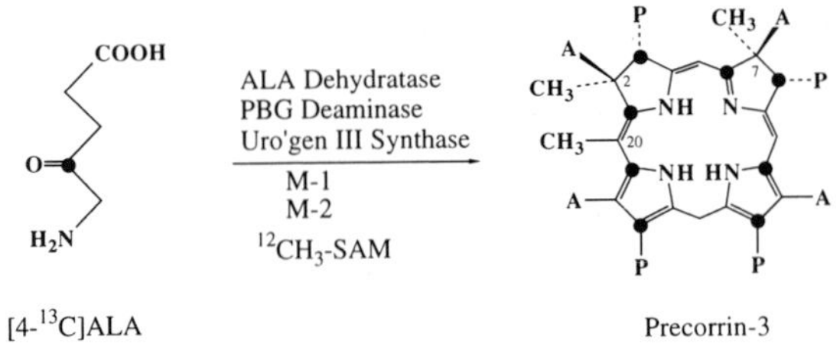

FIG. 7. Synthesis of precorrin-3x in a multi-enzyme one-flask system including M-2, the product of the *S. typhimurium cbiL* gene.

activity. The *Ps. denitrificans cobL* gene product has recently been reported (Blanche et al 1992a) to have two functions, methylation (of precorrin-6y at C-5 and C-15) and decarboxylation of the ring C acetate. SDS–PAGE and N-terminal sequence analysis revealed that two separate gene products in *S. typhimurium* (*cbiE* and *cbiT*) correspond to the *cobL* gene product; *cbiE* is homologous to the methyltransferase region and *cbiT* to the decarboxylase region, i.e., in the *Salmonella* genome the 5,15-methylase and decarboxylation activities are separated.

From its similarity to the *Ps. denitrificans cobI* gene product (31% nucleotide sequence identity, 71% conservation), the *cbiL* gene product was predicted to be the *S. typhimurium* precorrin-2 methyltransferase (M-2). The expressed protein could be used, although less efficiently than the CobI enzyme, in a multi-enzyme one-flask synthesis of precorrin-3 from the building block, ALA, as shown in Fig. 7, both in the NMR sample tube and preparatively (on the multi-milligram scale), by adding the five overexpressed enzymes to the substrate ALA in the presence of SAM (Warren et al 1992). The structure of precorrin-3, a dipyrrocorphin with the constitution shown (Fig. 7), revealed subtle differences in the ^{13}C NMR spectrum reflecting the influence of the new methyl group at C-20 on the conjugated system, which results in a preponderance of the tautomer whose electronic array is prepared for the next *C*-methylation step and which is directly observed via ^{13}C-enrichment on NMR spectroscopy.

As a guide to the expected order of insertion of the remaining methyl groups on the periphery of this last intermediate, precorrin-3, we recall the earlier pulse-labelling experiments in which the substrates, uro'gen III and precorrins 2 and 3 (as Factors II and III respectively), were incubated with SAM for several hours with a cell-free extract capable of synthesizing cobyrinic acid before being pulse-labelled with [[13]CH₃]SAM. An examination of the different intensities of the [[13]CH₃]methyl resonances in the resultant cobester identified the sequence of methyl group insertion (in three independent studies: Scott 1990 and references therein, Uzar et al 1987, Blanche et al 1990) as *C*-methylation at C-17, followed by C-12 (now known to occur first at C-11, see below) then at C-1, C-5 and C-15. This differentiation between C-5 and C-15 was found in *Pr. shermanii*

(Scott 1990), whereas the reverse order, C-15 before C-5 was found in *C. tetanomorphum* extracts (Uzar et al 1987). In the third study (*Ps. denitrificans*, Blanche et al 1990) the C-5/C-15 distinction could not be made. However, the welcome agreement that C-17 is the first site of alkylation on the precorrin-3 template and that this is followed by *C*-methylation at C-11 then C-1 imposes certain restrictions on the type of structure expected for the missing intermediates precorrins-4, 5 and 6, corresponding to methyl insertion at C-17, C-11 and C-1, respectively. Inspection of the stereochemistry of the C-17 methyl group in vitamin B_{12} reveals that, of the five methyl groups on quaternary carbon centres in the vitamin, only that at C-17 is β-oriented. This stereochemistry was rationalized as long ago as 1976 (Scott 1978) when we suggested that methylation actually occurred at C-19, followed by ring opening, reclosure (by an allowed 16π electron process) to give the β-oriented C-19-methyl, and finally two [1,5]-sigmatropic shifts, to terminate at C-17 with β-orientation. Just as in *Ps. denitrificans*, where the methyl group at C-11 migrates later to C-12, the fourth methyl group (at C-17) might have moved from another position, in which case precorrin-4 might be either a C-19 or a C-18 β-methylated structure.

In order to begin the task of unveiling the remaining precorrin intermediates, we incubated precorrin-3 in turn with SAM and each of the putative, overexpressed methyltransferases, CbiE, CbiH, CbiF and CbiL, with surprising results. The only *C*-methylation observed was that catalysed by ORF-7 (the *cbiF* gene product), and turned out to be methylation at C-11! The new isolate, identified as the kinetic product by direct observation by NMR spectroscopy (Fig. 8), is a modified corphin bearing a fourth methyl group at C-11. Since biochemical conversion to cobyrinic acid has not yet been demonstrated, we have named the new substance compound 4x (Roessner et al 1992, Ozaki et al 1993); its structure recalls one of the four possible structures proposed for Factor S_3 (Müller et al 1986, 1987). The spectroscopic data for the Factor S_3 isolated from *Pr. shermanii* (Müller et al 1986, 1987) could not distinguish between methylation at C-1 and C-11, but it is now clear that the correct isomer is the one in which $11(\equiv16)$ α-methylation has taken place on the uro'gen I template. The new structural proposal for Factor S_3 was nicely confirmed by enzymic synthesis, this time using [4-^{13}C]ALA as substrate in a one-flask reaction mixture containing ALA dehydratase, deaminase ($\rightarrow$uro'gen I), CysG, CbiF, SAM and $ZnCl_2$ (Fig. 9). In the absence of the last of these enzymes (the *cbiF* gene product) 2,7,12-trimethylpyrrocorphin (Fig. 3) accumulates, but when the fourth enzyme (CbiF) was added, a new signal appeared at δ79 p.p.m. heralding the insertion of a fourth (α-) methyl group at the C-16 position (Fig. 9). The resultant zinc complex (as the octamethyl ester) was identical in every respect to factor S_3 isolated earlier from *Pr. shermanii* (Ozaki et al 1993). It only remains to define the absolute stereochemistry of the new chiral center at C-16. In further support of the concept of the control exerted by the coordinating metal, we have synthesized Factor S_1 by insertion of zinc into the

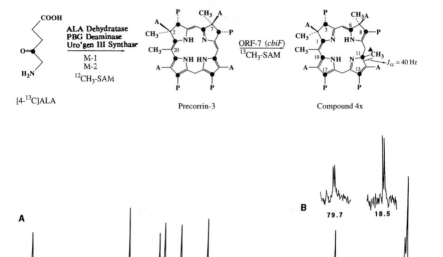

FIG. 8.   Multienzyme synthesis of compound 4x with the *cbiF* gene product and SAM.
The ^{13}C NMR spectrum of 4x (**A**) reveals 3 sp^2 (C-8, C-13 and C-17) but only one sp^3
propionate terminus (C-3) and a signal at $\delta79.7$ p.p.m. typical of an sp^3 carbon (C-11)
adjacent to nitrogen. The inset (**B**) shows the coupling ($Jcc = 40$ Hz) of the new ^{13}CH$_3$
($\delta18.5$) to the C-11 signal at $\delta79.7$ p.p.m.

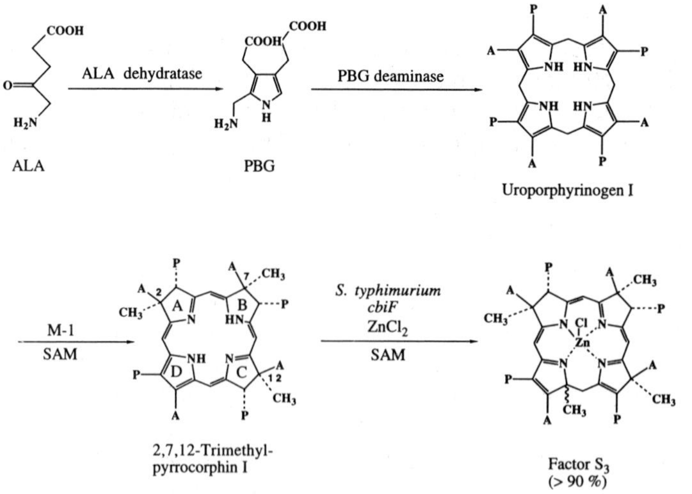

FIG. 9.   Multienzyme synthesis of Factor S$_3$.

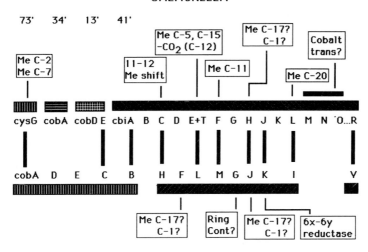

FIG. 10.   The location and functions of some of the known genes for B_{12} biosynthesis
in *Salmonella typhimurium* (Thibaut et al 1990a, Crouzet et al 1990, Blanche et al 1992b,
Warren et al 1992) and *Pseudomonas denitrificans* (Warren et al 1992, Thibaut et al
1990b,c, 1992, Blanche et al 1992a,b). In *S. typhimurium* the genes map at 14′ (*cobD*,
cobE; addition of aminopropanol), 34′ (*cobA*; adenosylation), 41′ (*cbiA-R*; cobinamide
biosynthesis), and 73′ (*cysG*; uro'gen III methyltransferase). Homologies between the
S. typhimurium and *Ps. denitrificans* gene products are shown. See the text for a
discussion of the functions of the gene products.

trimethylpyrrocorphin I (Fig. 9) in the presence of M-1 and SAM, producing
the 2,7,12,17-tetramethyl zinc corphinate identical to S_1 (J. B. Spencer,
unpublished results). These new findings (Roessner et al 1992, Ozaki et al 1993)
suggest a lack of substrate specificity in *C*-methylation by the methyltransferases
and provide a welcome confirmation of the function of the *cbiF* gene product
as a methyltransferase which attacks the α-position of pyrrole rings of
pyrrocorphins, whether type III or type I; they also open the door to synthetic
chemistry based on *C*-methylation at electron-rich centres.

In the recently described *Ps. denitrificans* mutant (Thibaut et al 1990a,b,c,
Crouzet et al 1990, Blanche et al 1992a) harbouring the eight genes necessary
for the synthesis of the metal-free precorrin-6x (Thibaut et al 1990b,c) (and hence
hydrogenobyrinic acid), the absence of cobalt suggests that zinc or magnesium
can be substituted for cobalt in order to achieve regio- and stereospecificity of
C-methylation and ring contraction the same as that found in the cobalt series.
Because the coordinating metal obviously exercises strict regio-control over this
type of process, our metal-free substrates must now be revisited as their

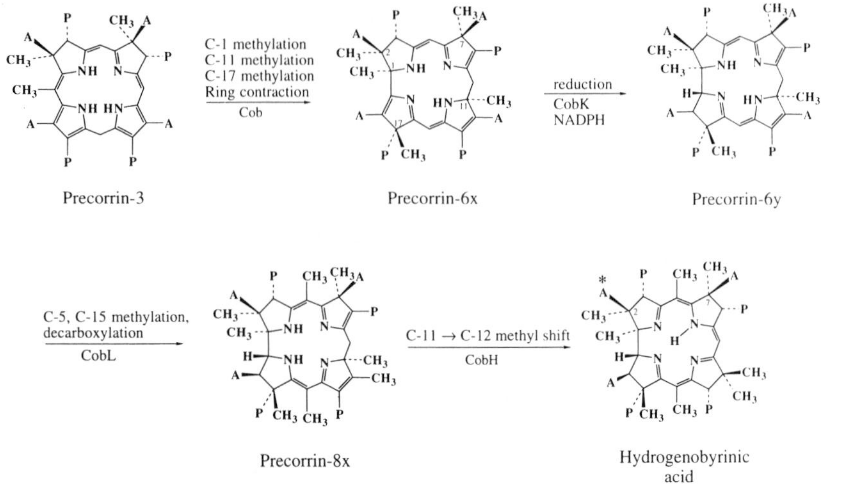

FIG. 11. Precorrins-6x, 6y and 8x, recently discovered metal-free precursors of hydrogenobyrinic acid in *Ps. dentrificans* expression systems.

complexes with cobalt and zinc to determine the effect of metal on the specificity of *C*-methylation in corrin biosynthesis.

Comparison of *Salmonella* and *Ps. denitrificans* genes for corrin synthesis

We are now in a position to compare the homologies and functions of the cloned genes of *Salmonella* and *Pseudomonas* (Fig. 10). Comparisons of amino acid sequences predicted from the nucleotide sequences of the *Ps. denitrificans* and *S. typhimurium* genes revealed similarity between: *cobH* and *cbiC*; *cobI* and *cbiL*; *cobJ* and *cbiH*; *cobK* and *cbiJ*; *cobL* and *cbiE + T*; and the *cobM* and *cbiF* gene products. No homologues were found for the *cobF*, *cobG*, *cbiD*, or *cbiG* gene products (Fig. 10).

The *Ps. denitrificans* enzymes responsible for reduction of precorrin-6x to precorrin-6y (CobK), the bismethylation and decarboxylation of 6y to 8x (CobL), and the rearrangement of the latter to hydrogenobyrinic acid (CobH) have recently been described (Thibaut et al 1990a, 1992, Crouzet et al 1990, Blanche et al 1992a) (Fig. 11). The remaining genes (*cobF*, *G*, *J* and *M*) probably encode enzymes responsible for the synthesis of precorrin-6x from precorrin-3, a process which must involve several labile intermediates derived from SAM-dependent *C*-methylations at C-1, C11, C-17, and ring contraction–deacylation (from C-20 and its attached methyl). None of these intermediates has been isolated from lysates of the recombinant *Ps. denitrificans* strain; they are presumably destroyed by the procedures used for the isolation of precorrin-6x (oxidation–esterification).

Notice that we have still to find the C-17 and C-1 methylases, the ring contractase–deacylase and cobalt insertase enzymes. The structure recently

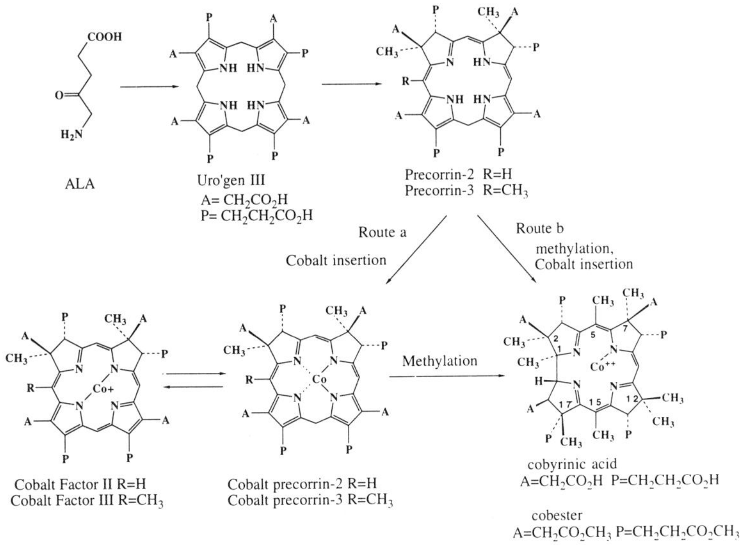

FIG. 12. Temporal resolution of cobalt insertion (into precorrin-2 and precorrin-3; route a) in *Pr. shermanii*.

suggested for precorrin-8x formed by C-5, C-15 methylation and decarboxylation in ring C and the gene (*cobH*) encoding the mutase for the subsequent C-11→C-12 transposition (Fig. 11) are important clues (Blanche et al 1992b, Thibaut 1992) to the mechanism of corrin biosynthesis, but it must be borne in mind that none of these species bears cobalt and that the final product in the eight-gene overexpression system in *Ps. denitrificans* is hydrogenobyrinic acid (cobalt-free cobyrinic acid). The absence of cobalt in the *Ps. denitrificans* metabolites precorrin-6x→8x is in accord with the fact that in this organism, amidation of cobyrinic acid takes place before cobalt insertion (Blanche et al 1992a). In contrast, hydrogenobyrinic acid is not a precursor of cobyrinic acid in *Pr. shermanii* (Podschun & Müller 1985), and, even in the genetically altered *Ps. denitrificans* strain, cobalt can be inserted into the product, hydrogenobyrinic acid, only by non-enzymic means.

Although it would seem inconceivable to organic chemists (and evolutionary theorists) that two distinct pathways should run in parallel to such a complex target as vitamin B_{12} (or even cobyrinic acid), the circumstantial evidence now suggests that the enzymes of the aerobic and anaerobic pathways are making and handling subtly different substrates—one set cobalt-free, the other coordinated with cobalt.

Cobalt insertion: experiments with *Pr. shermanii*

In collaboration with G. Müller (Müller et al 1991) we have shown that in *Pr. shermanii* extracts, cobalt is inserted into precorrin-2 and precorrin-3; the

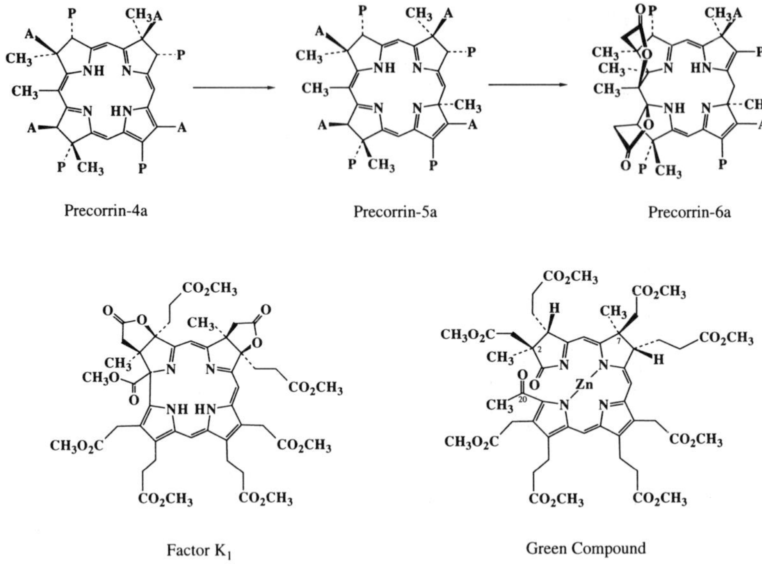

Precorrin-4a   Precorrin-5a   Precorrin-6a

Factor K₁   Green Compound

FIG. 13.   Proposed structures of the missing intermediates, precorrins-4a, 5a and 6a, and the *Pr. shermanii* metabolites K$_1$ and the 'green compound' (see *End note*).

resultant metal complexes serve as substrates for cobyrinic acid synthesis in this organism (Fig. 12).

It was also shown that cobalt insertion in *Pr. shermanii* is an enzymic process when cobalt–Factor III was isolated after incubation of Factor III with a *Pr. shermanii* homogenate (P. J. Santander, unpublished work). Thus, there is no longer any doubt that in *Pr. shermanii* (and presumably also in *Salmonella*) cobalt is inserted much earlier in the pathway than in *Ps. denitrificans*. In fact, CysG (M-1) can also catalyse the insertion of cobalt into Factor II (Spencer et al 1993).

Ring contraction: temporal resolution of a unique isotope exchange at a carbonyl group on the way to B$_{12}$

The structure of precorrin-6x dictates that, in *Ps. denitrificans*, ring contraction takes place between precorrin-3 and precorrin-6x, i.e., before the last two methylations (C-5 and C-15) and decarboxylation (C-12 acetate). No evidence is yet available (see *End note*) for the intermediacy of species corresponding to precorrins-4a, 5a and 6a, which should have the structures shown in Fig. 13 as their cobalt complexes. However, in *Pr. shermanii* it has been possible to show by NMR spectroscopy that a unique carbonyl group at C-27 (the ring A carboxyamide group of B$_{12}$) undergoes exchange in H$_2$O and that this group

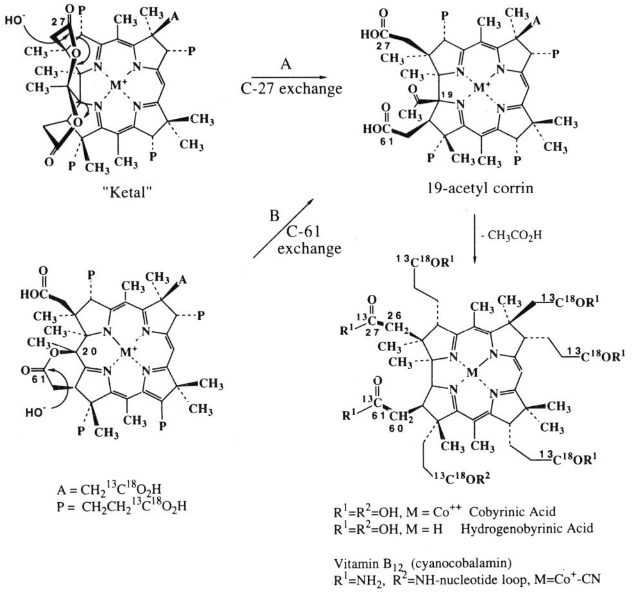

FIG. 14. Proposed mechanisms for $^{18}O/^{16}O$ exchange during ring contraction in *Pr. shermanii*.

has been involved earlier in the mechanism of ring contraction following *C*-methylation at C-17, C-11 and C-1 (Fig. 14). Several unusual ring-contracted metabolites and oxidation products (including Factor K_1 and the 'green compound' shown in Fig. 13) including C-1- (and/or C-19-) substituted dehydro *corrins* containing methyl groups at C-2 and C-7 have been isolated from *Pr. shermanii* (G. Müller, A. I. Scott et al, unpublished work); these structures suggest strongly that precorrin-3 has undergone 'early' ring contraction.

In corrin biosynthesis, it is now clear that the ring A acetate (C-27) serves a scaffolding role at least once before and/or during the ring contraction step, and is released by hydrolysis at C-27, whereas the ring D acetate, which may still be involved in formation of the ketal for the C-19 acetyl system, neither suffers direct hydrolysis nor participates in the hydrolytic step.

Evolution of the porphyrin–corrin pathway: why type III?

There seems no doubt that there are dual or parallel pathways for corrin synthesis in the primitive anaerobes and (anaerobic) photosynthetic and aerobic microorganisms (Fig. 15). The 'preregistration' (Eschenmoser 1988) of a prebiotic corrin template may well have diverted the subsequent biochemical machinery to a type III rather than a type I world, because the genes for uro'gen III synthesis were needed for the subsequent ring contraction step which requires

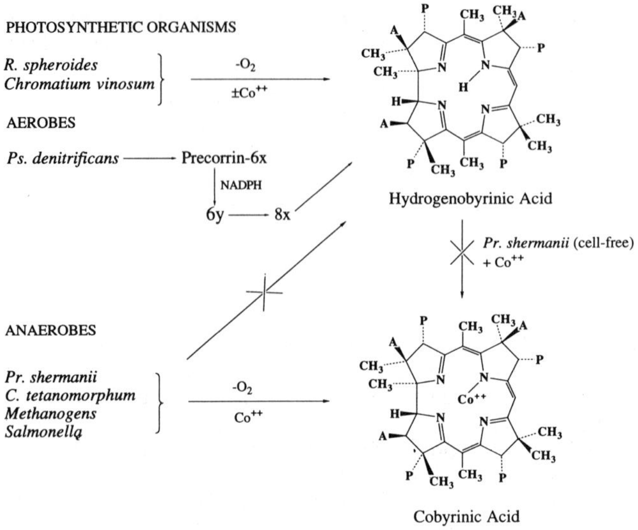

FIG. 15. Parallel, oxidative and metal-free routes to corrins in bacteria.

juxtaposition of the two adjacent acetate side chains (Eschenmoser 1988) of uro'gen III. Thus, when haem, chlorophyll and the cytochrome pigments became necessary cofactors for oxygen transport, photosynthesis and O_2 activation, the natural choice was uro'gen III, because this template was already in place as a necessary component of corrin synthesis under genetic control. The intriguing questions raised by the existence of these different pathways will be difficult to answer directly, but even within the context of 'modern' B_{12} biosynthesis, the duality of pathways (with and without O_2) suggest a vestige of evolutionary preference which now appears to be lateral rather than vertical.

Summary

We have shown how the B_{12} biosynthetic pathways differ between anaerobic bacteria, in which early insertion of cobalt takes place, and the aerobes (such as *Ps. denitrificans*), in which the metal ion does not enter the macrocycle until well after the ring contraction and substrate methylation steps are complete. In addition, the specificities of the methyltransferases encoded by *cysG* and *cbiF* appear to be quite unrestricted, such that they can be used to prepare several *C*-methylated type I pyrrocorphins and corphins (including Factor S_3).

Acknowledgements

A special word of thanks is due to my senior collaborators: Professors G. Müller (Stuttgart), P. M. Jordan (London), M. Kajiwara (Tokyo) and J. Roth (Utah). I am indebted to my present group of highly motivated, young colleagues whose recent discoveries form the main theme of this account: Drs P. J. Santander, R. Danso-Danquah, M. D. Gonzalez, M. J. Warren, K. Iida, C. Pichon, T. Xue, P. Nayar and J. B. Spencer, Ms B. P. Atshaves, Ms K. Campbell, Ms S. Estrada, Messrs N. Anousis, C. Min, S. Ozaki, J. Park and M. Holderman. Their research efforts in the areas of molecular biology, enzymology and NMR spectroscopy were guided at all times by Drs Charles Roessner, Neal Stolowich and Howard Williams whilst Mr Y. Gao provided untiring help with NMR instrumentation. We thank NIH, NSF, the Robert A. Welch Foundation, SERC (UK) and NATO for generous financial support.

References

Battersby AR, Fröbel K, Hammerschmidt F, Jones C 1982 Biosynthesis of vitamin B$_{12}$: isolation of dihydrosirohydrochlorin, a biosynthetic intermediate. Structural studies and incorporation experiments. J Chem Soc Chem Commun, p 455–457

Blanche F, Debussche L, Thibaut D, Crouzet J, Cameron B 1989 Purification of S-adenosyl-L-methionine:uroporphyrinogen III methyltransferase from *Pseudomonas denitrificans*. J Bacteriol 171:4222–4231

Blanche F, Thibaut D, Frechet D et al 1990 Hydrogenobyrinic acid: isolation, biosynthesis and function. Angew Chem Int Ed Engl 29:884–886

Blanche F, Thibaut D, Famechon A, Debussche L, Cameron B, Crouzet J 1992a Precorrin-6x reductase from *Pseudomonas denitrificans*: purification and characterization of the enzyme and identification of the structural gene. J Bacteriol 174:1036–1042

Blanche F, Famechon A, Thibaut D, Debussche L, Cameron B, Crouzet J 1992b Biosynthesis of vitamin B$_{12}$ in *Pseudomonas denitrificans*: the biosynthetic sequence from precorrin-6y to precorrin-8x is catalyzed by the *cobL* gene product. J Bacteriol 174:1050–1052

Brunt RD, Leeper FJ, Grgurina I, Battersby AR 1989 Biosynthesis of vitamin B$_{12}$: synthesis of ($\pm$)5^{13}C Faktor 1 methyl ester. Determination of the oxidative state of precorrin-1. J Chem Soc Chem Commun, p 428–431

Cameron B, Briggs K, Pridmore S, Brefort G, Crouzet J 1989 Cloning and analysis of genes involved in coenzyme B$_{12}$ biosynthesis in *Pseudomonas denitrificans*. J Bacteriol 171:547–557

Crouzet J, Blanche F, Cameron B, Thibaut D, Debussche L 1990 Biochemical and genetic studies on vitamin B$_{12}$. In: Baldwin TO, Raushel FM, Scott AI (eds) Chemical aspects of enzyme biotechnology: fundamentals. Plenum, New York, p 299–315

Eigen M, Lindemann BF, Tietze M, Winkler-Oswatifsche R, Dress A, Van Halseler A 1989 How old is the genetic code? Statistical geometry of tRNA provides an answer. Science 244:673–679

Eschenmoser A 1988 Vitamin B$_{12}$: experiments concerning the origin of its molecular structure. Angew Chem Int Ed Engl 27:5–39 (Angew Chem 100:5–39)

Georgopapadakou NH, Scott AI 1977 On B$_{12}$ biosynthesis and evolution. J Theor Biol 69:381–384

Jeter RM, Roth JR 1987 Cobalamin (vitamin B$_{12}$) biosynthetic genes of *Salmonella typhimurium*. J Bacteriol 169:3187–3198

Leeper FJ 1985 The biosynthesis of porphyrins, chlorophylls, and vitamin B$_{12}$. Nat Prod Rep 2:561–580

Leeper FJ 1987 The biosynthesis of porphyrins, chlorophylls, and vitamin B$_{12}$. Nat Prod Rep 4:441–469

Leeper FJ 1989 The biosynthesis of porphyrins, chlorophylls, and vitamin B$_{12}$. Nat Prod Rep 6:171–203

Mason SF 1991 Chemical evolution: origins of the elements, molecules, and living systems. Oxford University Press, Oxford

Müller G, Schmiedl J, Schneider E et al 1986 The structure of factor S_3, a metabolite of *Propionibacterium shermanii* derived from uroporphyrinogen I. J Am Chem Soc 108:7875–7877

Müller G, Schmiedl J, Savidis L et al 1987 Factor S_1, a natural corphin from *Propionibacterium shermanii*. J Am Chem Soc 109:6902–6904

Müller G, Zipfel F, Hlineny K et al 1991 Timing of cobalt insertion in vitamin B_{12} biosynthesis. J Am Chem Soc 113:9893–9895

Ozaki S-I, Roessner CA, Stolowich NJ et al 1993 Multienzyme synthesis and structure of factor S_3. J Am Chem Soc 115:7935–7938

Podschun TE, Müller G 1985 Hydrogenobyrinsaüre und vitamin B_{12}. Angew Chem 97:63–64

Roessner CA, Warren MJ, Santander PJ et al 1992 Expression of 9 *Salmonella typhimurium* enzymes for cobinamide synthesis. Identification of the 11-methyl and 20-methyl transferases of corrin biosynthesis. FEBS (Fed Eur Biochem Soc) Lett 301:73–78

Schopf W (ed) 1983 Earth's earliest biosphere. Princeton University Press, Princeton, NJ

Scott AI 1978 Biosynthesis of vitamin B_{12}. In search of the porphyrin–corrin connection. Acc Chem Res 11:29–36

Scott AI 1990 Mechanistic and evolutionary aspects of vitamin B_{12} biosynthesis. Acc Chem Res 23:308–317

Scott AI, Williams HJ, Stolowich NJ et al 1989 The structure of sirohydrochlorin I, a dimethylisobacteriochlorin derived from uro-porphyrinogen I. J Chem Soc Chem Commun, p 522–525

Scott AI, Warren MJ, Roessner CA, Stolowich NJ, Santander PJ 1990 Development of an 'overmethylation' strategy for corrin synthesis. Multi-enzyme preparation of pyrrocorphins. J Chem Soc Chem Commun, p 593–597

Spencer JB, Stolowich NJ, Roessner CA, Scott AI 1993 The *E. coli cysG* gene encodes the multifunctional protein, sirohaem synthase. FEBS (Fed Eur Biochem Soc) Lett 335:57–60

Thibaut D, Couder M, Crouzet J, Debussche L, Cameron B, Blanche F 1990a Assay and purification of *S*-adenosyl-L-methionine:precorrin-2 methyltransferase from *Pseudomonas denitrificans*. J Bacteriol 172:6245–6251

Thibaut D, Debussche L, Blanche F 1990b Biosynthesis of vitamin B_{12}: isolation of precorrin-6x, a metal-free precursor of the corrin macrocycle retaining five *S*-adenosylmethionine-derived peripheral methyl groups. Proc Natl Acad Sci USA 87:8795–8797

Thibaut D, Blanche F, Debussche L, Leeper FJ, Battersby AR 1990c Biosynthesis of vitamin B_{12}: structure of precorrin-6x octamethyl ester. Proc Natl Acad Sci USA 87:8800–8804

Thibaut D, Couder M, Famechon A et al 1992 The final step in the biosynthesis of hydrogenobyrinic acid is catalyzed by the *cobH* gene product with precorrin-8x as the substrate. J Bacteriol 174:1043–1049

Uzar HC, Battersby AR, Carpenter TA, Leeper FJ 1987 Biosynthesis of porphyrins and related macrocycles. Part 28. Development of a pulse labelling method to determine the C-methylation sequence for vitamin B_{12}. J Chem Soc Perkin Trans I, p 1689–1696

Waditschatka R 1985 Die Porphyrinogen ⇄ Pyrrocorphin-Tautomerisierung. Dissertation no 7707, ETH, Zürich, Switzerland

Warren MJ, Roessner CA, Santander PJ, Scott AI 1990a The *Escherichia coli cysG* gene encodes *S*-adenosylmethionine-dependent uroporphyrinogen III methylase. Biochem J 265:725–729

Warren MJ, Gonzalez MD, Williams HJ, Stolowich NJ, Scott AI 1990b Uroporphyrinogen-III methylase catalyses the enzymatic synthesis of sirohydrochlorins II and IV by a clockwise mechanism. J Am Chem Soc 112:5343–5345

Warren MJ, Roessner CA, Ozaki S, Stolowich NJ, Santander PJ, Scott AI 1992 Enzymatic synthesis and structure of precorrin-3, a trimethyldipyrrocorphin intermediate in vitamin B$_{12}$ biosynthesis. Biochemistry 31:603–609

End note. In a short time after the Ciba Foundation symposium was held most of the concepts and predictions about the 'missing' intermediates, precorrins-4a, 5a, etc. (shown as hypothetical structures in Fig. 13) have been swept away by a flood of new information. Work from the French groups (Thibaut et al 1993, Debussche et al 1993) and from our laboratory (Scott et al 1993, Spencer et al 1993, Min et al 1993) has revealed the structures of three new intermediates formed by the *cob* genes *cobG*, precorrin-3x, *cobJ*, precorrin-4 and *cobM*, precorrin-5; see structures below. These exciting developments have radically altered our views about the B$_{12}$ pathway.

Precorrin-3   Precorrin-3x   Precorrin-4

Debussche L, Thibaut D, Danzer M et al 1993 Biosynthesis of vitamin B$_{12}$: structure of precorrin-3B, the trimethylated substrate of the enzyme catalysing ring contraction. J Chem Soc Chem Commun, p 1100–1103

Min C, Atshaves BP, Roessner CA, Stolowich NJ, Spencer JB, Scott AI 1993 Isolation, structure and genetically engineered synthesis of precorrin-5, the pentamethylated intermediate of vitamin B$_{12}$ biosynthesis. J Am Chem Soc 115:10380–10381

Scott AI, Roessner CA, Stolowich NJ, Spencer JB, Min C, Osaki S-I 1993 Biosynthesis of vitamin B$_{12}$: discovery of the enzymes for oxidative ring contraction and insertion of the fourth methyl group. FEBS (Fed Eur Biochem Soc) Lett 331:105–108

Spencer JB, Stolowich NJ, Roessner CA, Scott AI 1993 Biosynthesis of vitamin B$_{12}$: ring contraction is preceded by insertion of molecular oxygen. J Am Chem Soc 331: 115:11610–11611

Thibaut D, Debussche L, Fréchet D, Herman F, Vuilhorgne M, Blanche F 1993 Biosynthesis of vitamin B$_{12}$: the structure of factor IV, the oxidized form of precorrin-4. J Chem Soc Chem Commun, p 513–515

DISCUSSION

Castelfranco: The *cysG* gene encodes a single protein, as I understand it, with several different activities. Have you tried to separate out these activities?

Scott: When Martin Warren truncated the gene, the protein produced did not work. If you cut off the front end, the remainder will catalyse only the methyl transfer. We don't yet know whether the N-terminal part will catalyse the dehydrogenation and metal chelation on its own.

Castelfranco: Does the cobalt insertase from *E. coli*, CysG, require ATP?

Scott: No, it doesn't, unlike magnesium chelatase and the *Ps. denitrificans* cobaltochelatase (CobN, S, T).

Warren: Insertion of cobalt into precorrin-2 and precorrin-3 can occur at the reduced level.

Scott: Not enzymically; oxidation up to Factor II is necessary before cobalt is inserted.

Warren: Cobalt does go in non-enzymically at the reduced level, doesn't it?

Scott: Cobalt can go into Factor II non-enzymically, but at a rate at least a thousand times slower than the catalytic rate.

A. Smith: Metal ions such as cobalt are inserted by CysG from *E. coli*, which doesn't make vitamin B_{12}. How sure are you that the CysG enzyme from *Salmonella typhimurium* will do the same thing, insert cobalt and make B_{12}?

Scott: The circumstantial evidence is John Roth's demonstration (Goldman & Roth 1993) that the 66 *S. typhimurium cysG* mutants (*cysG* maps at 72′) won't make sirohaem or B_{12}. We have *cysG* from *S. typhimurium*, but we didn't think it was worth studying because of its high similarity to *E. coli cysG*. Perhaps we should do the experiments.

Warren: There is 90% amino acid sequence identity between *S. typhimurium* CysG and *E. coli* CysG.

Rebeiz: Is there any evidence that the enzymes operate as a multienzyme complex in solution? The enzymes are probably regulated in the cell, but what happens when they're in solution in a test-tube?

Scott: I don't think they operate as a complex. You can put two in, for example, and go to a certain point then add the third, and so on.

Beale: Does *E. coli* have any of the other genes, even though it doesn't make B_{12}?

Stamford: As far as I know, the latest genetic maps present no loci for cobalamin synthetic genes.

Beale: Can you pick up homologous sequences by Southern hybridization?

Warren: John Roth has looked very hard but has found nothing (personal communication).

Castelfranco: *S*-Adenosylmethionine appears to be a key cofactor both in vitamin B_{12} synthesis and in chlorophyll synthesis. *S*-Adenosylmethionine is extremely polar and positively charged, and as such I would not expect it to move freely between compartments within cells. There must be some intracellular carriers; these would have an extremely important role in regulation of these reactions. This is only speculation, but I feel sure it is true in the chloroplast envelope. From what I've heard here, I would suspect this might be the case in these prokaryotes also.

Thauer: Prokaryotes generally don't have compartments.

Castelfranco: Is that really true?

Thauer: There is only one accepted exception, the cyanobacteria, which contain separate compartments surrounded by the thylakoid membranes.

Akhtar: The enzymes are cytoplasmic in *S. typhimurium*.

Castelfranco: Is *S*-adenosylmethionine also made in the cytoplasm?

Akhtar: I assume so.

Castelfranco: Then its movement would not be a problem.

Stamford: *S*-Adenosylmethionine may well have a role in regulation of cobalamin biosynthesis. *S*-Adenosylmethionine-dependent methyltransferases are normally inhibited by their product, *S*-adenosylhomocysteine. SUMT from *Ps. denitrificans* is particularly susceptible to this inhibition. The K_m for *S*-adenosylmethionine for that enzyme is around 6.3 μM. The K_i for S-adenosylhomocysteine is around 0.32 μM, a concentration 20-fold lower than the K_m for the substrate, which might create a bottle-neck on that pathway.

Akhtar: Compartmentalization becomes an important issue when you transfer cytoplasmic *S*-adenosylmethionine into the membrane. I have the feeling that B$_{12}$ might be a rather simple-minded system in this respect. Is there any evidence that any of the enzymes are not truly soluble?

Stamford: All the methylases, from *Ps. denitrificans* at least, seem to be soluble.

Warren: There are some sequences in the *Salmonella* cobIII cluster which look as though the proteins encoded might have membrane-bound domains.

Scott: But these are not biosynthetic enzymes necessary for cobyrinic acid synthesis.

Beale: The B$_{12}$ biosynthetic intermediates are also soluble. Chlorophyll biosynthetic proteins may be membrane-associated because the intermediates are insoluble.

Scott: Could any of the more biologically oriented amongst us comment on why a molecule as complicated as B$_{12}$ should have a metal-free and a metal-containing pathway? It's such a complicated product that it seems odd that there should be parallel pathways.

A. Smith: The only thing that occurred to me, which is why I asked you about CysG (p 303), is that ferrochelatase will insert zinc into protoporphyrin. Were you doing your assays in the cell or in a cell-free system?

Scott: We just had the enzyme.

A. Smith: It is possible that *in vivo* it doesn't actually work like that.

Scott: As I mentioned in my paper, we know from our experiments with *Pr. shermanii* and from experiments with cobalt done at Cambridge that cobalt is inserted early in the anaerobes studied so far.

Battersby: The work done at Cambridge on the timing of cobalt insertion in *Pr. shermanii* used pulse labelling. Unlabelled sirohydrochlorin, which yields precorrin-2 *in situ*, was incubated with the cell-free *Pr. shermanii* enzyme system with a very small amount of high specific activity ^{60}Co ions. After 10 minutes, a large amount of unlabelled Co(II) ions were added, greatly lowering the specific activity of the cobalt in the medium, together with [*methyl*-^{14}C,^{13}C]SAM. The cobyrinic acid isolated after 16 hours incubation was found to have a molar specific activity of its cobalt almost the same (93%) as that of the initial ^{60}Co. This means that essentially all of the cobalt of the final cobyrinic acid is inserted during the first 10 minutes of incubation. In addition, the ^{14}C activity of the cobester prepared from this cobyrinic acid, and especially its ^{13}C NMR spectrum, showed clearly that all of the last five methyl groups, including that at C-17, are inserted *after* cobalt insertion. Thus, cobalt insertion occurs at or close to precorrin-3 and before methylation at C-17. (R. A. Vishwakarma, S. Balachandran, N. P. J. Stamford, S. Monaghan, S. Prelle, F. J. Leeper & A. R. Battersby, unpublished work 1987–1992.)

Scott: I should stress, in a tribute to Gerhard Müller, that these cobalt experiments are quite difficult to get right.

Beale: Does the existence of the two pathways correlate at all with the degree of anaerobiosis?

Scott: Photosynthetic bacteria, e.g., *Rhodobacter sphaeroides*, and the aerobic *Ps. denitrificans* all finish up with hydrogenobyrinic acid. The cobalt is inserted later, into the ring A,B diamide.

Beale: Rhodobacter is not a strict anaerobe, but a facultative anaerobe.

Scott: That's right. There's no enzyme I know of that will convert cobalt-free cobyrinic acid, hydrogenobyrinic acid, into cobyrinic acid—in other words, there's no cobalt 'insertase' at that point. Cell-free *Pr. shermanii* accumulates cobyrinic acid, with cobalt already inside. Cobalt-free cobyrinic acid is not metallated in *Pr. shermanii*, or in any other organism, as far as I know.

Beale: It could be that facultative anaerobes need to avoid a certain combination of metal redox states, or that their pathway avoids the involvement of an enzyme that can catalyse the insertion of the wrong metal.

Battersby: Professor Scott, how firm is the structure of Factor K$_1$? Was the material ^{13}C-labelled, or is the structure just a possible one based on ^{1}H NMR without NOE studies?

Scott: We have mass spectroscopy data and the number of *meso* protons from the ^{1}H NMR structure. We will need ^{13}C-labelling to complete the structure.

Eschenmoser: This structure should not be taken lightly. It is a dramatic structure, because it points to the possibility that the enzymic ring contraction chemistry on the higher oxidation level could also occur spontaneously.

Scott: There might even be a methyl ketone rather than a carboxylic acid at C-1 or C-19. Another metabolite (B$_2$) doesn't have a pendent group at C-1 or C-19. These structures need to be confirmed but, constitutionally, there is no doubt that they represent early ring contraction, possibly at the precorrin-2 level.

Chang: Why does the methylation occur at C-11 first, with the methyl group then shifting to C-12?

Arigoni: Methylation at C-11 triggers off in a reasonable way the cleavage of the acetyl group (Scheme 1).

Battersby: Notice also that this view of the probable mechanism for the removal of the acetyl group from C-1, with which I entirely agree, generates an extended enamine which provides the perfect reactivity for C-1 methylation.

Scott: That means that C-11 methylation could be concomitant with deacylation.

Akhtar: One could argue that the anhydride moiety generated from the acetyl group and the acetate side chain might spontaneously hydrolyse non-enzymically. However, Nature rarely leaves even such a trivial event to non-enzymic chance. Is there an enzyme which will hydrolyse the anhydride group? Such a process would need to be enzyme-catalysed for the OH group to be put specifically on one side of the anhydride bond.

Arigoni: Generation of the anhydride intermediate can be resorted to in the explanation of the formation of doubly ^{18}O-labelled acetate species (p 281). In this scheme, if the first oxygen transfer occurs during generation of the *C*-acetyl group, the ratio of singly to doubly labelled species in the released acetate will depend on the regiospecicity of the OH$^-$ attack during hydrolysis of the anhydride group (cf., Scheme 1). Our evidence for the formation of doubly

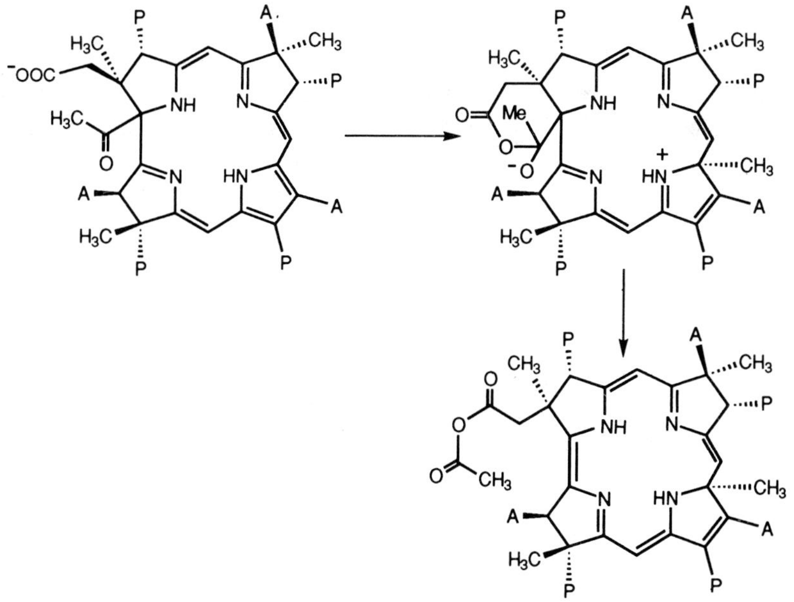

Scheme 1 (*Arigoni*)

[18]O-labelled acetate is clearly insufficient and we are still working on the problem. The original work of Kurumaya et al (1989) suggests that more than one oxygen atom is released from the ring A acetic acid chain during formation of vitamin B_{12}.

References

Goldman BS, Roth JR 1993 Genetic structure and regulation of the *cysG* gene in *Salmonella typhimurium*. J Bacteriol 175:1457–1466
Kurumaya K, Okazaki T, Kajiwara M 1989 Studies on the biosynthesis of corrinoids and porhyrinoids. I. The labeling of oxygen of vitamin B_{12}. Chem & Pharm Bull (Tokyo) 37:1151–1154

B₁₂: reminiscences and afterthoughts

Albert Eschenmoser

Laboratorium für Organische Chemie, ETH-Zentrum, Universitätstrasse 16, CH-8092 Zürich, Switzerland

Abstract. This paper describes some of the chemistry of synthetic corrinoids and corphinoids carried out at the ETH (Swiss Federal Institute of Technology) after the completion of the non-enzymic synthesis of vitamin B₁₂ and attempts to delineate the interplay between mechanistic hypotheses, model studies and the experimental research in B₁₂ biosynthesis which had spurred this work. The afterthoughts deal with why and how the work on corrinoids eventually led to an experimental involvement in the problem of a chemical aetiology of the structure of natural nucleic acids.

1994 The biosynthesis of the tetrapyrrole pigments. Wiley, Chichester (Ciba Foundation Symposium 180) p 309–332

I should like to introduce my reminiscences by showing the majestic Piz Palü in the Engadine, viewed from the Diavolezza (Fig. 1), which is symbolic of this symposium in many ways, not least of Sir Alan Battersby and his research achievements in the field of porphinoid biosynthesis. The image can also represent the relationship between research on B₁₂ biosynthesis and the type of chemistry which we have done in this context and on which I am intending to reminisce here. Those people on the snowless and comfortably accessible foreground look up to the beautiful snow-capped mountain, over the icy rocks of which the true mountaineers fight their way up to the top. The people in the foreground propose and discuss the best pathways by which to reach the top; after all, from their position, they have an excellent panoramic view of the mountain and of the grooves and crests in its overall structure. However, beside model exercises, they can do little except watch with great interest and admiration how their friends and heroes up on the mountain find their way to the peak, and be surprised by the unexpected turns they sometimes take.

My reminiscences begin in 1964 (Fig. 2 and 3). I go that far back because this was the time when we had chemically synthesized the first corrin, and had thought that a ligand system, which we then named 'corphin', might be the central type of intermediate on the biosynthetic pathway to vitamin B₁₂. At that time, it seemed clear that the C-20 of porphyrinogen type III was the carbon which would end up as the carbon of the angular methyl group at position C-1

309

FIG. 1.   Piz Palü (3912m) in the Engadine, viewed from the Diavolezza. ©Photoglob, Zürich, Switzerland.

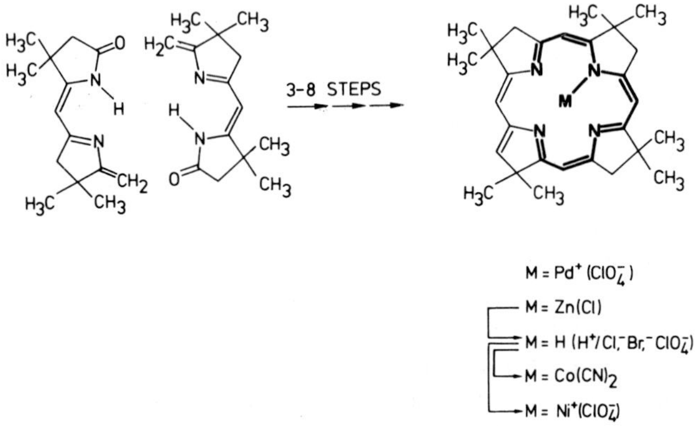

FIG. 2.   The first corphin synthesis (Johnson et al 1968, Müller et al 1973).

of B_{12}'s corrin system. A proton-catalysed *reductive* formation of a three-membered ring at the site of the A–D *meso* position followed by a further reduction to the corrin was our cherished mechanistic hypothesis (Fig. 3). In order to demonstrate the feasibility of such a process, we synthesized an octamethyl corphin (Johnson et al 1968), but failed, despite intense effort, to find a reductive corphin → corrin ring contraction (Müller et al 1973). In 1972 ^{13}C NMR spectroscopy revealed that the methyl group at C-1 of B_{12}'s corrin system stemmed from methionine and that the C-20 *meso* carbon was eliminated somewhere along the biosynthetic pathway (Brown et al 1972, Scott et al 1972). The reductive corphin → corrin ring contraction hypothesis was dead; the corphin ligand system, however, survived (Eschenmoser 1986).

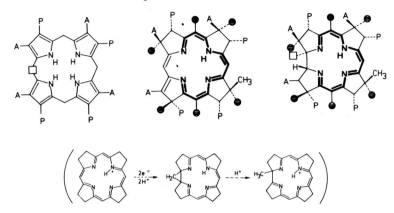

FIG. 3. Our early hypothesis on B$_{12}$ biosynthesis, involving the proton-catalysed reductive formation of a three-membered ring at the A–D *meso* position followed by further reduction to the corrin.

Now I come, of course, to the non-enzymic synthesis of B$_{12}$. Our version involved a photochemical A→D cyclization step (Fig. 4) (Eschenmoser 1968, Yamada et al 1969, Eschenmoser 1970, 1971, 1974, Eschenmoser & Wintner 1977), which was conceived in the wake of Woodward and Hoffmann's rules on orbital symmetry which included a new concept of sigmatropic rearrangements (Woodward 1967), a development that in itself had grown out of Woodward's research on creating the A–D ring junction in the Harvard component for the Harvard–ETH version of the synthesis of B$_{12}$ (Woodward 1968, 1971, 1973). In this photochemical A→D ring closure, a ring D hydrogen atom moves to the methylidene carbon of ring A in the triplet excited state of the chromophore π system (populated via triplet sensitization by the excited chromophore of the cyclization product) and forms a diradical whose singlet analogue, an ylide, valence-tautomerizes to the corrin structure (Eschenmoser 1976, Neier 1978).

An extended investigation followed this synthesis, spurred by questions such as: could this powerful type of A→D ring closure play a role in the biosynthesis of vitamin B$_{12}$; and, are there dark versions of the photochemical A→D secocorrin → corrin cycloisomerization (the biosynthesis does not require light)? Figure 5 summarizes what emerged from that research, all carried out in model systems. The dark simulation of the photochemical process by an electrochemical oxidation–reduction sequence, the oxido-reductive version of the A/D secocorrin → corrin cycloisomerization, was realized by Bernhard Kräutler as part of his PhD: in the dark, the photo-induced excitation of a chromophore electron is replaced by an oxidative removal of that electron and the electron's de-excitation by a reductive return of it, both processes being induced electrochemically (Kräutler et al 1976). The purely reductive version, starting from a Δ^{18}-dehydrosecocorrin complex, came in 1975 (Pfaltz et al 1975),

312

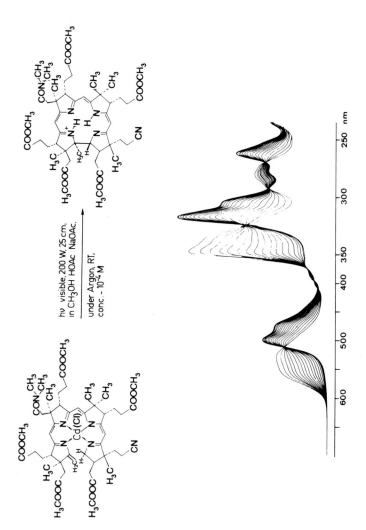

FIG. 4.   The photochemical cycloisomerization of A–D-secocorrin to corrin in B₁₂ biosynthesis (Fuhrer 1973).

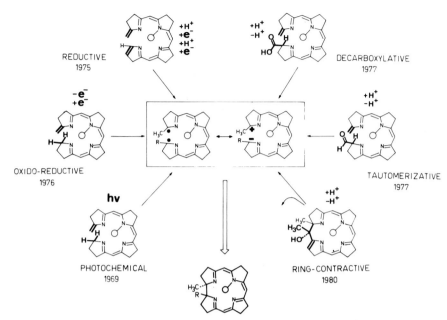

FIG. 5. The search for potentially biomimetic 1, 19-secocorrin → corrin cyclizations: photochemical (Yamada et al 1969); oxido-reductive (Kräutler et al 1976), the dark simulation of the photochemical process; reductive, starting from a Δ ^{18}-dehydrosecocorrin complex (Pfaltz et al 1975); decarboxylative (Pfaltz et al 1977); tautomerizative (Pfaltz et al 1977); ring-contractive (Rasetti et al 1981a,b).

followed by a decarboxylative (Pfaltz et al 1977), a tautomerizative (Pfaltz et al 1977) and, finally, in 1981, a ring-contractive version (Rasetti et al 1981a,b) to which we shall return below.

What originally had appeared to be the toughest of all problems in a chemical B$_{12}$ synthesis—the molecule's A–D ring junction—now emerged as a structural region which can be seen to form itself with surprising ease and in many different ways, once one approaches the problem in the correct way. Let us, at this point, remember Alan Johnson, who had been the first in this field to use an A→D ring closure strategy, when he prepared a corrole, an aromatic octadehydro derivative of the corrin nucleus (Johnson & Kay 1964, 1965). In parallel with our work on the 'dark versions' referred to above, we and others showed that corrinoids are, in fact, formed easily at all possible chromophore oxidation levels from the octadehydro level up to the level of the corrin system itself (Ofner et al 1981) (Fig. 6).

We originally had embarked on our 'post-B$_{12}$ work' with the intention of gathering chemical information that could assist research on B$_{12}$ biosynthesis, but the results of this work brought about a drastic change in our attitude towards

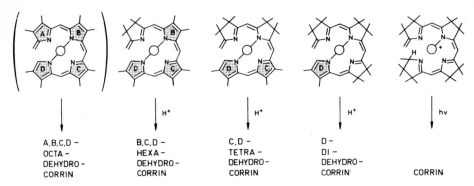

FIG. 6. A–D secocorrinoid → corrinoid cyclizations on different oxidative levels (Yamada et al 1969, Ofner et al 1981).

the structure of vitamin B_{12} (Eschenmoser 1976). Although this molecule had originally represented—to us, as well as to everybody else—the prototype of a complex and synthetically difficult natural product structure, as a result of our change of attitude it lost its aura of complexity and became viewed as a structure that is *superficially* complex, but intrinsically elementary. Elementary, but in what sense?

The crucial event on the way to a biosynthetic pathway's birth is a preceding accidental formation of the molecule which will become the biomolecule. For a molecule to be *selected* as a biomolecule, it has first to be there. Such processes of 'preformation' are determined by the environment—prebiotic or biotic—and, to an extent, by the intrinsic chemical properties of the molecules involved and of their precursors. Although we have no access to the former influences, the latter are susceptible to analysis. If they have had to be preformed accidentally, biomolecules—especially the oldest ones, those indispensable to life—should be intrinsically elementary structures, elementary according to a criterion of the potential for constitutional self-assembly, that is, assembly of a molecular structure from precursors without external instruction.

Bearing B_{12}'s postulated intrinsic structural simplicity in mind, we extended the kind of synthetic analysis which we had carried out on the A–D junction to other structural features of the B_{12} molecule (Eschenmoser 1988). So, we found that the corphin chromophore (which is, electronically, closely related to the corrin chromophore) is the thermodynamically most stable hexahydroporphinoid chromophore among the possible tautomers of a porphyrinogen, provided that the system is complexed by a central divalent metal ion such as magnesium, zinc, nickel or cobalt. Conditions which impose such a metal ion onto a porphyrinogen bring about its conversion, via a complicated but self-directing series of tautomerizations, into a pyrrocorphinate or (in the presence of protons) a corphinate complex (Figs. 7 and 8). In other words, the

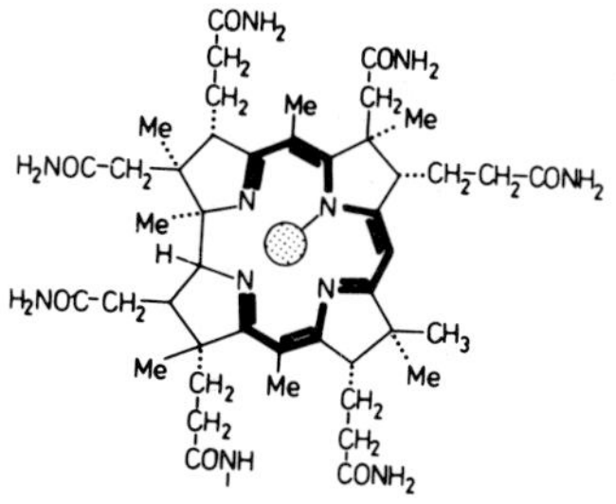

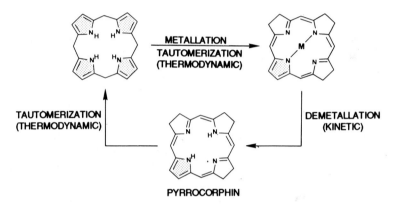

FIG. 7. The chromophore double bond arrangement in B$_{12}$. The corphinoid (corrinoid)-type of chromophore is a thermodynamically favoured arrangement of the double bonds in hexahydroporphinoid–metal complexes. The shaded sphere represents a divalent metal cation such as Mg^{2+}, Zn^{2+}, Ni^{2+} or Co^{2+}.

FIG. 8. Tautomerization of a porphyrinogen to a pyrrocorphin (Eschenmoser 1986, 1988).

corrinoid chromophore contains the thermodynamically most stable arrangements of the double bond system within a hexahydroporphinoid metal complex.

Once one moves along the track suggested by a postulate like the one referred to above, one starts to ask questions which one would otherwise not ask. One of the more successful of such questions was, could the site of attachment of the nucleotide chain to the ring D propionic side chain be the thermodynamically preferred site among the four apparently equivalent side chain sites? Amazingly, this turned out to be true: the outcome of our experiments carried out in this context (Fig. 9) can be rationalized only in terms of this conclusion (Eschenmoser 1988, Kreppelt 1991). If one takes the heptakis(cyanomethyl) ester of cobyrinic acid, whose seven carboxyl functions are all mildly activated towards nucleophilic attack without any differentation, and leaves it to react with the free nucleotide component (an inner ammonium phosphate salt) under appropriate conditions, the nucleotide chain attaches itself regiospecifically to the propionic site at ring D to form vitamin B$_{12}$ and only traces, if any, of B$_{12}$ isomers (Fig. 9). The

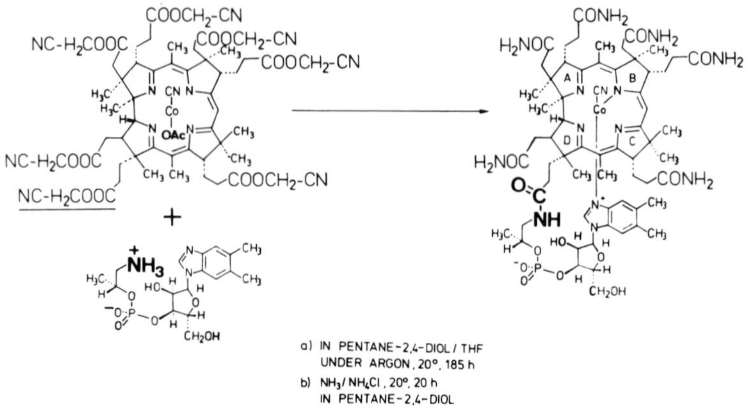

a) IN PENTANE-2,4-DIOL / THF
   UNDER ARGON, 20°, 185 h
b) NH₃/ NH₄Cl, 20°, 20 h
   IN PENTANE-2,4-DIOL

FIG. 9.   The site of attachment of the nucleotide loop at ring D in vitamin B_{12} is the thermodynamically preferred site among the seven possible isomeric attachments. Under appropriate conditions the free nucleotide component attaches to the heptakis-(cyanomethyl) ester of cobyrinic acid regiospecifically, without instruction, at the ring D propionic acid site to form vitamin B_{12} (Eschenmoser 1988, Kreppelt 1991).

nucleotide chain selectively finds its natural attachment site without external instruction: this specific structural feature of the vitamin B_{12} molecule is in fact able to 'assemble itself'.

Let us go a little further back in our reminiscences. When we were planning the total synthesis of vitamin B_{12} in 1960, it was clear both to R. B. Woodward, and to us at the ETH that the target had to be cobyric acid (a hexa-amide with the ring D propionic carboxyl group in the free form), because this naturally occuring B_{12} derivative had already served as the starting material in a partial synthesis of the vitamin (Friedrich et al 1960). This meant that any synthetic plan had to take care of the problem of keeping the ring D propionic acid function differentiated constitutionally from all the other carboxyl functions throughout the synthesis. It was inconceivable for us to consider the possibility of a regioselective attachment of the nucleotide loop to an undifferentiated ring D carboxyl function at that time, and eleven years later on 'Black Friday' (the day on which it was discovered that all we had available of a synthetic intermediate on the way to cobyric acid had lost its ring D differentiation due to an unforeseen accidental over-reaction), and another year later, when both cobyric acid syntheses were accomplished and the question arose as to whether the known cobyric acid → vitamin B_{12} transformation should be repeated or whether a new one should be developed. This was the time of pure 'constructionist' thinking in vitamin B_{12} synthesis; no trace of the idea had yet emerged that the structure of B_{12} might have an elementary nature, an idea which is required for 'selectionist' questions to be asked.

Why a corrin and not a corphin? Why did Nature choose the ring-contracted corrin skeleton for its cobalt complex cofactor in preference to the simpler, more directly accessible, corphin ligand which, after all, has a chromophore system more or less equivalent electronically? In the course of our work we had recognized another revealing thermodynamic preference: the corrinoid ligand skeleton offers a better spatial coordination fit to metal ions such as nickel(II) and cobalt(II) than the corphinoid skeleton (Fig.10). This was evident from X-ray analyses (with Christoph Kratky, Graz) of a series of corphinoid (mostly) nickel complexes (Kratky et al 1985, Eschenmoser 1986). The ligand periphery in these complexes is ruffled (Fig. 10), a property that reveals the coordination hole of the ligand system to be too large for the central metal ion. The juxtaposition of the X-ray crystallographic structures of two nickel(II) complexes with identical ligand π systems in Fig. 10, one with a corrinoid ligand skeleton and the other with a corphinoid ligand skeleton, illustrates the perfect fit of the nickel ion in the corrinoid ligand's coordination hole. The conclusion that this difference parallels the relative thermodynamic stability of the two classes of metal complex is corroborated by the observation (see below) that hydrocorphinoid complexes rearrange to corrinoid complexes when their constitution offers a pathway to do so (Rasetti et al 1981a,b). Furthermore, nickel(II) corphinates can be demetallated, whereas nickel(II) corrinates cannot under similar conditions (Lewis et al 1983).

What are the consequences of these structural and thermodynamic differences between corrinoid and corphinoid metal complexes for the metal ion's propensity to coordinate axial ligands? Figure 11 summarizes the expectations: axial coordination in corphinates should be stronger than in corrinates. In fact, nickel(II) corrinates are known to behave uniformly as square planar complexes, in contrast to corresponding corphinates, which are penta- (or hexa)-coordinated

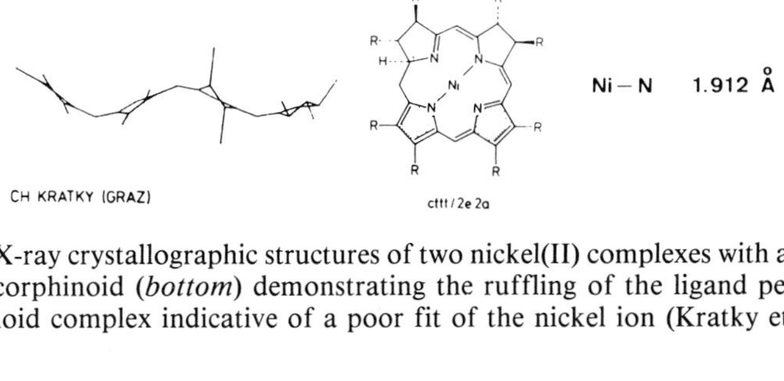

FIG. 10. X-ray crystallographic structures of two nickel(II) complexes with a corrinoid (*top*) or a corphinoid (*bottom*) demonstrating the ruffling of the ligand periphery in the corphinoid complex indicative of a poor fit of the nickel ion (Kratky et al 1985).

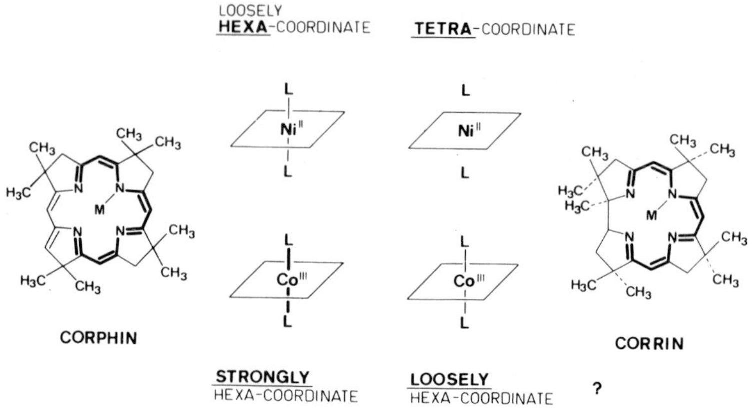

FIG. 11. The expected relative properties of axial coordination in corphinate and corrinate complexes.

in nucleophilic solvents. This residual axial reactivity of nickel(II) when coordinated in corphinoid ligands is believed to be an important aspect of the specific properties of cofactor F_{430} (Kratky et al 1984). But what about the cobalt complexes? Experimentally, we have done far less than this question requires. Figure 12 shows a result (Zimmermann 1989): the rate of solvolysis of the dicyano-cobalt(III) complex to the corresponding aquo-cyano-cobalt(III) complex is much greater in the corrin than in the corphin series. As expected, cobalt(III) corrinates undergo faster ligand exchange than cobalt(III) corphinates.

As well as the corrin chromophore, the corrin skeleton and the nucleotide loop, the crown of peripheral methyl groups is also a characteristic feature of the structure of vitamin B_{12}. Most of those methyl groups are placed at positions which, teleonomically, may be interpreted as having (or having had) the function of stabilizing the corrin chromophore against tautomerization or dehydrogenation to pyrrolic derivatives. In our studies on corphinoids we observed that pyrrocorphinates (Fig. 13, top left corner) can be easily and regioselectively C-protonated at the pyrrole ring to give the corresponding corphinates. What we deduced from this was found to be true: the role of the protons can indeed be taken by electrophilic methyl groups (e.g., CH_3I). We carried out a systematic study of the regioselectivity rules controlling such non-enzymic C-methylations in hexahydroporphinoid systems; the conclusions from these studies (Fig. 13) are pleasing from a theoretical point of view, useful from the standpoint of chemical synthesis and, last but not least, provoking in view of the constitutional course of the sequence of enzymic C-methylations in vitamin B_{12} biosynthesis (Uzar & Battersby 1982, 1985, Scott et al 1984). Figure 14 gives

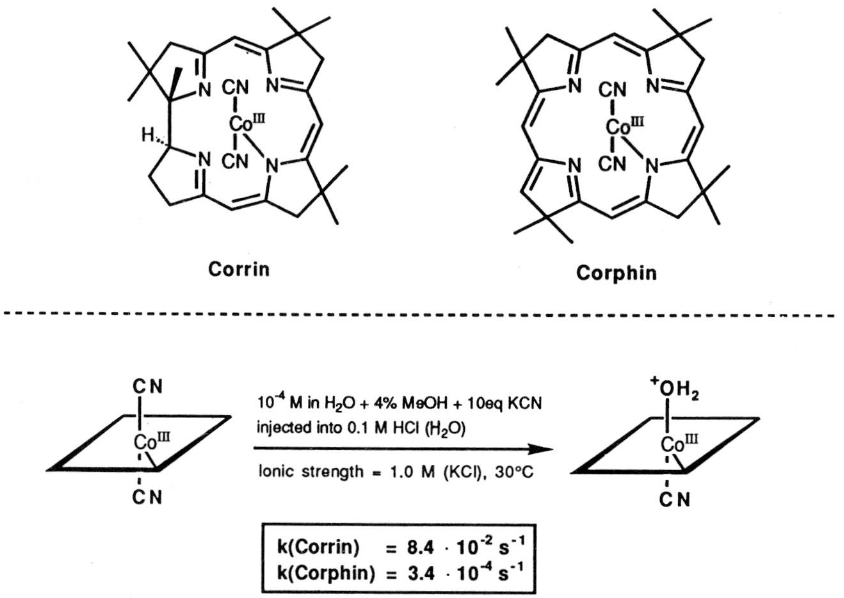

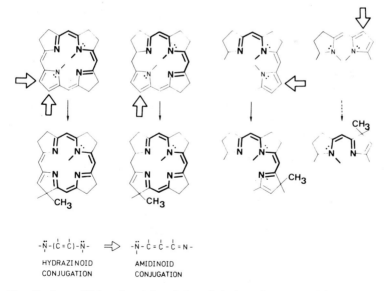

Corrin   **Corphin**

$$10^{-4} \text{ M in H}_2\text{O} + 4\% \text{ MeOH} + 10\text{eq KCN}$$
injected into 0.1 M HCl (H$_2$O)

Ionic strength = 1.0 M (KCl), 30°C

| k(Corrin) | = $8.4 \cdot 10^{-2} \text{ s}^{-1}$ |
| k(Corphin) | = $3.4 \cdot 10^{-4} \text{ s}^{-1}$ |

FIG. 12. Relative rates of ligand exchange in a dicyano-cobalt(III)–corrinate and a dicyano-cobalt(III)–corphinate. The rate of solvolysis of the dicyano-cobalt(III) complex to the aquo-cyano-cobalt(III) complex is much greater in the corrin than in the corphin (Zimmerman 1989).

$$-\ddot{N}-(\overset{|}{C}=\overset{|}{C})-\ddot{N}- \quad \Longrightarrow \quad -\ddot{N}-\overset{|}{C}=\overset{|}{C}-\overset{|}{C}=N-$$

HYDRAZINOID   AMIDINOID
CONJUGATION   CONJUGATION

FIG. 13. Regiospecificity of peripheral *C*-methylation of pyrrocorphinates and related corphinoids.

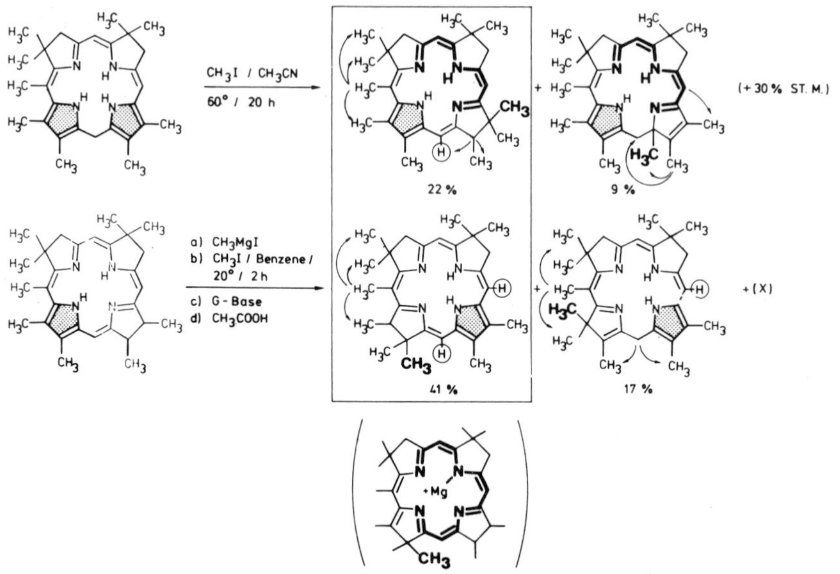

FIG. 14.   Examples of peripheral *C*-methylations of corphinoids (Eschenmoser 1986, 1988).

examples of such methylations, which are remarkably reminiscent of what can happen under enzymic control (Scott et al 1990).

Let us now return to the topic of ring contraction. At the time of the chemical B_{12} synthesis, this topic was at the heart of the B_{12} problem at Harvard as well as at the ETH, and it is still exactly that in today's research on B_{12} biosynthesis. Over a span of three decades, generations of chemical hypotheses on the question of how Nature might achieve the corrin ring contraction came and went again, each time blown away by still another surprising discovery made 'over there, on the Piz Palü'. The story had started with the discovery that the C-1 methyl group stems from methionine and not from the C-20 carbon (Brown et al 1972, Scott et al 1972). This finding blew away the hypothesis of a reductive corphin → corrin ring contraction via a three-membered ring. This body-blow was followed by the dramatic discovery that the *meso* carbon (C-20) becomes methylated and not a C_1 but a C_2 fragment is expelled in the ring contraction process (Battersby et al 1979, Müller et al 1979). This did away with the hypothesis of a A→D secocorrin→corrin cycloisomerization of a 19-formyl secocorrin derivative (Fig. 5), a hypothesis which had been shown in a model study to represent a smooth and chemically appealing pathway to the corrin system—appealing, because, with its expulsion of a C_1 fragment as formic acid, it embodied the attractive concept of a porphyrinogen → corrin transformation requiring neither reduction nor oxidation. This concept—nicely, so it seemed—survived the step to the next generation of chemical model, the

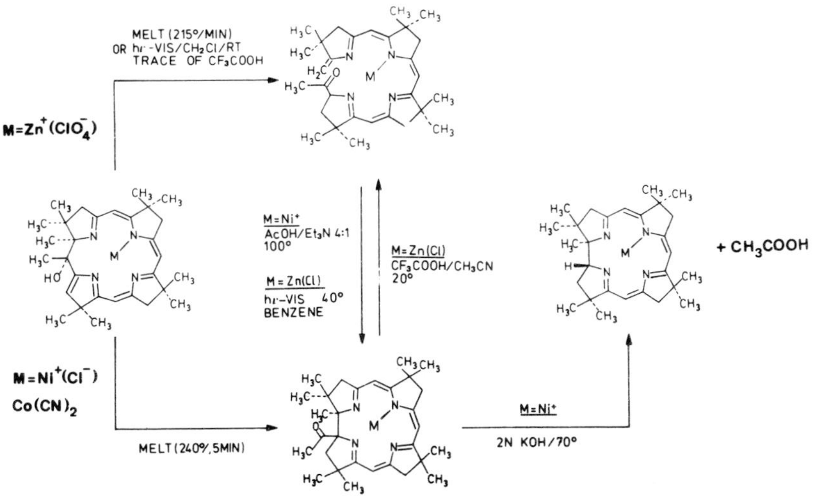

FIG. 15. Model experiments on the rearrangement of dihydrocorphinol to corrin showing the ring contraction and the expulsion of a C$_2$ fragment (Rasetti et al 1981a, b).

hydrocorphinol → corrin ring contraction (Figs. 5 and 15) that operates with the expulsion of the required C$_2$ fragment as acetic (instead of formic) acid (Mombelli et al 1981, Battersby et al 1981) to form a corrin intermediate containing an acetyl group at the ring D corner. This hypothesis, again well documented in model studies, proved remarkably resilient, and underwent an extension to the dilactone version (Fig. 16), in which the carboxyl function of the ring A acetic acid side chain was assigned the important function of assisting the functionalization of the C-20 *meso* position and, furthermore, to constitute in its lactone form a clamp that would keep the system from dissociating between rings A and D during the contraction process (Eschenmoser 1988). This extension originated from a search for an elementary non-enzymic reaction sequence that not only would perform a metal- and proton-assisted conversion of uroporphyrinogen III into a (proton instead of methyl) analogue of cobyrinic acid, but also would do so by selecting specifically the A–D *meso* position of uroporphyrinogen (III) for the ring contraction, because this position differentiates itself from the three other *meso* positions by being the only one which is flanked by two acetic acid side chains (Fig. 17).

We failed to demonstrate such a process in a model system; it probably would have been necessary to search for it in uroporphyrinogen III itself. None the less, such a non-enzymic uroporphyrinogen → corrin transformation still awaits discovery, but—and what a big but—it will very probably have to be induced by *oxidation*. The most recent and, once again, dramatic discoveries 'on Piz Palü' (Thibaut et al 1993, Debussche et al 1993; Battersby 1994, this volume) now

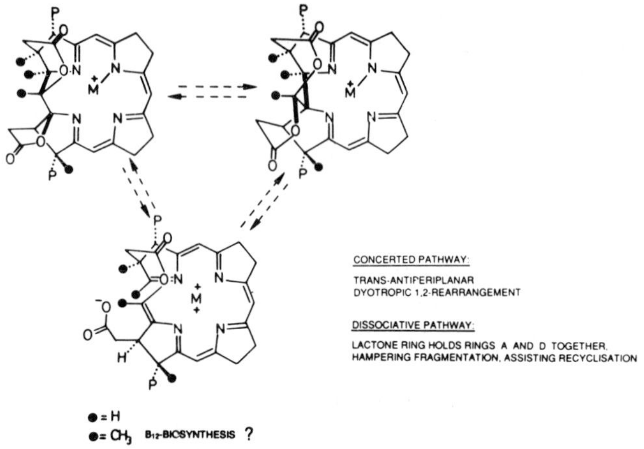

FIG. 16. The dilactone hypothesis in its original form. The carboxyl function of the ring A acetic acid side chain assists in the functionalization of the C-20 *meso* position and in the lactone form constitutes a clamp to prevent dissociation between rings A and D during the ring contraction process.

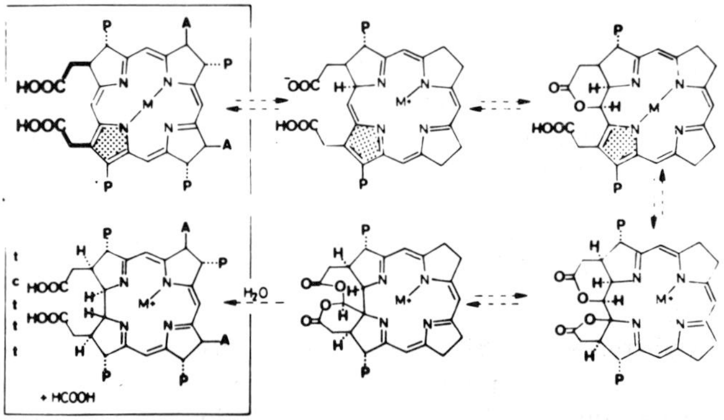

FIG. 17. Our original working hypothesis for the transformation of a urocorphin to a urocorrin, proposed to rationalize why the B_{12} ligand is ring-contracted at the A–D *meso* position and not at one of the other *meso* positions.

unambiguously show that the postulated constancy of oxidation level in B_{12} biosynthesis is incorrect (Ichinose et al 1993). Nature indeed does something we thought it would not do if it could avoid it, namely oxidize and then reduce again. However, once you look behind the structures of the intermediates occurring along the oxidative ring contraction pathway (e.g., the acetyl group at the ring A and not the ring D, corner) and analyse the type of reactions which can lead to and from them, you find yourself being taught once again an extraordinary lesson on chemical reactivity; the hydrocorphinol → corrin ring contraction pathway may have been 'good chemistry', but the opportunities for a ring contraction the system creates for itself by going up one step in oxidation level are extraordinary, offering easier, smoother and more elementary, 'better' chemistry for ring contraction. To have to recognize this *after* the molecular biologists have shown that it is the case is a reminder of an old lesson: appealing postulates may make you blind to alternatives.

The pathway from 5-aminolaevulinic acid to vitamin B$_{12}$ has a good chance of being—once it is known fully—the most extraordinary chapter of chemistry that has had to be learned from the biosynthesis of a biomolecule, and the story of that learning process will be an exceptional documentation of how bioorganic research in the second half of our century has been, and had to be, done.

'On Natural Products Synthesis: from the Synthesis of Vitamin B$_{12}$ to the Question Concerning the Origin of its Molecular Structure' is (an English translation of) the title of a lecture I gave in 1978 that was published in 1982 (Eschenmoser 1982) which, in retrospect, describes fairly accurately the general direction which our work has taken over the past 15 or so years, towards the chemical aetiology of biomolecules. The chemistry of the biosynthesis of biologically fundamental molecules leads to the problem of their origin and, eventually, to the problem of biogenesis. At its root, the problem of biogenesis is one that belongs primarily to bioorganic chemistry—actually, to synthetic bioorganic chemistry; it will be the task of this branch of science to provide the empirical and conceptual framework of an experimentally documented view of life's origin in terms of molecular structure. The widespread opinion that 'we shall never know', although plausible for obvious reasons, remains nothing more than an opinion. The issue is to recognize aspects of the problem which are 'time-independent' and accessible to experiment. Since the advent of Stanley Miller's classic experiment in 1953 (Miller 1953), so-called prebiotic chemistry (Miller & Orgel 1974) has been trying to do this and has in fact collected important relevant knowledge. However, one of the recent conclusions from that particular field of activity is the urgent call for other approaches.

One other approach starts from asking questions of the sort we had met when trying to understand the structure of vitamin B$_{12}$: why a corrin and not a corphin? There is a close connection between that question and the one that became the determinant of the research we do today: why pentose and not hexose nucleic acids? (Eschenmoser 1991, Eschenmoser & Loewenthal 1992, Eschenmoser & Dobler 1992). The formal similarity and conceptual identity of the two questions make the path from one to the other appear a rather direct one, but it wasn't. That trail led via the chemistry of α-aminonitriles, purines and pteridines (Ksander et al 1987) to the study of aziridin-2-carbonitrile (Drenkard et al 1990, Wagner et al 1990) (later oxirane-carbonitrile) and the aldolization chemistry of glycolaldehyde phosphate and on to the chemistry of hexose 2,4,6-triphosphates and pentose 2,4-diphosphates (Müller et al 1990).

We ask such questions, and derive experiments from them, on the basis of the concept that we can approach the problem of a biomolecule's origin by studying the chemistry of potential structural alternatives, molecular structures that—according to chemical reasoning—could be imagined to have offered themselves as potential biomolecules but were rejected by Nature in favour of the one which actually became the biomolecule. The task is to think of such alternative structures, to synthesize them chemically, to study their chemistry and to systematically compare relevant properties with those of the actual biomolecule. Of course, the choice of the alternatives is crucial, and must take into account a molecular structure's potential for constitutional self-assembly. Because the chemical reasoning leading to such a choice is derived from a structural hypothesis about the biomolecule's origin, and because the outcome of the comparison of relevant properties (relevant with respect to function) feeds back on that hypothesis, the concept connects experimental observations of (time-independent) chemical properties with chemical hypotheses about a biomolecule's origin in a potentially iterative way. If the concept were to be applied to life's replicators, the nucleic acids, that iterative network of connections might teach us something about the origin of life itself.

It was the work on sugar phosphates, our observation of the great ease by which hexose 2,4,6-triphosphates and pentose 2,4-diphosphates are formed, that had led us to ask why Nature chose pentose and not hexose sugar building blocks for her nucleic acids. The first experimental study derived from this question was a model study on the chemistry of homo-DNA (Fig. 18); it gave us an unexpectedly rich harvest of results, some of them highly surprising and directly relevant to our understanding of the structure and the structural behaviour of natural DNA, revealing the fundamental importance of the five-membered sugar ring in the natural nucleic acid's structure (DNA as well as RNA) (Hunziker et al 1993). The effect of insertion of a single methylene group into the sugar ring of DNA on its overall structure is illustrated in Fig. 19 (Roth et al 1993). Not only are the structures of these duplexes drastically different—quasi-linear versus helical—but homo-DNA duplexes are also consistently thermodynamically more stable relative to the single strands than corresponding DNA duplexes. It's not so much the purine–pyrimidine base pairing that makes double-stranded DNA helical, but the five-membered nature of the sugar ring.

The concept propounded above calls for comparison of DNA (or RNA) with structural alternatives that are built of potentially natural, fully hydroxylated hexose building blocks and not of a dideoxy sugar which can have had hardly any chance of being a prebiological natural product. The comparisons which have been achieved so far are highly revealing: both allopyranosyl-(6′→4′) and altropyranosyl-(6′→4′) oligonucleotides show drastically weakened purine–pyrimidine pairing, and (adenine/uracil-containing) glucopyranosyl-(6′→4′)

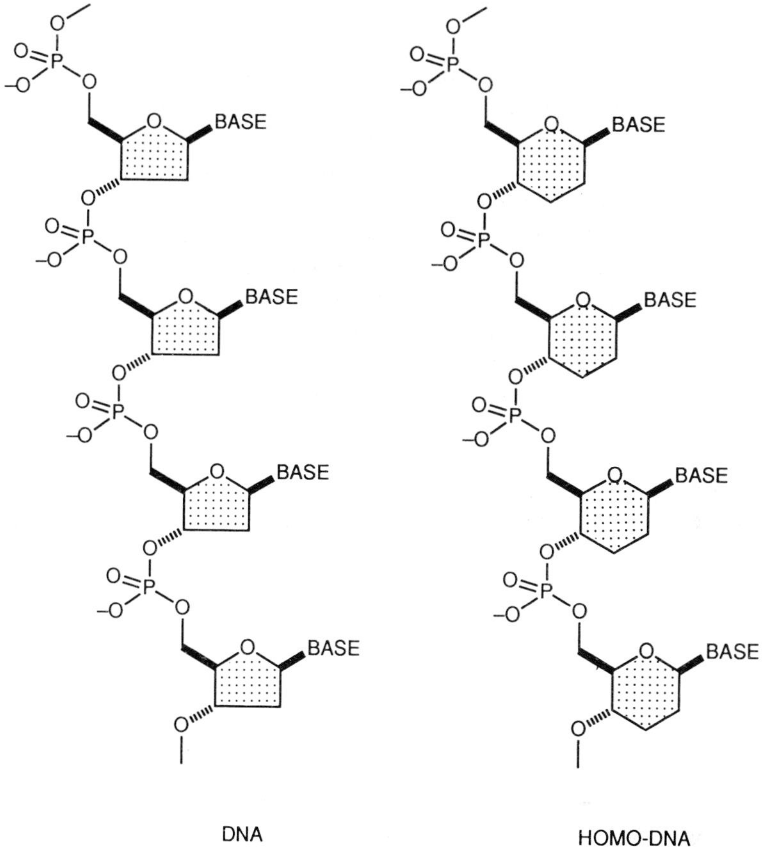

FIG. 18.   The constitution and configuration of homo-DNA (*right*) in comparison with that of natural DNA (*left*).

oligonucleotides do not pair at all (Fischer 1992, Groebke 1993, Diederichsen 1993, Helg 1994). All our results can be interpreted consistently in terms of structural handicaps, steric in nature, that hamper base pairing in fully hydroxylated hexopyranosyl-(6′→4′) oligonucleotides. Our observations and extrapolations from these results point towards the conclusion that it is unlikely that an oligonucleotide system with pairing properties comparable to those of RNA (or DNA) will be found in the hexose series. Much more clearly than we could have hoped at the outset, our experiments promise an answer to the 'why not' question we posed: that answer may well take the remarkably simple form, 'too many atoms in a hexose'.

Encouraged by this, we have extended our studies to the pentopyranosyl series (Eschenmoser 1993) and discovered a constitutional isomer of RNA, an

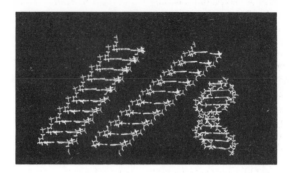

FIG. 19.   NMR structural models of a homo-DNA duplex (*left*, ddGlc[A₅–T₅] duplex in two possible conformations) in comparison with the corresponding natural DNA duplex (*right*, d[A₅–T₅]).

oligonucleotide system with the phosphodiester link between the 4′ and the 2′ positions of the ribopyranose ring, showing a purine–pyrimidine pairing that is stronger and at least as selective as (if not more than) it is in natural RNA (Pitsch et al 1993). Needless to say, a great many experiments on the chemistry of this 'p-RNA' are awaiting us. Knowledge of the existence of this RNA isomer may be an indispensable step on the way to comprehending the origin of RNA. Let us remember what actually had led to that step—it was the vitamin B_{12} (A→D) ring closure problem, both in synthesis and in biosynthesis.

Acknowledgements

The work here was supported by the Swiss National Science Foundation, by Ciba-Geigy AG, Basel and by Firmenich & Cie., Geneva.

References

Battersby AR 1994 New intermediates in the B_{12} pathway. In: The biosynthesis of the tetrapyrrole pigments. Wiley, Chichester (Ciba Found Symp 180) p 267–284

Battersby AR, Matcham WJ, McDonald E et al 1979 Biosynthesis of vitamin B_{12}: structure of the trimethylisobacteriochlorin from *Propionibacterium shermanii*. J Chem Soc Chem Commun, p 185–186

Battersby AR, Bushell MJ, Jones C, Lewis NG, Pfenninger A 1981 Biosynthesis of vitamin B_{12}: identity of fragment extruded during ring contraction to the corrin macrocycle. Proc Natl Acad Sci USA 78:13–15

Brown CE, Katz JJ, Shemin D 1972 Biosynthesis of vitamin B_{12}: study by ^{13}C magnetic resonance spectroscopy. Proc Natl Acad Sci USA 69:2585

Debussche L, Thibaut D, Danzer M et al 1993 Biosynthesis of vitamin B_{12}: structure of precorrin-3B, the trimethylated substrate of the enzyme catalysing ring contraction. J Chem Soc Chem Commun, p 1100–1103

Diederichsen U 1993 A. Hypoxanthin-Basenpaarungen in Homo-DNA-Oligonucleotiden. B. Zur Frage des Paarungsverhaltens von Glucopyranosyl-Oligonucleotiden. PhD thesis, ETH no. 10122, Zürich, Switzerland

Drenkard S, Ferris J, Eschenmoser A 1990 Chemie von α-Aminonitrilen. Aziridin-2-carbonitril: photochemische Bildung aus 2-Aminopropennitril. Helv Chim Acta 73:1373–1390

Eschenmoser A 1968 Current aspects of corrinoid synthesis. Proc RA Welch Found Conf Chem Res 12:9–47

Eschenmoser A 1970 Roads to corrins. Q Rev 24:366–415

Eschenmoser A 1971 Studies on organic synthesis. Pure Appl Chem (suppl) 2:69–106

Eschenmoser A 1974 Organische Naturstoffsynthese heute. Vitamin B$_{12}$ als Beispiel. Naturwissenschaften 61:513–525

Eschenmoser A 1976 Post-B$_{12}$ problems in corrin synthesis. Chem Soc Rev 5:377–410

Eschenmoser A 1982 Über organische Naturstoffsynthese: von der Synthese des Vitamin B$_{12}$ zur Frage nach dem Ursprung der Corrinstruktur. Nova Acta Leopold Neue Folge 55:5–47

Eschenmoser A 1986 Chemistry of corphinoids. Ann NY Acad Sci 471:108–129

Eschenmoser A 1988 Vitamin B$_{12}$: experiments concerning the origin of its molecular structure. Angew Chem Int Ed Engl 27:5–39

Eschenmoser A 1991 Warum Pentose- und nicht Hexose-Nucleinsäuren? Nachr Chem Tech Lab 39:795–806

Eschenmoser A 1993 Hexose nucleic acids. Pure Appl Chem 65:1179–1188

Eschenmoser A, Dobler M 1992 Warum Pentose- und nicht Hexose-Nucleinsäuren? Einleitung und Problemstellung, Konformationsanalyse für Oligonucleotid-Ketten aus 2',3'-Dideoxyglucopyranosyl-Bausteinen ('Homo-DNS') sowie Betrachtungen zur Konformation von A- und B-DNS. Helv Chim Acta 75:218–259

Eschenmoser A, Loewenthal E 1992 Chemistry of potentially prebiological natural products. Chem Soc Rev 21:1–16

Eschenmoser A, Wintner CE 1977 Natural product synthesis and vitamin B$_{12}$. Science 196:1410–1420

Fischer RW 1992 Allopyranosyl-Nukleinsäure: Synthese, Paarungseigenschaften und Struktur von Adenin-/Uracil-haltigen Oligonukleotiden. PhD thesis, ETH no. 9971, Zürich, Switzerland

Friedrich W, Gross G, Bernhauer K, Zeller P 1960 Partialsynthese von Vitamin B$_{12}$. Helv Chim Acta 43:704–712

Fuhrer W 1973 Totalsynthese von Vitamin B$_{12}$: der photochemische Weg. PhD thesis, ETH no. 51058, Zürich, Switzerland

Groebke K 1993 Über Purin-Purin-Paarungen bei Hexosepyranose-Nukleinsäuren. PhD thesis, ETH no. 10149, Zürich, Switzerland

Helg A 1994 Allopyranosyl-Nukleinsäure: Synthese, Paarungseigenschaften und Struktur von Guanin/Cystosin-haltigen Oligonukleotiden. PhD thesis, ETH no. 10464, Zürich, Switzerland

Hunziker J, Roth H-J, Böhringer M et al 1993 Warum Pentose-und nicht Hexose-Nucleinsäuren? Oligo(2',3'-dideoxy-β-D-glucopyranosyl)nucleotide ('Homo-DNS'): Paarungseigenschaften. Helv Chim Acta 76:259–352

Ichinose K, Kodera M, Leeper FJ, Battersby AR 1993 Proof that the biosynthesis of vitamin B$_{12}$ involves a reduction step in an anaerobic as well as an aerobic organism. J Chem Soc Chem Commun, p 515–517

Johnson AP, Wehrli P, Fletcher R, Eschenmoser A 1968 Corphin, a corrinoid-porphinoid ligand system. Angew Chem Int Ed Engl 7:623–625

Johnson AW, Kay IT 1964 The pentadehydrocorrin (corrole) ring system. Proc Chem Soc, p 89–90

Johnson AW, Kay IT 1965 Corroles. I. Synthesis. J Chem Soc, p 1620–1629

Kratky C, Fässler A, Pfaltz A, Kräutler B, Jaun B, Eschenmoser A 1984 Chemistry of corphinoids: structural properties of corphinoid nickel(II) complexes related to coenzyme F430. J Chem Soc Chem Commun, p 1368–1371

Kratky C, Waditschatka R, Angst C et al 1985 Die Sattelkonformation der hydroporphinoiden Nickel(II)-Komplexe: Struktur, Ursprung und stereochemische Konsequenzen. Helv Chim Acta 68:1312–1337

Kräutler B, Pfaltz A, Nordmann R, Hodgson KO, Dunitz JD, Eschenmoser A 1976 Versuche zur Redox-Simulation der photochemischen A/D-Secocorrin → Corrin-Cycloisomerisierung. Elektrochemische Oxydation von Nickel(II)-1-methyliden-2,2,7,7,12,12-hexamethyl-15-cyan-1,19-secocorrinat-perchlorat. Helv Chim Acta 59:924–937

Kreppelt F 1991 Regioselektive Rekonstituierung von Vitamin B_{12} durch Nukleotidierung von Cobyrinsäure-heptakis(cyanmethyl)ester. PhD thesis, ETH no. 9458, Zürich, Switzerland

Ksander G, Bold G, Lattmann R et al 1987 Chemie von α-Aminonitrilen. Einleitung und Wege zu Uroporphyrinogen-octanitrilen. Helv Chim Acta 70:1115–1172

Lewis NJ, Pfaltz A, Eschenmoser A 1983 Acid-catalyzed demetalation of nickel-hydrocorphin and cobalt-corrin complexes with 1,3-propanedithiol. Angew Chem Int Ed Engl 22:735–736

Miller SL 1953 A production of amino acids under possible primitive earth conditions. Science 117:528–529

Miller SL, Orgel LE 1974 The origins of life on earth. Prentice-Hall, Englewood Cliffs, NJ

Mombelli L, Nussbaumer C, Weber H, Müller G, Arigoni D 1981 Biosynthesis of vitamin B_{12}: nature of the volatile fragment generated during the formation of the corrin system. Proc Natl Acad Sci USA 78:9–10

Müller PM, Farooq S, Hardegger B, Salmond WS, Eschenmoser A 1973 Metal-free derivatives of the corphin ligand system. Angew Chem Int Ed Engl 12:914–916

Müller G, Gneuss KD, Kriemler H-P, Scott AI, Irwin AJ 1979 20-methylsirohydrochlorin: revised structure for a trimethylisobacteriochlorin intermediate (factor III) in the biosynthesis of vitamin B_{12}. J Am Chem Soc 101:3655–3657

Müller D, Pitsch S, Kittaka A, Wagner E, Wintner CE, Eschenmoser A 1990 Chemie von α-Aminonitrilen. Aldomerisierung von Glycolaldehyd-phosphat zu racemischen Hexose-2,4,6-triphosphaten und (in Gegenwart von Formaldehyd) racemischen Pentose-2,4-diphosphaten: rac-Allose-2,4,6-triphosphat und rac-Ribose-2,4-diphosphat sind die Reaktionshauptprodukte. Helv Chim Acta 73:1410–1468

Neier R 1978 Über den Mechanismus der photochemischen A/D-Secocorrin→Corrin Cycloisomerisierung. PhD thesis, ETH no. 6178, Zürich, Switzerland

Ofner S, Rasetti V, Zehnder B, Eschenmoser A 1981 Aufbau der Ligandsysteme des C,D-Tetradehydrocorrins und Isobakteriochlorins durch Sulfidkontraktion. Helv Chim Acta 64:1431–1443

Pfaltz A, Hardegger B, Müller PM, Farooq S, Kräutler B, Eschenmoser A 1975 Synthese und reduktive Cyclisierung eines Δ^{18}-Dehydro-A/D-secocorrinkomplexes. Helv Chim Acta 58:1444-1450

Pfaltz A, Bühler N, Neier R, Hirai K, Eschenmoser A 1977 Photochemische und nicht-photochemische A/D-Secocorrin → Corrin-Cyclisierungen bei 19-Carboxy- und 19-Formyl-1-methyliden-1,19-secocorrinaten. Decarboxylierbarkeit und Deformylierbarkeit von Nickel(II)-19-carboxy-bzw. 19-formyl-corrinaten. Helv Chim Acta 60:2653–2672

Pitsch S, Wendeborn S, Jaun B, Eschenmoser A 1993 Why pentose- and not hexose-nucleic acids? Pyranosyl-RNA ('p-RNA'). Helv Chim Acta 76:2161–2183

Rasetti V, Pfaltz A, Kratky C, Eschenmoser A 1981a Ring contraction of hydro-porphinoid to corrinoid complexes. Proc Natl Acad Sci USA 78:16–19

Rasetti V, Hilpert K, Fässler A, Pfaltz A, Eschenmoser A 1981b The dihydrocorphinol → corrin ring contraction: a potentially biomimetic mode of formation of the corrin structure. Angew Chem Int Ed Engl 20:1058–1060

Roth H-J, Leumann C, Eschenmoser A, Otting G, Billeter M, Wüthrich K 1994 Warum Pentose- und nicht Hexose-Nucleinsäuren? Oligo-(2′,3′-dideoxy-β-D-glucopyranosyl)-nucleotide ('Homo-DNS'): ^{1}H-, ^{13}C-, ^{31}P- und ^{15}N-NMR-spektroskopische Untersuchung des Paarungskomplexes von ddGlc(A-A-A-A-A-T-T-T-T-T). Helv Chim Acta 76:2701–2756

Scott AI, Townsend CA, Okada K, Kajiwara M, Whitman PJ, Cushley RJ 1972 Biosynthesis of corrinoids. Concerning the origin of the methyl groups in vitamin B$_{12}$. J Am Chem Soc 94:8267–8269

Scott AI, Mackenzie NE, Santander PJ et al 1984 Biosynthesis of vitamin B$_{12}$: timing of the methylation steps between urogen III and cobyrinic acid. Bioorg Chem 12:356–362

Scott AI, Warren MJ, Roessner CA, Stolowich NJ, Santander PJ 1990 Development of an 'overmethylation' strategy for corrin synthesis. Multi-enzyme preparation of pyrrocorphins. J Chem Soc Chem Commun, p 593–597

Thibaut D, Debussche L, Fréchet D, Herman F, Vuilhorgne M, Blanche F 1993 Biosynthesis of vitamin B$_{12}$: the structure of factor IV, the oxidized form of precorrin-4x. J Chem Soc Chem Commun, p 513–515

Uzar H, Battersby AR 1982 Biosynthesis of vitamin B$_{12}$: pulse labelling experiments to locate the fourth methylation site. J Chem Soc Chem Commun, p 1204–1206

Uzar H, Battersby AR 1985 Biosynthesis of vitamin B$_{12}$: order of the later C-methylation steps. J Chem Soc Chem Commun, p 585–588

Wagner E, Xiang Y-B, Baumann K, Gück J, Eschenmoser A 1990 Chemie von α-Aminonitrilen. Aziridin-2-carbonitril, ein Vorläufer von rac-O^3-Phosphoserinnitril und Glycolaldehyd-phosphat. Helv Chim Acta 73:1391–1409

Woodward RB 1967 The conservation of orbital symmetry. Chem Soc Spec Publ 21:217–249

Woodward RB 1968 Recent advances in the chemistry of natural products. Pure Appl Chem 17:519–547

Woodward RB 1971 Recent advances in the chemistry of natural products. Pure Appl Chem 25:283–304

Woodward RB 1973 The total synthesis of vitamin B$_{12}$. Pure Appl Chem 33:145–177

Yamada Y, Miljkovic D, Wehrli P et al 1969 A new type of corrin synthesis. Angew Chem Int Ed Engl 8:343–348

Zimmermann K 1989 Vergleichende Untersuchungen an Cobaltcorphin- und Cobaltcorrinkomplexen. PhD thesis, ETH no. 9038, Zürich, Switzerland

DISCUSSION

Arigoni: It would be nice to have some discussion of the philosophical content of Albert's presentation, on the value of looking at things in the way that he does, as well as a discussion of the details.

Scott: Albert is being far too modest in saying that his models were incorrect, because almost every one of us working in the biosynthetic field has depended so heavily on his intuition and insight into what is possible, chemically. Of course, the biochemistry that we are seeing now has to have a chemical rationale. Albert's models lead us in a direct way to a quick understanding of the newly observed biochemical events. His model chemistry and chromophores, involving the pyrrocorphins and the corphins, are about to play yet another role, because there is still a small black box to deal with.

Eschenmoser: There is hardly any other example of the biosynthesis of a natural product where a similar cascade of radically unexpected discoveries would have blown away so many hypotheses based on chemical reasoning as B_{12} biosynthesis did, demonstrating how awkwardly limited chemical reasoning can be when it is challenged to read from a unique natural product's structure the possible pathways of that structure's biosynthesis. But more than that, it seems to me that at least one of the hypotheses, the one based on the experimentally documented hydroporphinol→corrin ring contraction, eventually had an anticatalytic effect on the course of research, perhaps not so much on the bioorganic part, certainly not on the biological part, but definitely on the chemical part; it has simply cut off any further conceptual search for chemically feasible ring contraction pathways on higher oxidation levels, mainly as a consequence of its concurrence with that postulate of a constant oxidation level being maintained in B_{12} biosynthesis.

Battersby: The balance, Albert, was strongly positive—there is no question about that. Your ideas about extended amidine systems, about the preferred sites of methylation, about the effect of metal ion complexation, and so on, have all greatly helped our thinking. An enormous amount of wonderful chemistry came from your models. I don't agree with what you said about an anticatalytic effect at all.

Arigoni: One anti-catalytic effect stemmed from the identification of acetic acid as the volatile fragment released during formation of cobyrinic acid from, say, precorrin-3. On a book-keeping basis, this suggested that no oxidizing or reducing cofactors were involved anywhere in the biosynthetic scheme. Because, on top of that, Albert's model studies were compatible with this idea, we were all fairly convinced that it had to be that way. The later developments taught us a great lesson.

Scott: It wasn't until the explosion of molecular biology which made this recent work possible that we could test these ideas. All the formative mechanistic concepts are in place and must be obeyed in whatever we finally suggest for the mechanism of ring contraction.

Jordan: How did methylation at position 11 figure in some of your earlier considerations, when you were analysing all the possibilities?

Eschenmoser: If, and only if, the *meso* position C-10 of the ring C methylation substrate has the ring B double bond in the macrocyclic position C-9–C-10, the ring C methylation should—from the point of view of intrinsic reactivity— strongly prefer position C-12 because then the methylation extends the (strongly stabilizing) amidinic donor–acceptor conjugation from rings A and B to include

Eschenmoser's Corphin

Factor S$_1$

ring C. If, however, the substrate has the ring B double bond in position C-9–C-8, ring C is an isolated pyrrol π-system and methylation may occur—again from the point of view of intrinsic reactivity—at any of the four positions. Position C-11 would become one of two favoured methylation sites if the ring D double bond around C-16 were between C-16 and C-15.

Scott: Consider Albert's first synthetic zinc–corphinate and the structure of factor S$_1$; S$_1$ has turned out to be the first naturally occurring corphinate, and we consider the type I-derived pyrrole pigments to be natural products. The corphins were highly predictive and helpful to all of us. We would not have been able to get the structures of Factors S$_1$ and S$_3$ so quickly without Albert's work.

Battersby: There is another example. Factor IV is in fact a tetradehydrocorrin. The structural work by Blanche and Crouzet's groups on that substance was helped by Albert's previous synthesis of a tetradehydrocorrin with methyl groups all the way around the macrocycle. It had two methyls on each of the reduced rings, four on rings C and D and one at C-1, nine methyl groups in all. The match of the UV–visible absorption of that synthetic material, particularly in the protonated form, to Factor IV was very close.

Akhtar: In the last two or three decades the ETH has become like the Vatican—it tells us what is the correct behaviour and the correct scientific language. Whenever I use the term 'chemical' when I mean non-enzymic I am usually corrected immediately. Albert actually spoke of 'chemical synthesis'. I would like to stand up for the old way of speaking, because what Albert has shown us is yet another triumph of 185 years of chemistry. There is a flamboyance and beauty about non-enzymic, 'chemical' reactions, as we have just witnessed. No chemist ought to be ashamed of that. Enzymic chemistry has an elegance, but it is narrow-minded; test-tube chemistry has much greater vitality.

Eschenmoser: My emotions are in accord with yours, but there is an underlying problem that bothers me in the use of the terms biochemistry and chemistry.

Speaking of 'biochemistry and chemistry' in some contexts is like speaking of 'trees and plants'. Juxtaposing the terms biochemistry and organic chemistry would sound logical, but not biochemistry and chemistry. I therefore belong to the group who are sensitive to the terms enzymic and chemical.

Arigoni: The reason why I am finicky about the distinction between 'chemical' and 'non-enzymic' is that when people refer to non-enzymic reactions as being 'chemical' they give the impression that the enzymic reaction does not belong to chemistry.

Scott: In the overall base-pairing in homoDNA, is there a preponderance of Watson–Crick pairing or is there a mixture of Watson–Crick with Hoogsteen and other unusual pairings?

Eschenmoser: That depends on the sugar backbone. In homoDNA, whenever the opportunity exists for Watson–Crick pairing, there is Watson–Crick pairing. In allopyranosyloligonucleotides, however, the best pair, guanine–cytosine, is not a Watson–Crick pair but a reverse Hoogsteen pair (with a protonated cytosine).

Akhtar: Another type of DNA analogue, called PNA, has been made recently. PNA has a pseudo-protein backbone onto which the bases are attached. The work done on these has not been extensive, but PNAs seem to have caught biochemists' imagination. I was quite surprised to see here that, at least between A and T, the hydrogen bonds were perfect.

General discussion II

Aerobiosis and anaerobiosis

Thauer: It's always nice to speculate that an anaerobic organism such as *Clostridium* is old and that one such as *Escherichia coli*, being a facultative organism, is less old. However, consider the propionics. As a microbiologist, I view these as 'secondary' anaerobes, not as true anaerobes. Many of their metabolic features are like those of an aerobe which has lost the ability to use molecular oxygen. They are, for example, the only microorganisms which do not use molecular oxygen as their main electron acceptor that have a pyruvate dehydrogenase complex—a characteristic feature of aerobes—and they have catalase and other haem proteins. If you think there are different pathways in tetrapyrrole biosynthesis in different microorganisms, you would have to consider that the propionics, being Gram-positive eubacteria, are phylogenetically different from the Gram-negative pseudomonads and enterobacteria, rather than considering them as older than the others.

The enterobacteria, which include *E. coli* and *Salmonella*, are facultative anaerobes which grow equally well anaerobically and aerobically. They change their metabolism completely when they switch from one mode of growth to the other. More than 50 genes are turned on, and about the same number are turned off, when they change from aerobic to anaerobic conditions. A pathway under anaerobic conditions will not necessarily be the same under aerobic conditions. *E. coli*, for example, can synthesize ubiquinone under anaerobic conditions using a non-O_2-requiring pathway, but under aerobic conditions it uses an O_2-requiring pathway. Another example is the biosynthesis of pyridine nucleotides. The nicotinamide moiety is synthesized in some microorganisms, e.g., yeast, from tryptophan in aerobic conditions, and from aspartate and a triose phosphate under anaerobic conditions. Frequently in our discussions the question of where the oxygen in a molecule has come from has arisen. Under aerobic conditions a bacterium might use O_2 to introduce the oxygen whereas under anaerobic conditions it might use H_2O, with a completely different pathway. For example, pseudomonads can aerobically degrade many aromatic compounds using monooxygenases and dioxygenases. If they grow on nitrate anaerobically, they use a completely different pathway to degrade the same aromatic substrate. They degrade phenol under aerobic conditions by attack with molecular oxygen to produce catechol, which is further degraded to β-ketoadipate; in anaerobic conditions they carboxylate the phenol to *p*-hydroxybenzoate, which, after conversion to the CoA ester, is reduced to

333

benzoyl-CoA and from there on to cyclohexenecarboxyl-CoA. If you are not careful and work with cells grown under semi-anaerobic conditions, the cells might contain a mixture of enzymes from two different pathways. If one finds two pathways for tetrapyrrole biosynthesis, one should bear in mind that alternative pathways can operate under anaerobic and aerobic conditions. If you want to be really sure you are dealing with a completely anaerobic organism, you should select something such as a species of *Clostridium* that does not have cytochromes.

Battersby: What about *Clostridium tetanomorphum*?

Thauer: That would be a good organism to choose. You could also try methanogens, but the disadvantage is that genetic manipulations are not yet possible with either *C. tetanomorphum* or methanogens.

Castelfranco: You mentioned anaerobes which utilize nitrate as a terminal electron acceptor. Is that under aerobic or anaerobic conditions?

Thauer: Anaerobic.

Castelfranco: It's anaerobic in the sense that it does not require air, but there is an electron transport chain analogous to the one that goes to oxygen, but it goes only to nitrate. This is the distinction between true 'fermentative' behaviour and anaerobic respiration. I wonder whether these organisms that reduce nitrate or sulphate are something else, whether they have a third way of making a living.

Thauer: Molecular oxygen has two functions. One is to be the electron acceptor in the respiratory chain. In this function, in which O_2 is reduced to H_2O, O_2 can be substituted by other electron acceptors, such as NO_3^- or SO_4^-, which have sufficiently positive redox potentials. The second function is in oxygenase reactions, in which molecular oxygen is incorporated into organic or inorganic molecules. These reactions are, as far as we know, completely dependent on O_2.

Battersby: All the work in cell-free systems from *Pseudomonas denitrificans* by Blanche and Crouzet's groups was carried out anaerobically.

Hädener: A related question is whether or not *E. coli* is able to biosynthesize B_{12}. When we produced selenodeaminase (p 109), we used a methionine auxotroph in which the last step of the biosynthesis of methionine, the conversion of homocysteine to methionine, was blocked. In *E. coli* there are two genes encoding enzymes that can catalyse this reaction, and one of these, encoded by *metH*, is B_{12} dependent. We blocked the B_{12}-independent enzyme, encoded by *metE*, so the B_{12}-dependent enzyme should still be operating. This would not matter normally because *E. coli metE* mutants are grown aerobically in a minimal medium devoid of B_{12}, but in the presence of methionine (or selenomethionine). However, Cohen & Saint-Girons (1987) have speculated that *E. coli* might be able to synthesize B_{12} anaerobically, as *Salmonella typhimurium* is (Jeter et al 1984). We came across this when we tried to save money by adding only 10 mg/l of selenomethionine to our medium.

Selenodeaminase purified from such a fermentation was a mixture of protein that was almost fully substituted with selenomethionine (44%) and enzyme that was not substituted at all (56%) (Hädener et al 1993). There had been a lag phase in the fermentation following a normal initial phase. Some hours later cell growth resumed. This indicates that deaminase fully substituted with selenomethionine was produced during the initial phase as long as selenomethionine was available. Ordinary, sulphur-containing, methionine was obviously produced during the later phase although the medium did not contain vitamin B_{12}. Owing to the sensitivity of selenomethionine to oxidation, our fermentation protocol required partially anaerobic growth conditions which could favour *de novo* B_{12} biosynthesis. Endogenous production of methionine by the B_{12}-dependent gene product could then account for the large proportion of unsubstituted deaminase found in this fermentation. Alternatively, one might argue that revertants had taken over during the lag phase.

Thauer: Did you have a deletion in the non-B_{12}-dependent enzyme? You cannot have revertants with a deletion.

Hädener: The probability of there being revertants with this particular *metE* mutant is very low. We know that the lesion is a small deletion, but its extent is unknown (P. Oliver, unpublished work).

Beale: A third possibility is that there was a contaminating organism which was a B_{12} producer.

Scott: John Roth has looked very carefully for B_{12} with *E. coli* grown anaerobically. He has never detected B_{12} formation, but there could be special conditions under which it is possible.

Thauer: Haven't hybridization studies with DNA from *Salmonella typhimurium* been done? If the genes are in *E. coli*, they should be picked up by hybridization.

Akhtar: In polyketide biosynthesis there are many examples that would have given positive results in hybridization experiments, but this is of only phylogenetic importance.

Scott: The consensus at the moment is that *E. coli* doesn't make B_{12}, but it could be engineered to make it. The strangest thing of all *is* that *S. typhimurium* doesn't really need B_{12}.

Warren: As John Roth has pointed out, *S. typhimurium* devotes about 1% of its total genome to B_{12} biosynthesis, yet it makes only a few molecules per cell. He has been trying to work out what the selection pressure was for this, since *E. coli* would appear to have given up its ability to make B_{12} *de novo*.

Scott: *S. typhimurium* does have propanediol dehydratase, and therefore B_{12} coenzyme requirements, when grown aerobically.

Thauer: Bacteria have genes for functions which you don't see. The essential problem is that the way we grow the bugs in the laboratory is not the way they live naturally. If 1% of the *S. typhimurium* genome is devoted to B_{12} biosynthesis, I would think that there are conditions under which it is required— we just don't know what they are.

Beale: These conditions may not be intrinsic to the organism's growth, but instead may have something to do with mutually beneficial interactions with other organisms in its environment. No one would guess why soy-beans have globin genes unless they grew soy-beans symbiotically with *Rhizobium*, which utilizes leghaemoglobin to provide the microenvironment necessary for N_2 fixation.

Scott: We are all pretty well agreed that plants don't make B_{12}.

Beale: We are convinced that they don't make B_{12}, and most of us are convinced that they don't need B_{12} either.

Thauer: What about yeast?

Beale: I'm not sure.

Thauer: There is a lot of vitamin B_{12} in beer.

Scott: It's particularly high in Guinness, by the way. I remember a colleague who worked for Guinness telling me that it's really very good for you and that the company let him drink as much as he liked!

References

Cohen GN, Saint-Girons I 1987 Biosynthesis of threonine, lysine and methionine. In: Neidhardt FC (ed) *Escherichia coli* and *Salmonella typhimurium*. American Society for Microbiology, Washington, DC, p 429–444

Hädener A, Matzinger PK, Malashkevich VN et al 1993 Purification, characterization, crysallisation and X-ray analysis of selenomethionine-labelled hydroxymethylbilane synthase from *Escherichia coli*. Eur J Biochem 211:615–624

Jeter RM, Olivera BM, Roth JR 1984 *Salmonella typhimurium* synthesizes cobalamin (vitamin B_{12}) *de novo* under anaerobic growth conditions. J Bacteriol 159:206–213

Summing-up

Duilio Arigoni

Laboratorium für Organische Chemie, ETH-Zentrum, Universitätstrasse 16, CH-8092 Zürich, Switzerland

Scanning through the book of the Ciba Foundation symposium held in 1958 on the Biosynthesis of Sterols and Terpenes, I came across the following paragraph in Sir Robert Robinson's closing remarks (Ciba Foundation 1959, p 302): 'It must be remembered that the speculative side started the ball rolling and is now, I hope, to be replaced entirely by the experimental method. The experiments themselves and the result of them are providing sufficient clues for further work.' Indeed, the speakers at this meeting largely have refrained from indulging in speculations and have provided excellent proof for the now generally accepted belief that it is experimentation which leads the way in biosynthetic studies and, eventually, discloses for our delight the often cunning and always exquisitely rational way in which Nature deals with complex biosynthetic problems.

Let me comment briefly on a few selected themes which have emerged during this symposium.

I think one cannot over-emphasize the tremendous impact that NMR spectroscopy and later, and perhaps even more dramatically, techniques allowing the isolation of single genes and the expression of the encoded proteins have had on the development of the topic under discussion. In future, organic chemists interested in biosynthetic problems will have to cope with such techniques much as they have had to do in the past whenever new spectroscopic methods appeared on the scene. No doubt an efficient solution of problems with the complexity displayed by most of the molecules we have been talking about will require close cooperation of the organic chemists with their more biologically oriented partners who, in turn, will no doubt take advantage from a sustained contact with organic chemistry.

We also witnessed impressive progress in the area of purification and crystallization of biosynthetic enzymes. It is important to realize that the determination of the structure of these molecules by X-ray crystallography is, of course, only the beginning of a new story. Knowledge of the three-dimensional arrangement of the atoms in both the catalyst and the substrate provides crucially important information but tells us rather little about the dynamics of the reaction under investigation. Ideally, one would like to obtain a motion picture of the catalyst in action. This is certainly a formidable task, but there are hints that specifically mutated enzymes can be put to good use in tackling the problem.

In his presentation on haem d_1 Dr Chang raised the tantalizing question concerning the unknown link between the structure of the molecule and its biological activity. This is an excellent challenge for the organic chemist. In too many cases we know what the compounds look like and in which processes they participate, but we fail to ask the question as to why these specific structures rather than others were selected by evolution. I am convinced that there is more chemical wisdom engraved in the structure of such molecules than many organic chemists would dream of and with a discerning eye it should be possible to understand how, for example, the embedment of a nickel ion into the ligand system of the F_{430} cofactor imparts to the metal the specific properties which enable it to participate in the last step of methane biogenesis.

If one subscribes, as I do, to Albert Eschenmoser's philosophy, according to which biologically important molecules, or some close relatives, were first generated spontaneously in a prebiotic environment, one is then confronted with yet another chemical challenge, namely that of mimicking such prebiotic processes under appropriate laboratory conditions. Professor Eschenmoser has already provided many beautiful illustrations of this approach, but much remains to be done. Let us keep in mind, for example, that the apparently simple condensation of two units of 5-aminolaevulinic acid to give porphobilinogen cannot be controlled efficiently in the absence of the enzyme.

Finally, I was impressed by the phenomenon of heterogeneity detected in the biosynthesis of both chlorophyll and vitamin B_{12}. Eventually, somebody will have to investigate the biological significance of this heterogeneity, which, for the time being, appears to be something beyond the realm of understanding of the typical organic chemist.

This has been, in my opinion, a most enjoyable and fruitful meeting and I would like to express my thanks to the speakers for their presentations and to all discussants for their engagement and their contributions.

Last, but not least, we should thank Derek Chadwick and the members of the Ciba Foundation's staff for their impeccable organization of the meeting and for providing us with such warm and homely surroundings.

Reference

Ciba Foundation 1959 Biosynthesis of terpenes and sterols. Churchill, London

Index of contributors

Non-participating co-authors are indicated by asterisks. Entries in bold type indicate papers; other entries refer to discussion contributions.

Indexes compiled by Liza Weinkove

Subject index